1982

Basic
Applied
Logic

Basic Applied Logic

Kenton F. Machina
Illinois State University

Scott, Foresman and Company
Glenview, Illinois

Dallas, Tex.,   Oakland, N.J.,   Palo Alto, Cal.
Tucker, Ga.,   London, England

Acknowledgments

From an undated letter written in the late 1970s, soliciting funds for the Center for the Defense of Free Enterprise. Reprinted by permission.

From "The Invisible Wall" from *Dark Ghetto:* DILEMMAS OF SOCIAL POWER by Kenneth B. Clark. Copyright © 1965 by Kenneth B. Clark. Reprinted by permission of Harper & Row Publishers, Inc.

"The Theory That Doesn't Work." Reprinted by permission of FMC Environmental Research Laboratories, Princeton, New Jersey.

From *Managerial Accounting* by Ray H. Garrison. © Business Publications, Inc., 1976, p. 72. Reprinted by permission.

From the editorial page of the *Herald-Telephone,* September 19, 1969. Reprinted by permission. permission of Georges Borchardt, Inc.

From the editorial page of the Herald-Telephone, September 19, 1969. Reprinted by permission.

Russell Kirk, *Enemies of the Permanent Things.* New Rochelle, N.Y.: Arlington House, 1969.

From "A Phony Argument to Save the SST," *Louisville Courier-Journal,* January 4, 1971, editorial page. Copyright © 1971. The Courier-Journal. Reprinted with permission.

Excerpts from "Putting Up with the Ugly Duckling" by Burton Yale Pines in *Time,* Vol. 109, No. 12, March 21, 1977. Reprinted by permission from *Time,* The Weekly Newsmagazine; Copyright Time Inc. 1977.

From "Pre-teen alcohol dangers found" in *The Vidette,* April 26, 1977, p. 2. Reprinted by permission.

Excerpts from "Playing China Card Could Trap Us in a Triangle" by Alexander Yanov, *Los Angeles Times,* July 6, 1978. Reprinted by permission of the author.

Library of Congress Cataloging in Publication Data

Machina, Kenton F., 1942–
  Basic applied logic.

  Includes bibliographical references
  1. Logic. I. Title.
BC71.M24   160   81-9307
ISBN 0-673-15359-2   AACR2

Preface

This is a practical book, intended to help its users develop logical reasoning skills. It is designed for any introductory course that focuses on the development of those skills—whether that course takes an "informal" nonsymbolic approach or a more formal approach. Although it contains optional sections in which modern symbolic logic is carefully developed through first-order predicate calculus, the book maintains a practical orientation, emphasizing only those aspects of logical theory that bear on argument evaluation and argument construction. Thus, the book is well-suited for use in a wide range of introductory applied logic courses, introductory philosophy courses containing major units on logic, and courses in argumentation.

Basic Applied Logic is an accessible book, despite the seeming complexity of its subject. I have tried for an informal, helpful tone, and have taken pains to explain difficult points in detail so class time may be spent on the exercises and on student questions. Some extended passages drawn from various authors, and frequent shorter passages which are either drawn from "real-world" sources or sound as though they were, provide repeated examples of the applicability of logic, including formal logic, to argument evaluation. I have also included sections on important topics not usually found in introductory books, such as counterexample construction, supplying missing premisses, difficulties with symbolic logic approximation of English language connectives, and the logical evaluation of "ought" arguments. Finally, I don't believe an introductory book needs to be technically inaccurate merely because it is introductory. Accordingly, I don't try to pretend that the horseshoe captures the core meaning of "if . . . then," and I avoid claiming that the invalidity of the most detailed propositional logic diagram of an argument provides a proof of the ultimate invalidity of the argument.

Basic Applied Logic contains a large number of exercises, generally arranged in order of difficulty, appearing at the ends of the relatively short sixty-seven sections into which the text is divided. This arrangement provides the student with immediate reinforcement and the instructor with increased flexibility. The answers to some of these exercises are found at the back, along with a list of all the rules of deduction comprising the system of natural deduction taught in the optional symbolic logic sections. Answers to the remaining exercises are provided in an instructor's manual.

There is considerably more in this book than can be learned in a typical one-term beginning college logic course. In fact, if even most of the contents are taken seriously, there is enough here for a one-year course. That does not mean, however, that a one-term course should begin with page one and proceed through the first half of the book. There are a number of sensible routes through the book which emphasize different topics. The map found immediately after the Table of Contents displays the alternatives. For instance, the sections containing symbolic logic may be bypassed entirely in favor of Part III applications, or some of them may be completed before the course follows a chain of arrows leading to Part III. The instructor's manual contains specific suggestions for constructing various sorts of courses, from almost completely symbolic to completely nonsymbolic courses.

For me, the application of logic to argument analysis is a subject alive with challenges and potential, not a mere rote application of established rules. I believe that attitude will be apparent to anyone who works through this book, and that many such people will be encouraged to carry out that kind of application for the rich rewards it can bring.

It is impossible to acknowledge all those to whom I am intellectually indebted, especially because this is a textbook containing much that has become standard through the work of hundreds of people. For helping me to understand what has become standard, though, I am especially grateful to my teachers, David Kaplan, Donald Kalish, and Montgomery Furth. And for helping me to avoid errors and to think clearly about the manuscript, I am grateful to David Sanford and T. C. Carroll, Jr. Finally, for allowing me to try out my ideas, I acknowledge the several hundred students with whom various parts of this book have been used.

Kenton Machina

Contents

PART II. LOGICAL SYSTEMS 100

MAP OF ALTERNATE ROUTES
THROUGH THE TEXT

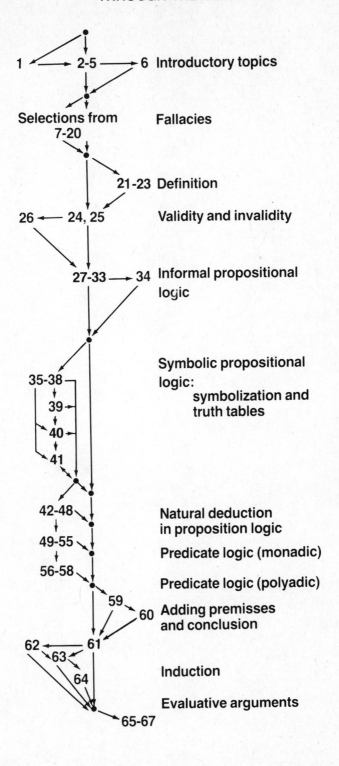

1 → 2-5 → 6 Introductory topics

Selections from   Fallacies
7-20

21-23 Definition

26 ← 24, 25   Validity and invalidity

27-33 → 34 Informal propositional
logic

Symbolic propositional
logic:
symbolization and
truth tables
35-38
39
40
41

42-48 Natural deduction
in proposition logic

49-55 Predicate logic (monadic)

56-58 Predicate logic (polyadic)

59
60 Adding premisses
and conclusion

62 ← 61
63
Induction
64

Evaluative arguments
65-67

Basic
Applied
Logic

To Connie, my wife, who put up with this project, for the most part, graciously.

PART ONE
Informal Analysis

1

The Nature of
Logic and Argument

1

Why the study of logic is useful

Everyone who has occasion to pick up this book sometimes engages in serious discussion about beer, baseball, profit-making, religion, or whatever. And in any serious discussion it's very likely that opposing opinions are offered, argued for, even yelled about. Most folks are not very good at deciding who, if anyone, is right in such a discussion. In fact, it's popular to escape from thinking in such situations by saying, "You have your point of view and I have mine; each of us has a right to his or her own views." Whatever the psychological benefits of this ploy, it remains a mere escape device, because there are ways of sorting through the reasons we have for our beliefs to see which of the reasons hold water and which do not.

It is not easy to find our way through the maze of reasons we have for our beliefs. If you do very much of it out loud, you're bound to get in trouble. It makes people uncomfortable, even angry, to have their beliefs challenged or to have their reasoning analyzed. Nevertheless, there is a payoff for people who can do this kind of analysis: they have a valuable tool for figuring out which opinion is best supported by all the available evidence.

This entire book is dedicated to helping you develop the intellectual skills needed to do the kind of unpopular analysis required to evaluate reasoning. If you have the courage, you can even use these skills on your own reasoning.

It shouldn't be too hard to see that the person who has developed skills like these has an advantage in almost any occupation or area of interest, especially if he or she uses the skills wisely without offending everyone in the neighborhood. Just think of the degree to which the level of political discussion in our country would be raised if the general public were well trained in the analysis of reasoning. Or consider the mistakes in business administration which could be avoided if administrators could skillfully sort out the arguments for and against marketing a new product, opening a new plant, or hiring a new manager. Even if you have the skills and others around you don't, some of these same advantages still are available to you, and will help you find out the truth about things that matter to you.

You already possess some of the skills I'm talking about, but almost everyone can use some practice to develop and extend their abilities in these areas in much the same way that even native speakers of English may profit from a course in English composition to sharpen their writing skills. For my part, I pledge to make this book as useful, interesting, clear, and technically correct as I can.

2

What you will study when you study logic

By now you already have some idea of what logic is all about. You realize that logic has something to do with evaluating the reasons people have for believing the things they believe. However, there are a great many different ways of evaluating people's reasons, and not all these evaluation techniques are applications of logic. For example, a clinical psychologist may carefully analyze the psychological factors which lead one of his or her clients to have various irrational beliefs about the world. The resulting analysis may show how the client's various psychological needs are met in part by certain peculiar beliefs he or she holds about other people. One could then say that the psychologist had discovered something about the reasons why the client believes what he believes. In this case, it appears that the "reasons" for the beliefs are the *causes* of the beliefs—that is, it appears that sometimes when we talk about the reasons why people believe things, we are talking about the causal factors which help to instill the belief or which help to keep the belief alive. However, logic is not concerned with the *causes* of beliefs, but rather with the *support* that can be offered to show the beliefs true or false.

Since logic, political science, psychology, history, rhetoric, advertising theory, and many other disciplines all study human beliefs from varying points of view, students sometimes become confused. In reading this logic book, one may discover that the favorable statements given by a famous person provide no evidence that a particular brand of dog food is good for dogs to eat. Yet, in reading a book on effective advertising techniques, one may find that testimonials from famous persons can be very persuasive. It seems that the logic book is telling you not to use such testimonials, but the advertising text is telling you that it is a good idea to use them. So it may initially seem as though logic contradicts advertising theory or the theory of public speaking. However, there need not be any contradiction. Recall that logic has to do with what counts as a good reason for believing something, while the example just given has to do with something else, namely with what kinds of reasons people commonly find persuasive. Very often, reasons which effectively persuade people to believe something are merely *causes* of belief but are not very *good* reasons for believing whatever is in question.

Although it is correct to say that logic studies the factors which make reasons genuinely good reasons for believing something, this characterization of the nature of logic should be clarified. Suppose that you happened to believe that Gerald Ford was the best president the United States ever had. Also suppose that you use that belief as a reason for also believing that he should have won election to the presidency in 1976, instead of Jimmy Carter. Logic, you have been told, studies whether or not your reasons are good ones. So, you expect that logic will tell you whether it is correct to believe that Gerald Ford was the best president the United States ever had. After all, your favorable belief about Ford is a reason, and it can hardly be a good reason if it is not a correct belief. Unfortunately, your expectation that logic will tell you about President Ford's suitability to be president will have to remain unfulfilled, for logic can tell you when a belief is correct only in very special cases. Logic alone cannot tell you whether Ford was an angel or a devil. What can logic do, then? It can reveal whether your first belief about Ford is related logically to the second belief about Ford. In other words, when logic addresses what constitutes a good reason or a bad reason for believing something, it will be telling you what kind of beliefs are appropriate to use as support in a given context for given purposes.

So far, I have been talking as though logic were the study of certain things about beliefs. However, logic also studies things that no one actually believes. For example, probably no one believes that George Washington, the first President of the U.S., now lives in Washington, D.C. Nor does anyone believe George was female in 1778. Nevertheless, logic can tell us that if these two things that no one believes are true, then yet another statement would be true: someone who was female in 1778 now lives in Washington, D.C.

My point is that logic is not restricted to the study of things people actually believe. Logic studies certain kinds of relationships between *any* truth or falsehood and any other truth or falsehood, whether or not anyone believes these truths or falsehoods.

I'll use the word "proposition" to refer to any truth or falsehood. *Logic can then be defined as the study of what makes the truth of one proposition good support or bad support for the truth of another proposition.*

The study of logic

2.1 State which of the following activities would definitely be a part of logic, and defend your answers by explaining how these activities fit the characterization of logic given above. In the case of those activities which might not fit the characterization, explain why they might not.
   a. Investigation of the cultural patterns of belief which make one nation or ethnic group different from another.
   b. Description of how hypnosis can induce beliefs in people.
   c. Study of the best ways of persuading a crowd that you know the way they can all become rich or healthy.
   d. Investigation of what conclusions can be properly drawn from a given set of data.
   e. Listing how beliefs have changed over the years.

2.2 a. Give two examples of possible causes for a belief in the existence of God, describing these causes in such a way that you make clear that they are not good reasons which tend in any way to show that the belief in God is true. Then do the same thing with two possible causes of a belief that there is no God.
   b. Describe one possible reason for believing in God which is actually relevant to showing that the belief in God is true.

3

Arguments

We have already seen that when someone gives evidence in support of some proposition, logic will be involved in evaluating whether that support is strong or weak. Sometimes we refer to the giving of evidence as *arguing* for something. Thus, logic will be concerned with arguments among other things. In fact, in this book we will concentrate entirely on the application of logic to arguments. My present purpose will be to clarify the meaning of *"argument"*.

Unfortunately, the word "argument" in English often is used to refer only to heated emotional debates in which little or no evidence is offered by either party to the debate. Naturally, logic doesn't have much application to that kind of argument. When I say that we will be concerned with arguments, I mean that we will study support and ask how strong it is. It doesn't matter whether this support is offered calmly or emotionally, or whether there is any disagreement involved. Hence, the following expresses an argument, for it offers support for the importance of boiling the containers properly:

When you can vegetables at home, you need to make certain that all the bacteria are killed after the containers are sealed, in order to prevent the food from spoiling. Thus it is important when canning to boil the sealed containers for the proper amount of time.

Some arguments consist of very many propositions that are related to each other in complex ways. You will want to get better at analyzing these complex relationships, for otherwise it will be easy to pull the wool over your eyes. An example of a fairly complex argument perhaps of some interest to you follows. We will have occasion to work with it from time to time in this book. For now, try to figure out what is being argued for, and what reasons are being given to support the author's point of view. The argument is excerpted from "The Suicide of the Sexes" by George Gilder:[1]

> The differences between the sexes are perhaps the most important condition of our lives. With the people we know best, in the moments most crucial in our lives together, sexual differences become all-absorbing. . . . These differences are embodied in a number of roles. The central ones are mother-father, husband-wife. . . .
>
> One of the best ways to enrage a young feminist . . . is to accuse her of having a maternal instinct. In a claim contrary to the evidence of all human history and anthropology—and to an increasing body of hormonal research—most of these women assert that females have no more innate disposition to nurture children than do men. . . . But whether instinctual or not, the maternal role originates in the fact that only the woman is necessarily present at birth and has an easily identifiable connection to the child—a tie on which society can depend. . . . The idea that the father is inherently equal to the mother within the family, or that he will necessarily be inclined to remain with it, is nonsense. The man must be made equal by culture. . . . For a man's body is full only of undefined energies. And all these energies need the guidance of culture. He is therefore deeply dependent on the structure of the society to define his role in it. Of all society's institutions that work this civilizing effect, marriage is perhaps the most important. . . . Marriage attaches men to families, the source of continuity, individuality, and order. . . .
>
> A job is thus a central part of the sexual constitution. It can affirm the masculine identity of its holder; it can make it possible for him to court women in a spirit of commitment; it can make it possible for him to be married and thereby integrated into a continuing community. . . .
>
> The feminist contention that women do not generally receive equal pay for equal work . . . should be considered in light of the . . . cost to the society of male unemployment. The unemployed male can contribute little to the society and will often disrupt it, while the unemployed woman may perform valuable work in creating and maintaining families. In effect, the system of discrimination . . . tells women that if they enter the marketplace they will probably receive less pay than men, not because they could do the job less well but because they have an alternative role of incomparable

[1]*Harper's Magazine,* July 1973, pp. 42–54.

value to the society as a whole. The man, on the other hand, is paid more, not because of his special virtue, but because of the key importance of taming his naturally disruptive energies. . . .

Nothing is so important to the sexual constitution as the creation and maintenance of families. And since the role of the male as principal provider is a crucial prop for the family, the society must support it one way or the other. Today, however, the burdens of childbearing no longer prevent women from performing the provider role; and if day care becomes widely available, it will be possible for a matriarchal social pattern to emerge. Under such conditions, however, the men will inevitably bolt. And this development . . . would probably require the . . . emergence of a police state to supervise the undisciplined men and a child care state to manage the children. Thus will the costs of sexual job equality be passed on to the public in vastly increased taxes. The present sexual constitution is cheaper.

Of course, the male responsibility can be enforced in many other ways, coercively or through religious and social pressures. . . . In modern American society, however, the "social pressures" on women for marriage and family are giving way to pressures for career advancement, while the social pressures on men are thrusting them toward . . . the delights of easy sex. . . . At this point, therefore, any serious . . . campaign for equal pay for equal work would be destructive. . . .

In general, the . . . social order will be best served if most men have a position of economic superiority over the relevant women in the community, and if in most jobs the sexes tend to be segregated by either level or function. These practices are seen as oppressive by some; but they make possible a society in which women can love and respect men and sustain durable families. They make possible a society in which men can love and respect women and treat them humanely.

This passage contains a great many claims that are controversial, but what makes it an *argument* is that the author offers support for the conclusions about what ought to be done in our society. Since the passage contains arguments in support of the author's position, we can do more than merely agree or disagree with him—we can analyze his reasoning to see if he adequately supports his conclusions.

By now you probably have a fairly clear idea of what I mean when I use the word "argument", but it would be nice to have a precise definition of the term as it will be used in this book. Sometimes it doesn't matter that the author of a particular passage wrote what he or she wrote in order to give support for something, and some arguments are so poor that they contain *no* actual support for anything and yet we will want to call them arguments. So, I will make a somewhat unorthodox move. I will allow that *any* set of propositions may be treated as an argument. That is, an *argument is any set of propositions where at least one proposition in the set is being treated as proposed support for the truth of some proposition in the set.* Note that according to this definition the decisive factor in what counts as an argument is how the propositions are *treated.* Also note that *proposed* support might not turn out to be real support. If you decide to treat a given set of propositions as an argument, then it automatically becomes an argument. However, your decision to treat a given passage in this way might be silly, unreasonable, or a waste of time. Naturally, we will want to treat passages as arguments only when there is good reason for doing so.

Even though you are free to treat *any* set of propositions as an argument without regard for the author's intentions, there are many occasions on which you will be trying to understand what a particular author actually intended. For instance, if you were asked to summarize Gilder's argument for discrimination against women, the summary should reflect Gilder's own intentions as well as possible as determined from a careful reading of his writing. In order to attribute an argument to a particular author, you need some evidence that the author intended to present the argument you are attributing to him or her.

Given this definition of "argument", you can see that only sets of propositions make up arguments. Since questions, exclamations of emotion (such as curses), and commands cannot be literally true or false, such linguistic items do not enter as constituents of arguments. If I give a reason why I am giving you an order, I am not arguing that my order is true, since orders are not the sort of thing that can be true (or false). The order is not a constituent proposition in any argument. However, I could argue that my order is reasonable. Then the proposition which enters into the argument is not the order itself but rather is the claim that the order is reasonable. That claim is the sort of thing which can be true or false. Similarly, I cannot give support for the truth of a question, since questions are not true or false, but I can argue that my question is legitimate or reasonable, where the proposition (that is, the truth or falsehood) in question is the claim of legitimacy or reasonability.

Arguments

3.1 From the sample argument given in this Section, pick out two or three propositions from the passage which you believe can reasonably be treated as being argued for by Gilder.

3.2 From the sample argument given in this Section, pick out three or four propositions for which Gilder gives no argument.

3.3 Is it reasonable to treat the following Sections of discourse as arguments? Why, or why not?

a. Last night's rain came so unexpectedly that we had all our windows open and the books on the shelf by the door got ruined. The next time there's even one cloud in the sky when I go to bed, I'm going to close the windows enough so it won't rain in.

b. The accident occurred when the northbound car swerved to avoid hitting a cat that had darted out into the highway. It swerved so suddenly that the driver lost control for a moment and the car went careening back and forth across the road. Finally, the right wheels went off onto the shoulder and the car smashed into the bridge railing.

c. I don't think I'll go shopping tonight, because if I go, I'll spend the money I've been saving for my friend's birthday present.

d. We ought to market thick crust pizza in this town, because our market research indicates that there is a strong demand for that type of product among potential customers.

4

Basic argument structure: premiss and conclusion

You now realize that in an argument there must be at least one proposition which is treated as proposed support for some proposition. Thus there are two basic components of an argument, the part which is treated as proposed support, and the part which that support is designed to uphold. Usually, the part which contains the proposed evidence or support consists of more than one proposition. However, no matter how many such propositions there are in any given argument, *any proposition which is treated as proposed support for the truth of some proposition in the argument is called a premiss.*

We also need a handy label for the other part of an argument—the part for which support is thought of as being proposed. Actually, I have already been using the word *"conclusion"* to refer to this part of an argument, and I will continue using that label for this part. To sum up, *when one proposition is treated as being offered in support of a second proposition, the second proposition is called a conclusion.*

To analyze an argument, first separate its premisses from its conclusions. Here's a simple case: suppose your roommate has complained that you were unfriendly today, and you reply angrily, "You have no room to talk; you were as grouchy as they come when you woke up yesterday". It's easy to treat this reply as an argument. Its conclusion would be that the roommate has no room to talk, and its premiss would be the rest of the reply. It will often be convenient to indicate the distinction between premiss and conclusion by writing arguments in the following fashion:

You were as grouchy as they come when you woke up yesterday.
You have no room to talk.

The line separates the premisses of the argument from its conclusion. All the premisses are written in a list above the horizontal line. When we write an argument this way, we are not indicating whether it is a good one or a bad one—we are only separating the premisses from the conclusions.

In more complex cases, such as the long passages by Gilder used as an illustration earlier, it is possible to have a great many premisses and quite a few conclusions. If you look back to Section 3 you will find Gilder arguing that the present social arrangement is cheaper than the arrangement which would have to be made if women were not discriminated against in employment. In order to support that conclusion, Gilder presents some premises:

P1. If a matriarchal social pattern emerges, the men in the society will "bolt"—they will become wild and predatory.

P2. If the men in the society bolt, a police state would be required in order to supervise the undisciplined men.

P3. If the women perform the role of provider for families, a child care state would be required in order to care for the children.

P4. If we have a police state and a child care state, taxes will be increased drastically.

The present sexual constitution is cheaper than the alternative which would emerge if women were not discriminated against in employment.

Examine the above argument carefully. (The premises are labelled with "P's.") Look how the premises seem to relate to one another, so that by putting them together, one can see how certain conclusions might follow. Ask yourself what it would be reasonable to believe if you truly accepted P1 through P4 as accurate.

Although the argument set out above represents part of Gilder's position, there are still quite a few pieces missing from it. Nevertheless, in the argument presented above, several premises work together to provide some support for the conclusion. Each of these premises is a premiss and not a conclusion, because none of these premisses is treated as providing support for any of the other premisses. For instance, I don't think it would be reasonable to suppose that P1 is to be treated as supporting P2. The premises are *related to* each other, but none is reasonably thought to provide *support for* the others. People often think that premisses which are all about the same subject, and which "hook together" logically in some vague way somehow ought to be treated as providing support for each other. But that is a mistake. (Of course, if you want to be unreasonable and stubborn about it, you can insist on treating *any* statement as proposed support for any other statement.)

We are now prepared to note one new thing about the above argument. Some of its premisses are argued for elsewhere in Gilder's article. P1 is controversial. And Gilder realizes that. So he provides support for the truth of P1 earlier in his essay. For example, Gilder claims that man's body is full of undefined energies, and that these energies need to be channeled by society if a man is to have a social role. For these and other reasons Gilder emphasizes the place of marriage in taming a man and making him a useful citizen.

Although a proposition like P1 serves as a *premiss* in the argument set out above, the very same proposition can be treated as the *conclusion* of some *other* argument at the same time. This sort of thing happens frequently. It leads to structures like the following:

P. Premiss

P. Premiss

C. Conclusion 1

P. Conclusion 1 (now thought of as a premiss)
P. Premiss
P. Premiss
_____
C. Conclusion 2

In the above structure, two initial premisses are treated as proposed support for Conclusion 1. This conclusion is then used as a premiss in the second argument to help support Conclusion 2. These two arguments form a chain directed toward establishing Conclusion 2.

If you look back to the definitions of "premiss" and "conclusion" you should be able to see that there is nothing wrong with supposing that one and the same proposition can play both the role of premiss and the role of conclusion. One and the same proposition can be treated as proposed support for some conclusion at the same time that it is treated as being proposed for support by other propositions. I like to call such propositions—the ones that play both roles in a chain of arguments—"intermediate conclusions". In Gilder's arguments, the proposition which was labeled "P1" is an intermediate conclusion. It is treated as proposed support for the conclusion that the present sexual constitution is cheaper than the alternative, and thus it is a premiss. Yet it is also treated as being supported by other premisses which appear earlier in Gilder's essay and thus it is also a conclusion. Despite the fact that it plays both roles, it is actually stated only once in the essay.

How do you tell what to treat as a premiss and what to treat as a conclusion? You ask the following question of each and every proposition under consideration: Is this proposition reasonably treated as proposed support for some claim? If so, then this proposition can reasonably be treated as a premiss. In addition, then ask the following question with respect to the very same proposition: Is this proposition reasonably treated as proposed for support by something else? If so, it should be a conclusion. Sometimes the answer to both questions will be "yes". If so, the proposition is both a premiss and a conclusion, and you are probably dealing with a chain of arguments.

Basic argument structure: premiss and conclusion

4.1 What is a premiss? What is a conclusion?

4.2 Can one proposition be both a premiss and a conclusion?

The passages in questions 4.3 through 4.5 have been adapted from various published writings. Each part of the passages is given a number for easy reference. Read the passages and answer the questions about them which follow.

4.3 *(1) The tie-up between "soft" marijuana and "hard" dope like heroin and cocaine is one major reason why the campaign to accept marijuana should be rejected, firmly and unequivocally. (2) A survey by a University of Wisconsin researcher among Madison, Wisconsin, high school students shows why. They found evidence of what law enforcement men have been saying,*

*that use of marijuana leads young men and women into trying the highly
dangerous drugs. (3) The survey found "an alarmingly high proportion of
those using pot are taking the more dangerous drugs such as LSD, speed,
hashish, and even opium."*[1]

a. This passage can reasonably be construed, *in its entirety,* as an argument for one main conclusion. If we interpret the passage that way, what
   does sentence (1) do?

   1. It states the conclusion of the whole argument: the campaign to accept
      marijuana should be rejected. It also states the reason which supports
      this conclusion.
   2. It states the conclusion of the whole argument, namely, that there is a
      tie-up between marijuana and hard dope, and it exhorts us not to allow
      marijuana to become legally acceptable.
   3. It serves as support for sentence (3).

   Explain why the answer you chose is better than the ones you rejected.
   (You might do this by explaining what goes wrong if you accept either
   one of the other answers.)

b. Still construing this passage in the same way, decide which of the following best describes the role of sentence (3).

   1. It states the conclusion of the whole argument.
   2. It supports the contention that the campaign to accept marijuana
      should be rejected, but it does not support the contention that the use
      of marijuana leads young men and women into trying highly dangerous
      drugs.
   3. It supports the sentences numbered (2), and these sentences in turn
      support the claim that there is a tie-up between marijuana and hard
      dope, which in turn supports the claim that the campaign to accept
      marijuana should be rejected.

   Explain why the answer you chose is better than the ones you rejected.

4.4 *(1) . . . America's anti-poverty program has been a sick farce in both North
and South. (2) In the South, it is clearly racism which prevents the poor
from running their own programs; in the North, it more often seems to be
politicking and bureaucracy. (3) But the results are not so different: (4) In
the North, non-whites make up 42 percent of all families in metropolitan
"poverty areas" and only 6 percent of families in areas classified as not
poor. . . . . (5) Behind it all is a federal government which cares far more
about winning the war on the Vietnamese than the war on poverty; (6) which
has put the war on poverty in the hands of self-serving politicians and bu-
reaucrats rather than the poor themselves; (7) which is unwilling to curb the
misuse of white power but quick to condemn black power.*[2]

a. Does the passage contain any support for sentence (2)? If so, what sentences do this supporting?

---

[1]From the editorial page of the *Herald-Telephone,* September 19, 1969. Bloomington , Ind.
[2]Stokley Carmichael, "Power and Racism," *Rhetoric of the Civil Rights Movement.* Haig and Ham-
ida Bosmagian, compilers (New York: Random House, 1969).

  b. Which of the following best describes the relationship between (1) and
     (5), (6), and (7)?
     1. (1) is offered as evidence for the truth of (5), (6), and (7).
     2. (1), (5), (6), and (7) all express criticisms of the federal government,
        but none of these sentences provide any support for any of the others.
     3. (5), (6), and (7) are offered as both explanation of why (1) came to be
        true and as support for the truth of (1), because Carmichael assumes
        that most people who believe (5), (6), and (7) are true will see that (1)
        could hardly help but come true under those circumstances.
     Explain and defend your choice of answer.
  c. What exactly do you think Carmichael is implying in sentence (5) when
     he refers to the war as "the war on the Vietnamese" rather than as "the
     Vietnam war"? (This is an example of a skillful use of a turn of phrase to
     make a point not directly related logically to the main point of the pas-
     sage. However, the occurrence of the phrase complicates the logical
     analysis of the passage, since Carmichael's particular view of the war
     could be false—in which case there would, strictly speaking, be no war
     on the Vietnamese—and yet it could be true that the federal government
     cares most about winning the war in Vietnam. A reasonable interpreter
     who disagreed with Carmichael's implied interpretation of the war would,
     in that case, mentally substitute the phrase "Vietnam war" for the phrase
     "war on the Vietnamese" in (5) and treat the logic of the passage accord-
     ingly, so as not to destroy the force of the passage as it relates to the war
     on poverty.)

4.5 *(1) The unabashed defender of traditional norms, and the unregenerate
    champion of prescriptive institutions—though they may have gained some
    ground in recent years—remain members of a Remnant. (2) To be conser-
    vative is to be a conservator—a guardian of old truths and old rights. (3)
    This rarely has been a popular office—not with the leaders of the crowd.*[3]

  a. Sentence (1) contains a Biblical term, "remnant", used to refer to the
     faithful few who remain sound in their convictions through trials and trib-
     ulations, thereby retaining the correct relation to God available to those
     with such convictions. Kirk takes this term and uses its connotations here,
     but he is not talking specifically about religious people. In this passage,
     do you think sentence (2) should be evaluated as:
     1. an accurate description of all those people commonly thought to be
        politically conservative,
     2. a definition of the sort of conservatism Kirk intends to talk about, per-
        haps capturing, in his view, the essence of true conservatism, or
     3. an explanation of how the words "conservative" and "conservator" are
        commonly used today?
     Defend your answer.

4.6 Rewrite each of the following arguments, labeling premisses with "P's" and
    drawing a horizontal line to separate the premisses from the conclusions. If
    one of the following arguments is a chain, write out each argument that
    makes up the chain, as was indicated earlier.

[3]Russell Kirk, *Enemies of the Permanent Things* (New Rochelle: New York: Arlington House,
1969).

a. The chancellor is not to be trusted. After all, he smokes a pipe, and pipe smoking (with tobacco) is an establishment habit.

b. The chancellor smokes a pipe, and therefore is not to be trusted since pipe smoking is an establishment habit. Another reason he is not to be trusted is that he is willing to compromise in a disagreement.

c. Pipes and pipe tobacco are heavily advertised in snobbish ads which run in magazines whose primary audience is upper middle class and upper class whites. Moreover, 85 percent of the American pipe smokers are in executive or professional occupations. It is therefore clear that pipe smoking is an establishment habit. The chancellor smokes a pipe, so he has an establishment habit. Thus, he is not to be trusted.

d. Pipes and pipe tobacco are heavily advertised in snobbish ads which run in magazines whose primary audience is upper middle class and upper class whites. Moreover, 85 percent of the American pipe smokers are in executive or professional occupations. It is therefore clear that pipe smoking is an establishment habit. The chancellor smokes a pipe, so he has an establishment habit. Anyone who has an establishment habit is not to be trusted. Furthermore, the chancellor is willing to compromise in a disagreement, and willingness to compromise is a sign of moral weakness. Thus, the chancellor is not to be trusted.

e. Proponents of reform in U.S. presidential election procedure argue that current procedures allow the election of a president getting less than a majority vote. They fail to mention that such an election can only happen if the candidate has broad support across the country. For the structure of the electoral college is such that a winning candidate must have actually carried many states. Under the reform plan as proposed, a minority president still could be elected. However, in this case there is no guarantee that the winner carry any states at all. But in order to have enough political power to actually govern the country after the election, the winner needs broad, widely based support. Thus, the old system provides greater assurance that the winner would be able to govern.

5

Some key words that give structure to arguments

By now it should be apparent that logical analysis of any argument depends heavily on understanding how the argument is put together. But as we have noted, there is often no single right way to look at the structure of an argument. Nevertheless no matter how difficult or arbitrary the decision regarding what to treat as a premiss and what to treat as a conclusion, that decision must be made before the logical analysis can be completed.

Fortunately, there are certain key words in English which can help us make these decisions reasonably. These key words relate the parts of arguments to one another, as in the following illustration drawn from a recent accounting text:[1]

> *If a processing center has work in process at the beginning or at the end of a period, . . . completed units alone will not accurately measure total output. THE REASON IS THAT part of the costs of the period will relate to the partially completed units in the ending inventory. These partially completed units will HAVE TO BE considered along with the fully completed units in measuring the period's output. HOWEVER, the partially completed units will be measured on an equivalent units basis. FOR EXAMPLE, 100 units 60 percent complete would be equivalent to 60 fully completed units. THEREFORE, the ending inventory in this case would be said to contain 60 equivalent units of production. These equivalent units of production would HAVE TO BE added to the fully completed units as a step toward computing the total output of the period.*

I have indicated key words and phrases by capitalizing them. Some of the words or phrases in capital letters indicate that a *conclusion* is being stated. These obviously include "therefore", but phrases such as "have to be", meaning "must be", often indicate a conclusion as well, suggesting that the reasons already stated make it necessary that the conclusion "has to be" true. Perhaps that is the indication in the above passage. On the other hand, the phrase, "the reason is that", naturally indicates a *premiss*. The phrase, "for example", often means that the material immediately following is *only* an example and is not really either an additional premiss or an additional conclusion. "However" often indicates an additional premiss, or an additional conclusion. (Thus, "however" is not very helpful in sorting out premisses from conclusions. Its usefulness comes from its ability to indicate that the material it sets off from the rest of the passage is somehow different from the immediately preceding material.) Using these clues, plus a little common sense, I arrive at the following arrangements of premisses and conclusions for the preceding above passage:

P1. Part of the costs of the period (that is, any period in which there is work in process at the end) will relate to the partially completed units in the ending inventory.

C1. These partially completed units have to be considered along with the fully completed units in measuring the period's output.

The idea behind this argument is that *because* the costs are related to ending inventory, the accountant needs to take note of the ending inventory. C1 is not used immediately. Instead, a new argument is given, as follows:

P2. Partially completed units will be measured on an equivalent units basis. For example, 100 units 60 percent complete would be equivalent to 60 fully completed units.

C2. The ending inventory in this case (that is, in the example) would be said to contain 60 equivalent units of production. (Other cases are handled in a similar way.)

[1]Garrison, Ray H., *Managerial Accounting* (Dallas: Business Publications, 1976), p. 72.

We can now put these two arguments together. The first one establishes that partially completed units have to be considered. The second one tells in what way they are to be considered. The purpose of the second argument is, so far as I can tell, merely to explain how to figure out what an equivalent unit of production is. I don't believe C2 is actually used as a premiss when we go on. Instead, now that C2 has shown what equivalent units of production really are like, we can use P2 over again with greater understanding of what it means. The next argument would look like this:

P3.   (Same as C1.)

P4.   (Repeat of P2 above.)

C3.   These equivalent units of production would have to be added to the fully completed units as a step toward computing the total output of the period.

This last argument makes sense because P4 tells us how to count partially completed units, while P3 tells us that partially completed units plus completed units have to be considered. If we want, we can now fit the very first sentence of the passage into the chain of argument, as follows:

P5.   (Same as C3.)

C4.   If a processing center has work in process at the beginning or at the end of a period, . . . completed units alone will not accurately measure total output.

While there are other ways of looking at the passage, I do think the above representation captures quite well the logical arrangement of the sentences in the passage, and it takes account of the key words which were put in capital letters earlier.

The preceding work may serve to illustrate how key words or phrases can help to separate premises from conclusions in a reasonable fashion. The useful key words or phrases used above were "the reason is that", "have to be", and "therefore". It is not hard to think of other phrases or words which serve much the same roles as these. Words such as "thus", "hence", "consequently", and "accordingly" often indicate that a conclusion is about to be stated. Words such as "since", "because", and "for" often indicate that a premiss is about to be stated. Other words which help to signal relationships of various kinds between whole sentences in a passage are words like "moreover", "nevertheless", "in addition", and "for instance", but these phrases do not seem to be connected with the distinction between premisses and conclusions. In fact, "moreover" and "nevertheless" often indicate that another whole argument is about to begin. Such an argument may be a second argument for the same conclusion the author has already argued for, or it may be an argument for one of his premisses.

Some key words that give structure to arguments

5.1 List the key words in the following passages which provide the clues to which sentences in the passages ought to be treated as playing the role of conclusion. Then rewrite each of the passages, separating them into their various component arguments, labeling each premiss and each conclusion with a "P" or a "C" and drawing a horizontal line between each premiss set and the conclusion that goes with that set as was done above for the argument from the accounting text. Do not expect that the sentences will necessarily appear in the right order within these passages as they are written below. Rather, go by the key words and common sense as to what is the point of each passage and each subpart of each passage.

a. *I may mention that the external ears of the common mouse are supplied in an extraordinary manner with nerves, so that they no doubt serve as tactile organs; hence the length can hardly be quite unimportant.*[1]

b. *The idea that the father is inherently equal to the mother within the family, or that he will necessarily be inclined to remain with it, is nonsense. The man must be made equal by culture . . . For a man's body is full only of undefined energies. And all these energies need the guidance of culture. He is therefore deeply dependent on the structure of the society to define his role in it.*[2]

c. Nothing is so important to the sexual constitution as the creation and maintenance of families. And since the role of the male as principal provider is a crucial prop for the family, the society must support it one way or the other.

d. *Since . . . whatever is equal or superior to the mind that possesses virtue and is in control does not make the mind a slave to lust; and since, because of its weakness, whatever is inferior to the mind cannot do this . . . therefore it follows that nothing can make the mind a companion of desire except its own will and free choice.*[3]

e. *There is much to be said for authorizing Concorde service into the New York City area, at least on an experimental basis. For one thing, banning it might be a futile attempt to block the inevitable. Supersonic travel, after all, is probably here to stay, if only because greater speed has always been the primary goal of transportation development . . . For another thing, a ban on the Concorde would betray the American tradition of welcoming rugged but fair competition in the marketplace. The staggering development and operating costs of the Concorde may make the*

---

[1]Charles Darwin, *The Origin of Species*, abridged and ed. by C. and W. Irving (New York: Frederick Ungar Publishing, 1956), p. 55.

[2]b. and c. are drawn from the selection by George Gilder in Section 2.

[3]St. Augustine, *On Free Choice of the Will*, tr. by A. S. Benjamin and L. H. Hackstaff, (Indianapolis: Library of Liberal Arts, 1964), p. 22.

plane one of history's landmark commercial disasters, but if Paris and London are willing to keep subsidizing it they are entitled to a chance to serve the U.S.'s major travel market . . . . Thus, before Concorde service at J.F.K. is ruled detrimental to the commonwealth, the big bird deserves at least a chance to demonstrate—in a carefully monitored test—that it is not quite the monster its critics contended.[4]

f. *Man's natural instinct moves him to live in civil society, for he cannot, if dwelling apart, provide himself with the necessary requirements of life, nor procure the means of developing his mental and moral faculties. . . . Civil authority, therefore, comes from nature, and the end [that is, the goal] of such authority is the preservation of order in the State. Consequently, if for the preservation of order capital punishment is necessary, the right to inflict capital punishment must be comprised in the civil authority. For nature never gives the end without the means. Capital punishment then is not legalized murder.*[5]

6

Sentences and propositions

According to our definition, arguments consist of propositions rather than sentences. I defined "proposition" as a truth or falsehood. What then, is the exact difference between a truth or a falsehood (a proposition) on the one hand and a sentence on the other? And what is the motive for saying arguments consist of propositions rather than sentences?

There are many different kinds of sentences. Questions, commands, and statements are all expressible by sentences. A moment's thought on the variety of sentences that exist should show you that not all sentences are of the sort that can be true or false—for example, a sentence that expresses a question is not the sort that can be true or false. Thus there is one quick and easy answer to the question why I do not say that arguments are made of sentences: arguments are made of things that are true or false, while not all sentences are that sort of thing.

The quick and easy answer above is not very revealing. It would not be hard to make up a special label for just those sentences which are either true or false. In fact, some logicians refer to such sentences as statement-making sentences. We might then say that arguments are made of statement-making sentences. But I don't want to say that. Why not? The answers to this question will be interesting and revealing.

---

[4]Burton Yale Pines, "Putting Up with the Ugly Duckling," *Time*, March 21, 1977, p. 41.
[5]John J. Ford, *Catholic Mind*, March 18, 1915. Reprinted in *Ethics and Metaethics*, ed. by R. Abelson (New York: St. Martin's Press, 1963), pp. 20–1.

First, we need to think more carefully about what a sentence is. Sentences may sometimes be conceived as strings of written words separated by spaces, or as strings of sound waves traveling through the air from speaker to hearer. If we think of sentences along those lines, each sentence has just one fairly definite location at a particular time. The sentences you are now reading, for example, are located at this time on the page in front of your eyes. Continuing with this conception of a sentence, we would have to say that the sentence

> Joe ran to the store.

is located just before the words you are reading right at this instant, and is a *different* sentence from the following sentence, which is located in a different place:

> Joe ran to the store.

The conception of a sentence which I have been describing makes sentences out to be physical phenomena, much like baseball bats, automobiles, and winter storms. On this conception, a sentence is destroyed when erased or when the sound waves which constitute it die out; sentences can sometimes be moved around like other objects, and the sentences in your book are not the same sentences as those found in other copies of the book.

However, there are also other common conceptions of what a sentence is, conceptions which are somewhat more abstract. In this respect sentences resemble automobiles. The idea of a sentence with which we have just been working made each sentence have its own location at any given time, just as we often think of each automobile as having its own location at any given time. However, we don't always think of automobiles in that way. Sometimes when we say things like

> Harry bought the same car Linda has after he heard good reports from her about the performance she was getting.

we do *not* mean Harry bought the same machine that Linda did, but rather that Harry bought the same *kind* of automobile Linda has. When we talk this way about automobiles, what counts as the same automobile is any automobile of the same *type* (perhaps the same make and model). And as you might have expected, we can talk in this way about sentences as well. If we adopt this more abstract way of talking about sentences, any two strings of words containing exactly the same words in the same order will be counted as just *one* sentence rather than two. On this conception of a sentence, a given sentence may appear in several places at once, and the sentences you are reading right now are thought of as the very same sentences as those appearing in other copies of this book.

Nevertheless, no matter which of the above conceptions of a sentence we use, one feature of a sentence remains: the notion of a sentence is so intimately connected to the notion of wording that it is not possible to change the wording of a sentence without thereby creating a *new* sentence. That is to say, one cannot have *one* sentence with alternative wordings. *A sentence is constituted by words, in a particular order.* When I say that I prefer to describe logic as dealing with propositions rather than sentences, it is because I believe that the notion of a truth or falsehood (the notion of a proposition) is *not* tied quite so tightly to wording, and that this is preferable for logic for two reasons.

The first reason I believe it is better to think of arguments as being composed of propositions rather than sentences is that people often construct arguments "in their heads", without writing or speaking, and we often want logic to deal with beliefs we attribute to people even though these beliefs have never been articulated by those people, even "to themselves". If we were to claim that logic deals only with sentences, I worry that unspoken arguments and unarticulated beliefs (perhaps held unconsciously) will be left out of the proper domain of logical investigation, for the ordinary conceptions of the nature of sentences would not allow sentences to exist where the wording had not been worked out. If there is no good way to say that unspoken arguments and unarticulated beliefs are composed of sentences with specific wordings, we may be better off saying that logic deals with propositions (truths and falsehoods) rather than sentences.

The second (and more important) reason why I think it better to talk about arguments as being composed of propositions is that the goodness or badness of an argument does not depend on the words which are used to express the argument. The quality of my argument does not depend on whether I speak broken or fluent English, whether I use good grammar, whether I intersperse German words into my sentences because I don't know the English word for what I mean. My premisses support my conclusion just as well or just as badly no matter what words I choose to express them. If I choose my mode of expression poorly, others may not be able to understand my argument simply because they do not know what I am saying, or because they will not pay attention to someone who speaks poorly. But my argument does not thereby lose any *logical* force whatsoever; it is still either good or bad. However, if we say that the argument consists of sentences rather than of propositions, we would then have to say that when the wording of the sentences changes in any way so does the argument. We would also have to say that a person who speaks broken English and therefore expresses himself poorly in English does not present the same arguments as someone who has exactly the same points to make but who does it in elegant English. We would have to say that the German-speaking person never gives the same arguments as an English-speaking person, even though the German sentences of the one person could be translated into the English sentences of the other. All these consequences of treating arguments as made of sentences rather than propositions seem to me to be undesirable. It seems much better to think of arguments as made up of propositions.

If my preference for propositions over sentences were to be made completely secure, I would have to explain more exactly what a proposition is. However, I'm not going to do that, for it would take us too far afield. Almost all logicians agree that arguments are best thought of as consisting of something like truths and falsehoods rather than what I have been calling sentences. But there is widespread disagreement about the nature of truths and falsehoods.

Sentences and propositions

6.1 On the first conception of the nature of a sentence, is the sentence "Joe ran home from the store" a different sentence from the sentence "JOE RAN HOME FROM THE STORE"? Explain.

6.2 Do the above sentences count as one or as two sentences according to the second, more abstract conception of sentences? Explain.

6.3 Do people speaking Spanish speak the same sentences as people speaking English? Explain. What about people who speak with a Bostonian accent? Do they speak the same sentences as those with a Western drawl?

6.4 Is the sentence "Bill hit Jane" the same sentence as "Jane was hit by Bill" on either conception of what a sentence is?

6.5 Construct a new sentence which probably counts as expressing the same proposition as the following sentence expresses: The truck was crushed by the train at 10 P.M.

6.6 Construct a simple argument using the sample sentence from 6.5 to express a premiss. Now rewrite the argument, substituting your answer to 6.5 for the sample sentence. Is the rewritten argument a different argument or the same argument according to the definition of "argument" we are using? Explain.

2

Distinguishing
Good Arguments from
Fallacious Ones

7

Fallacies

Up to this point we have not had much to do with *evaluation* of the arguments we have found. Now it is time to get to the main business of this study in logic: evaluation of arguments.

We will begin by looking at a few common types of errors people commit when arguing—failures to adequately support conclusions. On the surface many types of argumentation look as if they are logical when in fact they do not provide adequate support for their conclusions. Such types of argumentation are referred to as "fallacies". By working through the following sections on various fallacies, your ability to detect logical errors should be increased, and that meets our purposes. I should perhaps mention that there is no completely standard way of classifying fallacies, and I will feel free to adopt a scheme partially my own.

Although a fallacious argument appears to be logical when in fact it does not provide adequate support for its conclusion, I do not mean to label an argument fallacious merely when it has a false premiss. You may very well be inclined to say that an argument which has a false premiss does not provide adequate support for its conclusion, and thus that the falsity of the premiss guarantees the fallaciousness of the argument. But usually logic does not examine propositions for actual truth or falsehood, since normally there is no way that logic by itself can reveal whether something is true. Usually, it takes some kind of experience or background knowledge on your part, in addition to logic, in order to know when some proposition is true. Accordingly, I will try to describe each of the fallacies in such a way that the actual truth or falsehood of the premisses is irrelevant—the fallaciousness will come from the way premisses are related to conclusions, not from the falsehood of the premisses.

What then, do you say about an argument when you do happen to know that some of its premisses are false? Well, I think that what you say is just that some of its premisses are false. We don't need any special word to label such arguments. Naturally, if an argument does have a false premiss, then its conclusion might turn out false as well, no matter how well the premisses work together to lead up logically to the conclusion. For example, I might argue as follows:

> The moon is made of nice fresh lettuce.
> Rabbits love to eat fresh lettuce.
> _____
> Rabbits love to eat what the moon is made of.

This argument is well arranged internally, so that its premises lead logically to its conclusion. Yet it has a false premiss. The result in this case is a false conclusion. No special fallacy is committed by this argument.

Before we begin to list some fallacies, I want to block one possible misunderstanding. Since I refer to the fallacies as "errors", it would be easy for you to suppose I mean that it is always irrational to use an argument that exhibits one of these fallacies. But there very well may be occasions on which the use of a fallacious argument is quite rational, perhaps to slow down a discussion or to create confusion. It will still be fallacious—that is, the argument will not provide adequate support for its conclusion—despite the rationality of employing it. Hence, I am not arguing that you should never use fallacious arguments.

I believe the real benefit derivable from discussion of the fallacies is increased awareness of the sometimes subtle differences between arguing logically and arguing fallaciously. Accordingly, the following Sections will discuss not only fallacious arguments, but also the nonfallacious arguments which the fallacious ones imitate.

8

Attacking someone else's point of view (the ad hominem fallacy)

Some people say that love makes the world go 'round; I suppose that's true sometimes. Still, our poor world has seen enough disagreements, disputes, and outright battles which have shaped our outlooks, desires, and our environment. I'm sure you will say disagreement between people is one of the hard facts of life. And often when there is disagreement, arguments are given to show the other person's point of view is incorrect. Attacks on another person's point of view can be made in both logically proper and logically improper ways, and the goal of this Section is to clarify the difference. When a chemist reports new experiments which show that a currently accepted theory has something wrong with it because the new data is incompatible with accepted theory, then we have a case of a legitimate logically correct attack on others' views. Or, to take a more common example, when a college student points out that his friend's belief about the meeting time for class is probably mistaken since the belief conflicts with the printed schedule put out by the school, we again have a logically correct approach.

I am not saying that the chemist in my example is right, or that the printed class schedule is infallible. I am only claiming that we can recognize the procedures described above as rational logical ways of arguing against someone else's beliefs. The evidence presented in each case is relevant to the conclusion and it supports that conclusion adequately.

In contrast to the procedures described above there are other commonly used strategies for arguing against someone else's beliefs—strategies which are normally *fallacious* even though they may be psychologically effective. In outline form two of these fallacious strategies look like this:

(I)   Person P is bad in some respect.
      P believes that X is true.
      _____
      X is false, or silly, or worth no consideration.

(II) P says or believes that X is true.
      Person P is in some special circumstances that make it to P's advantage to say or believe X.
      _____
      X is false, or silly, or worth no consideration.

Here are some examples of the first of these two fallacious strategies:

(A)   Jones is always going around complaining that the welfare checks he gets are not enough for a person to live on. But we all know what a lazy dirty person he is, and that he can't manage to keep his apartment in order.
      _____
      So you can just forget what he says.

(B)   You never do the dishes when it's your turn, and now you have the nerve to complain that I didn't mow the lawn last Saturday.
      _____
      You have no basis for complaint.

(C)   You claim that we all ought to love one another, and that all people are brothers and sisters; but I saw your name on that neighborhood petition which is designed to keep black families from moving into the area. So it's clear that all your talk about love and brotherhood is wrong.

Each of these examples fits pattern (I) above because in each case the views held by someone else are rejected merely because the person holding these views has some supposedly negative characteristic. It is not necessary that the person under attack be said to be morally bad. It is enough, for instance, to merely point out that he or she is inconsistent, as in example (C) above.

When are these arguments of type (I) fallacious? Whenever the respect in which person P is said to be bad has little or nothing to do with the truth or falsehood of X—whenever the premises are irrelevant to the conclusion. Looking at the examples again, you can easily see that Jones' laziness and sloppiness has nothing to do with whether the welfare checks are sufficient. Similarly in (B) it is not hard to see that your failure to take your turn doing the dishes has nothing to do with whether you have grounds for complaint about the lawn mowing. And finally, in (C) we see that someone's inconsistency really ought not to show anything about whether the views they express are correct. (It might show that they are being a hypocrite, but that is another matter.)

Here is a fallacious example which fits the *second* strategy we outlined, strategy (II):

> (D) Jones, who is the president of General Motors, testified today before the Congressional committee on air quality. He argued that the automobile industry needs more time to develop nonpolluting engines, and the legal standards for the amount of exhaust pollutants allowed for next year's models should be relaxed. But of course it will cost G.M. millions of dollars to develop nonpolluting cars and to meet next year's standards.
>
> Therefore, we can dismiss Jones' testimony from consideration.

In this example, the idea is to reject out of hand what the other person has said merely on the grounds that the other person has something to gain by saying it. Such a procedure is fallacious because the benefit accruing to the other person is *irrelevant* to the correctness of his or her beliefs and statements.

Both these fallacious strategies have a Latin name ("ad hominem") and I believe you should try to remember it, since the name actually appears in essay writing, newspaper editorials, and other materials.

One particular type of fallacious argument strategy fitting loosely under strategy (I) deserves special mention. This is the "guilt by association" strategy. Here, one discredits a person by claiming that the person is in some way associated with other persons or points of view that are evil, silly, or otherwise bad. In outline form, the strategy looks like this:

> (Ia) Person P is associated with bad people, or holds views that are bad, or holds views that are associated with bad views. P also believes or says X.
>
> X is false, or silly, or worth no consideration.

Here is an example of this strategy at work:

> (E) Harry, who manages that restaurant on the corner where all the prostitutes hang out, says that our city's laws on gambling are unfair.
>
> But you don't have to pay any attention to the likes of Harry.

Here is a somewhat different kind of example:

(F)   Sally claims that the bars in town shouldn't be allowed to stay open so
       late, because it causes too much disturbance. But then that sort of
       reasoning goes right along with all sorts of other restrictive thinking,
       like claiming that stores should be closed on Sunday, or that books
       should be censored, or that X-rated movies should be banned.

       We'll not have to think one minute longer about Sally's argument.

I take it that in this last example, the intent of the speaker is to make Sally's
views about the bars look silly to a particular audience by associating Sally's
ideas with other ideas that the audience would find silly or distasteful. This example differs from the preceding one because in the latter example it is Sally's
ideas rather than Sally herself that are associated with something "bad", while
in the former example it was Harry himself, rather than his ideas, that was associated with something "bad". Both examples are equally fallacious, since
some sort of loose association between persons and ideas or other persons that
are "bad" is an inadequate basis for concluding that any views are incorrect or
unworthy of consideration.

I mentioned at the beginning of this Section that there are logically proper
ways of attacking other people's beliefs or statements. Notice that the premises
in the good arguments were *relevant* and showed the belief or statement in
question to be false or doubtful. The premises in the *fallacious* arguments were
not relevant to showing that sort of thing. The problem about ad hominem arguments is not in just the premises themselves—the problem is in the *relationship* (or lack of relationship) between premises and conclusions. We could
keep the premises the same in some of those fallacious arguments, but change
the conclusions and the result might be arguments that are not fallacious. Such
arguments occur frequently in court when the testimony of a witness is being
challenged. We get arguments like the following one:

(G)   Myrtle, who has testified that my client was at the scene of the fight, is
       a drunk and a chronic liar who enjoys the publicity she gets from tes-
       tifying.

       The jury should keep these special facts about Myrtle in mind when
       evaluating her testimony and comparing it to conflicting testimony
       given by other witnesses.

This example illustrates a general point. Certain kinds of references to a person's character, or special circumstances, may very well be relevant to drawing
conclusions about the degree of caution or suspicion with which one should
treat the statements made by that person. The key to deciding whether such an
argument is fallacious or not is the relevance test: ask yourself whether the
premises are truly relevant to the particular conclusion drawn from them.

**Attacking someone else's point of view
(the ad hominem fallacy)**

8.1 Make up your own example of an argument that fits the ad hominem pattern
(I). Is your argument fallacious? Why or why not?

8.2 Do the same thing for the ad hominem pattern (II).

8.3 Suppose that you are working for a company that has salespeople who need
to travel as a part of their jobs. You and one of the other people in the
company are talking to the boss about supplying cars to the sales force,
and you believe your company should lease the cars rather than buying
them. The other person believes the company should buy the cars. Give a
fallacious ad hominem argument for the conclusion that the other person is
mistaken. Give an argument that is not fallacious for the same conclusion,
making up the "facts" that you need for your premisses.

8.4 Using the following premisses, construct a fallacious ad hominem argument
by adding the appropriate conclusion to them. Can you come up with a
conclusion that could be drawn from these premisses without resulting in a
fallacious argument?

Jones works for Hanson Insurance Company as a salesperson. He believes
that it is a good thing to have insurance on your car.

8.5 What conclusion could be legitimately drawn from the premisses in example
(C) given earlier in this Section?

8.6 Suppose you say a number of vicious and unwarranted lies about the person who lives next door. Have you automatically committed the ad hominem
fallacy?

9

Asking the expert

It is one of the more important facts of life that each of us is very much dependent on other people for knowledge. We must rely on other people because we
don't have the time or the expertise to investigate everything ourselves. Trivial
examples of this dependency are easy to multiply: we look up the meaning of a
word in a dictionary written by other people; we read textbooks written by other
people. We find out what went on at the party last Saturday night by talking to
someone who was there. And on and on. If we didn't ever rely on the word of

others, we would have to learn everything from scratch on our own, and we wouldn't know much of anything.

It shouldn't come as much of a surprise to you to find I want to claim that there are both logically proper and logically improper ways to use testimony from other people in support of a conclusion. In outline form, the appeal to someone's testimony in an argument will look like this:

Person P says X.
X is true.

(Normally, something about person P will be added, to make P sound reliable.) This type of argument is more or less the opposite of the ad hominem pattern, where we *rejected* X because P said it. Just as with the ad hominem pattern, arguments fitting the preceding appeal to authority pattern will be fallacious when the premiss is irrelevant to the truth of the conclusion. Our problem then, is to describe when arguments fitting this pattern have premisses relevant to the truth of the conclusion.

The issue is simple: When is one entitled to trust the word of someone else as evidence for the truth about some matter? The answer seems equally simple: when one has good reason to think that the other person is reliable with respect to the matter. But how does one tell who is reliable? There's the rub. I can best clarify the difference between fallacious and nonfallacious appeals to authority by trying to characterize the situation when we have good reason to think someone reliable. That will be the focus for the remainder of this Section.

Let's get started by looking at some examples of arguments which fit the appeal to authority format:

The president says the economy will soon improve and unemployment will decline.
This improvement will take place as described.

Bubbles Bigsby, the famous sportscaster, says that Bugs Beer is great stuff.
Bugs Beer is great stuff.

As the old saying goes, "Like father, like son."
Joe won't turn out any better than his father did.

In the last argument the appeal is to a traditional saying, which is a kind of authority. So the authority to which appeal is made in an argument does not need to be a particular individual; the authority could be a whole group of individuals or it could be the traditions of the society in which the argument is offered. Now to decide whether you have good reason to think the authorities (or so-called authorities) cited in these arguments are reliable, you will have to appeal to whatever background information you can obtain. In other words, to decide on the fallaciousness or nonfallaciousness of each of these arguments requires you to figure out what you know about presidents, famous sportscasters, and old sayings. You are not looking for information which will automatically and definitely tell you whether the authority is right, but rather you are looking for evidence of reliability in the matters under discussion. The argument will be nonfallacious even when the authority is mistaken and the conclusion is in fact false, provided that you have good reason to think that the authority is reliable when you construct or analyze the argument. We are not looking for perfect authorities, but for people we can reasonably trust and people who are likely to be right. That will provide evidence that the conclusion of an argument is true. In this connection, the following points will be useful: (1) First, decide whether the authority is likely to be in a good position to give an informed or expert opinion about the subject. If we are dealing with an argument about biological processes, it will not be appropriate to cite as evidence the opinions of an expert accountant who knows little about biology. An argument which appeals to the authority of someone or some group of people who lack the evidence or training to give a reliable statement about the subject matter of the argument is fallacious. (2) Secondly (assuming that the authority passes the first test), look to see whether the authority has a reason to deliberately misrepresent or perhaps unconsciously deceive himself or herself about the matter in question. I believe my first sample appeal to authority argument—the one which uses the statement of the president about the economy—would normally be fallacious because of the factors just mentioned. Surely the president with all his or her economic advisors is in a position to be something of an expert about what will happen to the economy in the near future, even though the president may not be an economist. But we are all aware, I hope, that presidents do not always feel it possible to be completely open and honest about what they think will happen. In short, the political context in which the president operates makes the president's statement especially unreliable as evidence for the truth of the conclusion. Thus, even when the opinion cited in an argument is the opinion of a real expert who has carefully researched the subject, the appeal to authority will still be fallacious when the authority has reason to misrepresent or be self-deceived. (3) Finally, even if an appeal to authority is nonfallacious according to the preceding tests, the argument will still be fallacious unless you know the authority is in a position to know something on the subject which the arguer does not already know independently. Suppose two experts in advertising strategy are in disagreement about what style of ad will best sell a new automobile model. After some discussion one of the two says that he can prove he is right by citing the authority of an advertising executive who wrote a book a couple of years ago on advertising technique. I believe this appeal to authority would be fallacious if the two executives don't know the book author is better informed on the subject than they are. The situation would be different if two inexperienced and relatively ignorant people were disagreeing about the subject and then one of them came up with the testimony of an advertising executive. I don't think such an appeal to authority would be fallacious. (Of course even the expert can be wrong, but it is still reasonable to cite expert testimony when you know the expert is in a position to know more.)

It seems to me that if authorities can pass these three tests, it is reasonable to count on their testimony and appeals to their authority will not be fallacious. Otherwise, the appeals will be fallacious.

Asking the expert

9.1 Which of the following appeals to authority seem to you to be fallacious? Defend your position on each passage with a brief statement about why the appeal in that passage is or is not fallacious.

a. Jones argues that crime in his city will decrease next year because the mayor in his last news conference said it would.

b. Smith, who is the chief assistant to the mayor, argues that crime in his city will decrease next year because in a private meeting with his staff the mayor said that he had been studying crime reports, that he had talked to the local police, and that he had gathered as much evidence as possible before coming to the conclusion that crime will decrease in the city next year.

c. Jerry argues that Safe bath soap gives the most effective germ-fighting action compared to other leading brands, since Marla Kidding, the famous lady wrestler, said so on TV.

d. Lisa argues that Safe bath soap gives the most effective germ-fighting action compared to other leading brands, since Dr. Schenk, a famous bacteriologist, said so on TV in a commercial for Safe.

e. George argues that it is dangerous to the emotional development of children if homosexuals teach elementary grades, on the grounds that a famous female singer who also has raised several children says it is dangerous.

f. Hank argues that the difference in economic standing between the poorest 5 percent and the richest 5 percent of the population in our country is increasing significantly each year, since he read that is so in his sociology textbook.

9.2 Can an appeal to authority ever completely prove that one's position is correct, beyond the shadow of any doubt? If not, why are not all appeals to authority fallacious? If so, which appeals to authority can do the job?

9.3 Can any appeal to the authority of traditional views held in our society be nonfallacious? If so, how? If not, why not?

10

Justifying behavior by comparing it to others'

The last two Sections dealt with the assessment of others' opinions as being true or false. We turn now to certain arguments which attempt to justify one's own behavior or the behavior of groups one favors. As expected, there are both fallacious and legitimate ways of constructing such arguments. One of the most common fallacious arguments follows this general pattern:

> Person P_1 or group G_1 has behaved in a way that is similar to the way person P_2 or group G_2 has behaved.
> _____
> The behavior of P_2 (or G_2) is justifiable.

For example, doesn't the following argument sound familiar?

> There's nothing wrong with my forgetting to pick you up at the station. After all, you forgot to take my books back to the library last week.

In this argument, we see someone attempting to conclude that his own behavior is fine merely because it is similar to the behavior of someone else—presumably someone who has made a complaint against him. In fact, in its most common form the fallacious argument fitting the above pattern will be a response to criticism from person P_1 or group G_1, and person P_2 will be the speaker or group G_2 will be some group with which the speaker identifies. In that form, the author of the argument is defending himself or some group he likes from criticism by claiming the critics have done the same things themselves which they now criticize. Because that is the typical circumstance in which such arguments are put forward, this fallacy has become known as the "you too" fallacy (or more often by the Latin phrase "tu quoque", meaning "you too").

Once you see the structure of such arguments laid out bare as above, it is not hard to see that they are likely to be fallacious. After all, it is entirely possible that *both* P_1 and P_2 have been behaving badly; so that the similarity between their behaviors does nothing to prove P_2's behavior is justifiable. I cannot successfully justify police brutality in my own country by pointing out that police in other countries are just as brutal or more so. And perhaps most importantly, I cannot normally justify my own behavior by remarking that "everyone" does the same things I do.

Note that an argument of this sort remains fallacious even when the behavior in question *is* justifiable. For instance, we may agree that it is truly justifiable for parents to punish various genuine misdeeds of their child, but it is nevertheless fallacious to attempt to justify such punishing by arguing that other parents also punish their children, or that our society also punishes misdeeds of its members. The fallacy occurs in such arguments because the premises of these arguments do not support their conclusions, not because the behavior in question is unjustifiable.

As with most fallacies, this one mimics legitimate logically correct ways of arguing. For example, one can legitimately argue that P_2's behavior is justifiable by comparing it to P_1's behavior if one already knows that P_1's behavior is justifiable. In fact, this method of arguing is used extensively in our legal system, when a lawyer cites a precedent-setting case involving someone else whose behavior was found to be justifiable. This amounts to adding an extra premiss to the fallacious argument structure, which makes it legitimate:

> P_1's behavior was justifiable.
> P_1's behavior is similar to P_2's in every relevant respect.
> _____
> P_2's behavior is also justifiable.

You also could change an argument which commits the "you too" fallacy into a legitimate logically correct argument in another way. Instead of adding a premiss, as I did above, you could alter the conclusion in various ways. For instance, it seems to me to be quite reasonable to argue this way:

> You forgot to take my books back to the library last week even though you promised to do so. So you are not any better than I am even though I forgot to pick you up at the station.

This argument is not fully convincing. One would want to know how serious each of the mentioned incidents really was, and so on. Yet I think you can see that this type of conclusion is much more legitimate than the conclusion that forgetting to pick you up is justifiable.

The above illustrates only one type of conclusion which might be legitimately drawn from this sort of premiss. There are others. For instance, on some occasions it will be possible to legitimately conclude that P_1 appears to be hypocritical in condemning P_2 since P_1 consistently does the same things that he condemns P_2 for doing:

> Jones cheats on his income tax every year and falsifies the sales tax records of his business to avoid paying all the taxes he owes. So it is somewhat hypocritical of Jones to criticize those people who fail to report all their earnings in order to get higher welfare payments.

Note that the conclusion here does not fit the fallacy pattern.

On other occasions, what I am expected or allowed to do will truly depend on what others do. In such cases, arguments *fitting our pattern* may well be nonfallacious. For example,

> My whole community follows the conventions of the English language when speaking. Thus, it is justifiable for me to do so also.

Or again,

> It has become common and accepted practice for judges in this country to draw the line between inadmissible hearsay evidence and admissible reports of conversations according to the following guidelines: . . . . Therefore, in the absence of any legislative ruling to the contrary, it is legitimate for me as a judge to do the same.

This argument is rendered legitimate because the common practice of judges establishes a working principle by which all are expected to rule.

I do not have a complete catalog of conclusions which might be legitimate in this type of argument. But I *can* tell you that the basic test for legitimacy will always be the same: the premises must be *relevant* to establishing the conclusion if fallaciousness is to be avoided. The fallacious arguments fitting the general "you too" fallacy pattern fail to pass this relevancy test.

Justifying behavior by comparing it to others'

10.1 Change the conclusion in the final sample argument above (about Jones) to make the argument commit the "you too" fallacy.

10.2 Make up a sample argument chain which commits the "you too" fallacy in arguing for the ultimate conclusion that you ought not be punished for cheating on the last test you took in chemistry.

10.3 Find an example of the "you too" fallacy in a recent newspaper or magazine report, or describe one that was used on you lately.

10.4 Does the speeder in the following remarks made to the officer giving him a traffic citation commit the "you too" fallacy? Defend your answer.

You shouldn't give me a ticket. I was only going about five miles over the limit. That guy in the green Ford was doing about eighty and you didn't stop him. It's not fair!

10.5 Does the politician commit the "you too" fallacy in making the following remarks? Defend your answer.

Why should I be penalized for accepting a few gifts from people in exchange for helping them get their zoning problems solved? That sort of thing has been going on around here since time began. It's just understood by most people that special service from the zoning board has its price. Everyone knows that. My accusers are just playing politics.

10.6 Is it legitimate to conclude that P_1 *is not in any position to criticize* P_2 if P_1 has been behaving the same way P_2 has?

11

Applying general principles to specific cases

Is it true that all triangles have three sides? Of course. Strictly speaking, nothing counts as a triangle unless it has three sides. Then there is a true *general principle* that all triangles have three sides. Thus, the following argument is *not* fallacious.

All triangles have three sides.
George drew a triangle.

George drew something three-sided.

This is a simple example of the application of a general principle to a particular case, a case to which the general principle applies quite well. The result is a logically appropriate argument.

However, there are many fallacious examples of a similar type of argument, examples in which a general principle is applied to a special case which it really doesn't fit. Here is a classic illustration of how this can happen:

In the U.S. we have freedom of speech guaranteed by our Constitution.

It is legally permissible to suddenly yell, "Fire!" in a crowded theater in the U.S.

I presume you can see that this argument is not reasonable and that it is fallacious. But why is this argument fallacious even though it is similar in some ways to the argument about triangles which was not fallacious? Surely one important factor in the first argument is that the general principle stated as its first premiss has no exceptions. Absolutely every triangle has three sides. Thus when we apply the general principle to a specific case, we know that the case cannot be an exception to the general principle. On the other hand, the general principle occurring in the *second* argument is stated so broadly and in such general terms that it is easy to see it cannot mean that in absolutely every conceivable case of speech complete freedom is guaranteed. In a sense the principle allows exceptions. Complete freedom to say whatever one pleases, whenever one pleases to say it, at whatever volume, was never intended by the authors of the Constitution, and everyone is supposed to understand that. Just think what it would mean to have truly *complete* freedom of speech.

Any principle stated in very broad vague terms does not clearly specify the range of particular cases to which it applies. That means that there usually will be at least some special cases which at first sound as though they should be covered by the general principle but upon further examination are revealed to be exceptions to the general principle or perhaps doubtful cases as far as the general principle is concerned. Thus our principle about freedom of speech does not clearly specify the range of cases in which freedom of speech is granted nor is the definition of freedom of speech spelled out. Our nation has a considerable body of law intended to clarify the exact extent of our freedom of speech.

Not only is the freedom of speech principle stated broadly, but it is part of a set of principles which have to be taken together in order to understand the extent to which any one of them applies in a specific case. Thus, even if the freedom of speech principle were not broadly stated, it would still be possible to misapply the principle to a specific case merely by ignoring the proper role of the principle within the system of principles found in the Constitution. For example, the Constitution also guarantees freedom of religion. That guarantee would suggest the following argument is fallacious:

In the U.S. we have freedom of speech guaranteed by our Constitution.

It is legally permissible for the police, at the order of the mayor, to enter churches which the mayor dislikes and yell obscenities during the services in order to cause disruption.

The above argument suffers from several defects, one being its failure to take into account the freedom of religion principle when trying to apply the freedom of speech principle. In sum, when a general principle fits into a set of related general principles, it is possible to misapply it by ignoring the other general principles in the set.

I believe the real-life examples of this sort of fallacy usually occur when someone tries to get an audience to accept a point of view by appealing to some slogan or general principle which the audience will automatically accept without much thinking. The fallacy in the argument will perhaps be less likely detected if the general principle cited in the argument is one to which the audience has a great emotional attachment.

There is not a standard modern name for the fallacy we have been examining, so I will suggest one: the fallacy of misapplication of general principles. In outline form the arguments which commit the fallacy look like this:

(A general principle.)

(Application of the general principle to a special case for which the true application of the principle is doubtful or which may be an exception to the intended principle.)

Applying general principles to specific cases

11.1 Listed below are some general principles which could be used as premisses in arguments which exhibit the fallacy of misapplication of general principles. For each of these general principles make up such an argument. (You have to make up the conclusions and supply any additional premisses needed to make your examples complete.) Be sure your arguments are fallacious.
   a. The U.S. Constitution guarantees each citizen the right of freedom of speech.
   b. All men are created equal.
   c. You should treat other people in the same way you would like them to treat you.
   d. Honesty is the best policy.
   e. If it feels good, it's fine to do it.

11.2 If a general principle *legitimately* leads to a conclusion which you know to be false, that would be a reason to reject the principle, but not a reason to call the argument fallacious. Point out some of the *legitimate* consequences which actually follow logically from these principles and which tend to show something is wrong with these principles:
   a. Everyone who is on welfare is a chisler.
   b. Anyone who causes pain to another human being who never did any harm to him is acting immorally.

12

Inferring general principles from specific cases

In the previous Section we discussed problems associated with arguments in which general principles are applied to specific cases. We will now consider arguments which reverse that procedure, arguments in which general principles are concluded on the basis of the evidence offered by specific cases.

Surely one of the most fundamental human intellectual processes is generalizing. By this, I mean the process through which we come to believe certain general principles are true after we have experienced what we take to be specific instances of these general principles. For example, one may come to believe he will become sunburned after staying out in the sun for more than one hour in the spring before building up a tan, and that belief will probably be based on a few painful experiences. Here, the painful experiences are the instances of the general principle. We could lay out an argument which captures the reasoning involved in a case like this:

P1. The last three years when I stayed out in the sun for more than an hour before building up a tan in the spring; I got somewhat sunburned.

P2. One year earlier, I stayed out a little less than an hour under similar circumstances and I did not quite get burned.

C. If I stay out in the sun for more than an hour in the spring before building up a tan, I will get sunburned.

Such reasoning is used all the time to justify common beliefs as well as more complex scientific theories.

It is fairly obvious that most arguments which draw a general conclusion from specific examples can be questioned. One can ask how we know the specific examples cited in the premises of such arguments are good examples from which to generalize. It seems there is always the chance that the general conclusion drawn from specific examples could be mistaken, because the examples may not be typical.

The questions raised in the preceding paragraph are difficult to answer in a completely satisfactory way. In fact, I believe no one yet knows how to answer such questions fully and systematically. However, we can get started with a very tentative and partial set of pointers related to this issue:
(1) Some arguments which proceed from specific cases to a general conclusion are not vulnerable to the questions raised above. For example, in the argument

P1. Matt, Jeff, and Joe are the only people in this room.
P2. Matt is 18 years old, Jeff is 19 and Joe is 20.

C. Everyone in this room is under 21 years old.

We couldn't say the particular examples mentioned in P2 are perhaps not typical of the people in this room. (Naturally, it is still possible that some or all of P1, P2, and C are false. That is not the issue. The issue is whether the premisses, if true, provide good support for the truth of the conclusion.) Thus, in some special cases there can be no doubt about the general conclusion if the premisses are true. Unfortunately, the most interesting cases of generalization which you are likely to encounter are not like these.

(2)   It seems fairly clear that some arguments which do not fit into the special category described in the preceding paragraph are nevertheless quite "logical" while other such arguments are quite fallacious. For instance,

> P1.   I once knew a couple of fraternity members. Man,
> P2.   were they stuck up!
> _____
> C.   All fraternity members are stuck up.

This is logically so bad that one might be inclined to believe that its author has some emotional need to criticize fraternity members. On the other hand, when the Gallup or Harris poll argues that approximately 54 percent of the American people approve of the way the president handles foreign relations, that is a generalization based on interviews with only a relatively few people (about 1,500 people, in fact); yet such polls are remarkably accurate. There is some sort of important logical difference between the silly argument about fraternity men and the sensible argument about American opinion. An adequate logical theory will account for this difference.

The fallacy committed in the argument about fraternity men can well be termed the fallacy of hasty generalization. In outline form, the fallacy of hasty generalization often looks like this:

> (Premisses which include a description of some specific cases that would be instances of the generalization occurring in the conclusion, but not enough specific cases are mentioned to provide strong evidence for the truth of the generalization.)
> _____
> (A generalization of the specific cases mentioned in the premisses)

But another version of this same fallacy can be outlined as follows:

> (Premisses which include a description of some specific cases that would be instances of the generalization occurring in the conclusion, but the specific cases have been chosen in some special way which is likely to make them atypical, unusual, or nonrepresentative cases.)
> _____
> (A generalization of the specific cases mentioned in the premisses)

The first version of the fallacy differs from the second version only with respect to the type of premisses each contains. In the first version the conclusion is a generalization from too few cases. In the second version the conclusion is a generalization from cases which well may be unusual or atypical. The example argument about fraternity men seems to fit the first version. Here is an example of the second version:

In a recent study of graduate students enrolled in Ph.D. programs throughout the country, it was found that almost all of the 2,000 students of Italian extraction were highly intelligent.

We may conclude that the Italian people are a highly intelligent group.

Here we have a very specialized sample mentioned in the premises, even though the sample in this case was fairly large. A large sample is not automatically a good one. If the sample is drawn from the whole population in such a way as to systematically exclude some members of the population, then nothing about the whole population can be inferred legitimately from the characteristics of the sample, even if the sample is large. Statisticians would say the sample in such a case is *biased*.

There will not be a definite line between fallacious and nonfallacious generalization arguments. Rather, the arguments vary from terrible, to very strong. We shall try to use the label, "hasty generalization", only for those arguments that fall near the weak end of this spectrum of varying strengths.

The mathematical theory of probability and statistics comes closer than anything else now available to being the kind of theory we need for making the determination of fallaciousness or nonfallaciousness in this type of argument. However, the theory of probability and statistics can deal only with a small proportion of the arguments that fit the above outline form. Hence, we still are waiting for a theory that will help us determine which generalization arguments to buy and which to reject. Perhaps you will participate one day in the construction of the needed theory.

Inferring general principles from specific cases

12.1 What kind of premises would be needed for a nonfallacious generalization argument with the conclusion that the Italian people on the average are quite intelligent?

12.2 How many people do you suppose would have to be interviewed at random by opinion poll takers in order to get a sample large enough on which to base a nonfallacious generalization about the opinion of the adults in this state? (This question will require some knowledge of statistical theory to answer in detail, but there is a hint contained in this Section.)

12.3 Which of the following generalization arguments seem to commit the fallacy of hasty generalization? Defend your answers with a brief comment, both when you think the argument is fallacious and when you think it is not.

a. When I mentioned to my accounting teacher that I am taking a course in art history, she just sneered and shook her head. So I guess the study of art history is looked down on among the faculty here.

b. Both XYZ Corp. and Kool-Kar Co. started marketing a new car polish last year, and both have found the item to be highly profitable. So our profits would be improved if we put out a new car polish too.

c. The Smith family consists of Mr. and Mrs. Smith and their two daughters

Beth and Joanne. Mr. Smith, Mrs. Smith, Beth, and Joanne are each lawyers. Hence, all the members of the Smith family are lawyers.

d. Consider the millions of women in this country who are far more skilled than millions of men when it comes to caring for children in the home. This shows that women have a greater natural ability for child care than men do.

e. Recently, 400 randomly selected students on this campus were interviewed about their opinions on how the school could best help to conserve energy supplies. Over 63 percent of those interviewed stated that they believed that the biggest energy waste now occurs in the dormitories through student carelessness about leaving lights and appliances turned on when not in use. It's safe to say on the basis of this survey that 60 percent or more of the whole student body would believe the same thing.

f. Recently, 400 randomly selected students who were studying late in the library were asked whether their courses were intellectually exciting to them. Nearly 85 percent said "yes". We may conclude that the majority of students at this institution find their courses intellectually exciting.

13

Suppression of evidence

When there is inconclusive evidence available on both sides of an issue but an author presents only one side, the case for that side appears stronger than it actually is, and we can say that a fallacy has been committed. This sort of fallacy probably occurs more frequently than any other, and it is often extremely hard to detect unless you are very familiar with the argument's subject matter.

Television advertising presents some blatant examples of the suppression of evidence. (A great many devices are used in advertising—many of them unstated emotional appeals that have little or nothing to do with argument. I am not talking about such appeals. It is hard to evaluate the logic of something that is not even an argument. But in addition to the unstated nonargumentative appeals, advertising does contain actual arguments.) We can construe many ads as having roughly the same sort of conclusion, whether or not it is actually stated: the product or service being advertised is the best one for you to buy for the relevant purposes. Sometimes there are more specific conclusions stated in the ad itself. But in any case, suppression of evidence occurs frequently—in fact, almost universally. Bayer Aspirin informs us that hospital tests show that a high percentage of people can take Bayer without stomach upset. (This would be one of Bayer's premises.) What Bayer does not tell us is that the same tests apply to other brands of aspirin. (That is a piece of suppressed evidence.) Anacin tells us that their tablets contain two medically proven ingredients instead of just the one ingredient contained in aspirin, and that the pain reliever contained in their tablets is the first choice of doctors. What they do not tell us is that the two ingredients are aspirin and caffeine, that Anacin costs considerably more than ordinary aspirin, and the pain reliever is plain aspirin. Digel touts its special gas-relieving ingredient, and points out that as a result of this ingredient Digel will be more effective in relieving stomach discomfort in which gas is involved than mere antacids will be. Digel fails to mention that several other brands also contain similar ingredients and that some of these may be less expensive.

I believe the examples cited above constitute at least partial arguments, and they even present evidence that is relevant to establishing the superiority of the products mentioned. However, in each case important evidence that counts against the superiority of the product in question is not stated. Such arguments commit the fallacy of suppressing evidence.

Advertising is not the only realm in which suppression of evidence runs rampant. When someone is arguing for his own point of view, it is always tempting to present only the partial evidence which favors that point of view, leaving out anything that might be damaging to it. Any such argument that is inconclusive will be at least mildly fallacious, because in a case of uncertainty a proper assessment of the likelihood that the conclusion is true depends on weighing *all* the available evidence.

Despite the fact that this fallacy may seem fairly straightforward and easy to understand even if it is not always easy to detect in practice, there are some grave theoretical problems connected with it. Suppose there is some completely unknown fact which is relevant to a given argument. Naturally, since this fact is unknown to everyone, it is not mentioned in the premisses of the argument. Does that make the argument fallacious?

This problem arises because an argument which is not conclusive may be made weaker as more becomes known. Initially, it may seem that Joe committed the bank robbery, but as more facts come to light, the case against Joe may weaken until it no longer appears that he is guilty. Thus, so long as our argument is inconclusive, it remains possible that further evidence will make us want to change our conclusion. Does this mean that a nonfallacious argument has to contain even the unknown facts as premisses?

I think not. We do not want to require that in order to escape being fallacious an argument must contain *everything* known or unknown that is relevant to the truth of its conclusion. Rather, what we want to require is that the argument contain in its premisses either (1) enough information to conclusively establish its conclusion or (2) a complete account of all the *available* evidence that bears on the question of the truth or falsehood of the conclusion. If the first alternative is satisfied, no further evidence is needed, even if some is available, and no further evidence would make any difference to the acceptability of the argument. If the second alternative is satisfied, we shall not say the argument is fallacious, even if it should turn out later that some evidence which was not available at the time the argument was put forward shows that the conclusion is false. Thus, when an argument is not conclusive it is possible that the later addition of further premisses will make the conclusion appear more dubious.

If we apply these requirements to advertising we find that a nonfallacious argument in favor of buying a particular product would have to mention all the known negative features of the product as well as the known positive ones. Of course, we are not likely to find ads displaying that degree of candor on television sets or in newspapers. But that merely means that we can expect advertising to continue offering fallacious arguments.

On some occasions there simply will not be enough time for an arguer to present all the available evidence relevant to his conclusion. When that happens, the result will necessarily be a fallacious argument. Perhaps you think it's unfair to label incomplete arguments in such contexts fallacies since their incompleteness is in a way unavoidable. Nevertheless, the arguments in question do not include all the obtainable evidence and thus do not allow the reader or hearer to know the full available story. Thus, I will insist these arguments are fallacious for their conclusions are not warranted by their premises. It may just be that you can't avoid giving fallacious arguments which suppress some of the relevant evidence on some occasions.

There is, however, one way in which it seems possible to sometimes avoid suppressing evidence even while leaving out some relevant available evidence: when the evidence to be omitted is *already* available to the audience, then one might not need to actually repeat it all every time in order to avoid the fallacy. Much will depend on the particular context. For example, in a debate or in a trial a good deal of evidence may already have been presented for a particular position. Suppose you then have a chance to speak in favor of an opposing position. In such a context it seems you may assume that the evidence already presented is now available to your audience, and you may not need to present it all again in your own argument for the opposing conclusion. Sometimes it will be appropriate merely to make passing reference to the evidence already presented, as by saying "Despite all the reasons already presented, I wish to argue that. . . ." At other times it might not even be necessary to verbally acknowledge the opposing evidence which is already available; the context by itself may make the relevance of that evidence very clear to everyone concerned.

These considerations lead to the following final definition of the fallacy of suppression of evidence: *when the premises of an argument, if true, would not provide a completely conclusive case in favor of the conclusion of the argument, the argument must contain at least an implicit reference to all the available evidence relevant to the truth or falsehood of the conclusion in order to avoid committing the fallacy of suppression of evidence.*

Suppression of evidence

13.1 Take any subject about which you have some background information and produce a pair of detailed arguments relating to that subject. Construct each of your arguments so as to commit the fallacy of suppression of evidence. Let the first argument be for one conclusion, and the second for the opposite conclusion. Try to make each one look as convincing as possible by suppressing relevant evidence that would count in favor of the opposite conclusion.

13.2 Pick out two or three arguments from advertising or published political debate dealing with subjects that you know enough about for you to be able to point out how these arguments suppress relevant evidence. Summarize each argument and explain what evidence has been left out that might make the arguments weaker.

14

Lack of evidence

Normally, when we lack adequate evidence the reasonable procedure is to draw the conclusion that at least has the best support, or else to suspend judgment and draw no conclusion at all. Try to apply this point to the analysis of the following argument:

(A)   No one has ever been able to prove there is a God.

Thus, the reasonable thing to do is to believe there is not a God.

In this argument, nothing is said that would enable us to tell that the best support is for conclusion that there is no God. Thus, applying the general principle with which this paragraph began, the reasonable procedure is to suspend judgment about God unless there are other arguments we can look at to help us decide what to believe. However, argument (A) does not conclude that we ought to suspend judgment. Instead, it concludes that we ought to disbelieve in God. That makes argument (A) fallacious. In fact, argument (A) is just as illogical as its sister argument (B):

(B)   No one has ever been able to prove atheism is correct.

So, the best thing is to believe in God.

Both these arguments attempt to use the lack of evidence in favor of one proposition as if it were evidence in favor of the opposite proposition. Neither one of these arguments even comes close to showing its conclusion has more evidence in its favor than the opposite conclusion has.

The fallacy committed by arguments (A) and (B) is called the fallacy of arguing from ignorance. It has that name because in each of these cases the argument proceeds as though ignorance of the truth of one proposition counts as proof of the opposite proposition. In outline form the fallacy appears to look like this:

X has never been proved to be true.

X is false.

Or this:

X has never been proved to be false.

X is true.

Both forms are equally fallacious.

One qualification needs to be added immediately. Not all arguments fitting the above forms are truly fallacious. Consider the following argument, which fits the first form, but which does not seem to be fallacious:

(C) I've looked all through the drawer for my pen, and I found no sign of it.

The pen is not in the drawer.

In this example, the premiss essentially says that despite my best efforts I lack evidence of the pen's presence in the drawer. The conclusion says that the pen is not in the drawer, which means that it is false to say the pen is in the drawer. Thus, argument (C) can be seen as an instance of the first form of the fallacy of arguing from ignorance. Yet argument (C) surely is not fallacious.

We need to figure out why argument (C) is legitimate but (A) and (B) are fallacious. It seems to me the difference is this: the existence of God is a much more complicated matter than the existence of the pen in the drawer. We have good reason to believe that if the pen were in the drawer I would have found it after a good search, but we have little reason to believe that one can find a proof of God's existence so easily, even if there is one. Failure to turn up evidence in favor of X would seem to count against X only when X, if true, would leave its evidence freely scattered about the house, readily obtainable.

These considerations lead to the specification of background conditions which arguments fitting either of the above outlines must meet in order for these arguments to avoid being fallacious:

(Background condition which, if met, prevents an argument fitting the first form from being fallacious.)

X is the sort of proposition which probably would have been proved true already if it were true.

(Background condition which, if met, prevents an argument fitting the second form from being fallacious.)

X is the sort of proposition which probably would have been disproved already if it were false.

The argument about the pen meets the first of these conditions and is thus prevented from being fallacious, while the arguments about God fail to meet these conditions and are thus fallacious.

Obviously there will be room for dispute when determining the fallaciousness of a specific argument. You may think something is easy to gather evidence for, and someone else may think not. Nothing can be done about this. I never promised it would always be easy to tell the difference between a fallacious argument and a nonfallacious one.

Lack of evidence

14.1 The following arguments each might be fallacious, depending on the situation in which they are presented. Your task is to describe briefly a situation for each argument in which that argument would commit the fallacy of arguing from ignorance if the argument were to be put forward by someone in that situation. Also briefly describe a situation in which the argument would not be fallacious.

   a. You must not be pregnant, since you don't show any signs of pregnancy.

   b. I have found no reason to believe my brother has died. So, we can assume he's still alive.

   c. There's no evidence that you are telling the truth; hence, you are lying.

   d. There's no evidence that you are lying; hence, you are telling the truth.

14.2 Make up two examples of arguments which commit the fallacy of arguing from ignorance, supplying any background information needed to insure that these arguments are indeed fallacious.

14.3 Suppose you have a casual acquaintance named Leslie. Since you have not had a great deal of contact with Leslie, you would not be in a position to know anything about her which she has kept fairly private. Under such circumstances, the following argument from ignorance would be fallacious:

Leslie has not proved herself to be an untrustworthy person.

Therefore, she is not untrustworthy

Since this argument is fallacious, you should not be convinced by it. Does that mean you should think Leslie is untrustworthy? Why not? How does this fit in with the things said in this Section?

15

Ruling out alternatives

When you are trying to make a choice between going to a party and studying for an examination, you might present some small arguments to yourself, arguments in which you would weigh the advantages and disadvantages of either course of action. Ultimately, you might rule out one of the possible alternatives, leaving the other alternative as the one you would choose. We could then summarize the results of your thinking in a scheme like this:

I will either go to the party or I will study.
But I will not go to the party.

I will study.

In this case, the argument scheme has the following basic pattern:

A will happen or B will happen.

But A will not happen.

B will happen.

This pattern of argument should seem very "logical" to you. If its premisses are true, surely the conclusion is also correct. (Of course, the premisses might be in fact false, but that is another matter with which we are not now concerned.)

Patterns like this are so common to our everyday thinking that we can follow them easily. This leads to the occurrence of yet one more fallacy, the fallacy of *false dilemma,* which has the following outline form:

A is not true.

B is true.

Taken by itself I think that this outline would appear ridiculous. But if the context of the argument *falsely suggests* that there are only two alternatives, A and B, and then someone comes along and gives the argument that A must be ruled out and therefore we can conclude that B is the proper alternative, that person may appear to be arguing logically.

I think an example or two will help. Suppose you are an accountant in a manufacturing firm which has not been able to produce enough of its product to fill all orders promptly without having to work the assembly line people overtime. The workers are complaining and some have quit because of the long hours. This has been going on for a couple of months, and there has been talk about how the company should expand its production facilities by buying or renting more space and installing more equipment. You are called in to compute the effect of the expansion on the company's profits and compare that effect to the effect that would be achieved if the company merely refused to take more orders than it could handle with the present use of present equipment. After several days of calculating, you come up with an interesting conclusion: it is not true that expanding production facilities will increase profits. (This could happen if the expansion would be too expensive.) On the basis of all this, you might then argue as follows in a meeting of the management:

It is not true that expanding production facilities is the best thing to do financially.

The best thing to do, financially, is to refuse to take more orders than can be handled at present.

Do you see how this argument fits the outline given above? Do you also see how this argument seems reasonable, given the context in which it was said? (You should. If you don't, read it all again.)

This argument commits the fallacy of false dilemma. Why? Because there are viable alternatives that have not be considered by anyone in this story. What are these alternatives? You should consider using the present production facilities more fully, by operating a second shift or a third shift. Then you wouldn't have the problem of individual workers having to work such long hours. Of course, I have no idea whether this alternative would work out better than the ones which were considered, but it is at least worth investigating. What is wrong with the argument as stated is just that not all the reasonable alternatives were considered. One alternative was ruled out by your computations, and you jumped to the conclusion that the other alternative which everyone had been thinking of was the best one.

Here is another example. At many colleges and universities, students majoring in a science, a social science, or one of the humanities are required to take courses in a foreign language in order to get the most popular degree. Students have periodically objected to this requirement, usually on the grounds that it serves no legitimate educational purpose. The foreign language faculty members, often fearing for their jobs, have responded with a variety of arguments to the effect that the study of foreign languages can be useful in various ways. They also suggest that if the requirement were abandoned high schools would no longer be as inclined to offer foreign language instruction, and that would be bad. All these arguments are too lengthy to reproduce here. However, I do want to look closely at the last few steps in one possible argument related to this dispute.

Let us suppose that after lengthy investigation and arguments Jones has come to the conclusion that the satisfaction of the foreign language degree requirement at his institution serves no useful purpose for almost all the students affected. We can then picture Jones as going on with a chain of arguments as follows:

P. Satisfaction of the foreign language degree requirements at this institution serves no useful purpose for almost all the students affected by them.

C. Thus, we should not keep the present requirements. (That is, it is not true that we should keep the present requirements.)

P. (Same as C above.)

C. We should not have a foreign language degree requirement.

Look carefully at the last argument in the chain. It fits the pattern for false dilemma, and it has left an important alternative out of consideration, the possibility that the foreign language requirement could be changed to make it better rather than being abolished altogether. Perhaps you can think of another alternative. In any case, since there is at least one reasonable alternative that has been left out, we have here an occurrence of the fallacy of false dilemma.

If we reflect for a moment about what these examples show, I think we can see that arguments of the form

A is not true.

B is true.

seem reasonable even when fallacious because of a context in which it *appears* that A and B are the only alternatives.

The above account of the fallacy needs modification, however, in order to allow for variations that do not quite fit the standard form. One such modification is suggested by the fact that the form

A is true or B is true or C is true.

A is not true.

B is not true.
_____

C is true.

is nonfallacious. Corresponding to this reasonable form, though, there will be a fallacious type of argument:

A is not true.

B is not true.
_____

C is true.

This form will be fallacious when there are in fact more alternatives than A, B, and C. Other modifications of the outline for the fallacy of false dilemma are also possible, but they will be variations on the same basic theme.

_____

Ruling out alternatives

15.1 In each of the following situations describe an alternative that has been left out of consideration in the argument chains put forward in the situation.
  a. Situation: Judy's boyfriend, John, doesn't often suggest that they go anywhere together since he doesn't have very much money. Judy is considering breaking up with him in order to have a more exciting life.
  Judy's argument: If I break up with John, I'm really going to miss him because I like him very much. He's nice to be with, even though we don't go out as much as I'd like. If I start going out with other men, I might not find one as nice as John. So, I guess I won't break up with him. Unfortunately, that means that things will go on as usual and I won't get to go out much.
  b. Situation: George owns a small store, and does not do very much business. His prices are set somewhat higher than other stores in town, but he is only making $700 a month profit for himself as it is. He is considering the possibility of moving his store to a more central location, in the hopes of increasing the number of people who would come into the store.
  George's argument: My present store is worth $20,000 less than the one I would have to buy if I were to relocate, and I don't have the money to make up the difference. So I can't afford to move to that place. But it's the only place available right now, other than my current location.

Hence, I won't move now. That means that I'm going to have to settle for only $700 per month profit from my store, at least for the time being.

15.2 Pick out the argument from each chain in 15.1 which commits the fallacy of false dilemma, and write out that argument, exhibiting its premises and conclusion. Write it out in such a way that it is clear that it fits the outline for a false dilemma argument. (You may have to reword the argument a little bit in order to make it fit the outline.)

15.3 Some arguments that fit the outline,

> A is not true.
> _____
> B is true.

are *not* instances of the fallacy of false dilemma. What is the factor which determines whether an argument of this form is an instance of the fallacy or not?

16

Begging the question

As you know, to beg is to ask for charity, for a handout. We normally think of a handout as something material, some object, such as money. But it is also possible to beg for other things, such as mercy or love. Now suppose that you were involved in a dispute with someone over a question of importance to you, and your opponent in the dispute asked you to give in and agree with his position just to be cooperative. Your opponent would be asking you for a kind of handout, a gift; the gift requested is your agreement. In such a case we might say your opponent had "begged" for you to just "give" him the disputed matter, that is, the "question" under discussion. Although this way of phrasing things sounds a bit awkward, "begging the question" really does amount to something like begging someone to give their agreement on a disputed question just as a gift.

How exactly does someone go about this "begging"? Recall that we are interested in fallacies committed by *arguments*. Thus, if someone came right out and asked you to agree with them, just as a favor, *that* would not be a fallacy, because a *request* is not an argument. So if begging the question is to be a *fallacy*, we will have to identify some manner of *arguing* which amounts to begging an opponent to give in, rather than earning the opponent's consent through giving good evidence. It is not hard to find examples of arguments which can be said to do this.

Suppose that Mr. Softheart has just learned, to his horror, that the local animal shelter has put approximately 3,000 stray dogs and cats to death each month for the last year. He confronts the director of the shelter and is told that there is nothing else to do with all those animals, but he remains unconvinced.

In an attempt to get the county board to change their policy he argues that the shelter's killing of the healthy animals is immoral, and for that reason should not be allowed. We can represent his argument as follows:

> P1.   The killing of healthy animals at the shelter is immoral.
>
> C1.   This killing should not be allowed by the county board.

As could be expected, one of the board members objects that she sees nothing immoral in eliminating unwanted animals that would merely cost the taxpayers money to support if they were not killed. But Softheart is not so easily silenced. He responds with another argument, this time in support of P1:

> P2.   The killing of healthy animals at the shelter is nothing short of outright murder. And as we all will surely recognize,
>
> P3.   murder is just plain wrong.
>
> C2.   (Same as P1)

We now have a chain of arguments, where the second argument really belongs in front of the first one. The ultimate conclusion of the chain is C1.

I believe that the question has been begged in this chain of arguments in the part of the argument chain designed to support C2. What does Mr. Softheart mean when he says that the killing at the shelter is *murder?* Doesn't he probably mean to say that the killing is *wrong?* That's why P3 seems so innocent, so obviously correct. If "murder" means "wrongful killing", then of course murder is wrong, as P3 states. If I am right about all this, then Softheart begs the question when he asserts P2, because he is just saying that the killing at the shelter is wrong, using different words from those used in P1, but without giving us any more reason to believe him now than we had before. Since the context of the argument clearly called for him to *prove* the killing is immoral, he is not entitled to merely *assume* that it is immoral, as he seems to do in P2. If we wanted to be difficult, we could say that Mr. Softheart's argument chain might just as well be written as follows:

> P2   The killing of healthy animals at the shelter is immoral.
>
> P3.   Immoral killing is immoral.
>
> C2.   The killing of healthy animals at the shelter is immoral.
>
> P1.   (Same as C2)
>
> C1.   This killing should not be allowed by the board.

However, not all arguments which can reasonably be thought to beg the question are actually circular in the way our example was. The basic idea of question-begging consists of asking the audience to accept the conclusion of the argument without real proof, as a gift to the arguer. This basic idea does not require that a question-begging argument be circular, because this idea is much broader than the idea of circular argument. In fact, one *could* say that *any* bad argument begs the question, since any and all bad arguments fail to give good support for their conclusions, thus in effect they ask the audience to accept the conclusion without good support.

Although any and all bad arguments might be said to beg the question, one does not usually speak of question-begging quite so broadly. We want a somewhat more precise account of begging the question which will identify it as a type of fallacy.

Let me try to analyze the ordinary concept of begging the question as I have seen it used. The first condition which must be met for question-begging has to do with the purpose of the argument in context. The context in which an argument is put forward establishes what job the argument is designed to do, and question-begging arguments all have the same job within their contexts: to display to the audience some rational grounds for accepting their conclusions. Softheart's argument was supposed to display to the county board rational grounds for concluding that the killing at the animal shelter should not be allowed. Even if the audience already accepts the conclusion, the point of an argument which begs the question still has to be showing the audience some rational grounds for accepting the conclusion. I don't think an argument which has some other purpose can beg the question.

Secondly, for each such argument the context and the nature of the conclusion will indicate in at least some vague way what kinds of premises would be sufficiently *independent* of the conclusion to count as possible grounds for accepting the conclusion. This notion of independence is crucial to the determination of question-begging, for a question-begging argument will be one which has the purpose described in the previous paragraph, and which fails to meet that purpose because it fails to provide premises which are sufficiently independent of its conclusion. An example of this lack of independence is found in Softheart's argument for C2. P2 is not sufficiently independent of C2 to count as *grounds* for accepting C2. When one seeks grounds for accepting a conclusion, one does not want that conclusion itself presented in the guise of grounds. Rather, one wants something more independent of the conclusion, something that one could accept on its own, for its own reasons.

The notion of *independence* required here is extremely difficult to make precise, because it varies tremendously from context to context, from audience to audience. I believe it is at least partially dependent on the psychology of the audience. Perhaps independence amounts just to this: a premiss is independent of the conclusion on a particular occasion for a particular audience if that audience sees that (within the context) it has grounds for accepting the *premiss* and these grounds do not include the *conclusion.* At least some members of the county board in our example did not see that they had grounds to accept the premiss that killing the animals is murder, unless they already accepted the conclusion that the killing is immoral; thus, for them the *premiss* that killing the animals is murder did not count as being independent of the *conclusion* that killing them is immoral.

In addition to the required purpose and lack of independence, question-begging arguments must have a third feature, namely, that the premiss which fails to be independent must also be one that genuinely leads in the direction of the conclusion if the premiss is accepted. An irrelevant or useless premiss which fails to be independent of the conclusion does not seem to me to make the argument beg the question. The premiss must be important to the argument before its lack of independence makes a difference.

Summing up, we get the following definition of begging the question: *an*

*argument begs the question when it contains an important premiss which in the
context fails to be sufficiently independent of the conclusion and the purpose of
the argument is to display in the context rational grounds for accepting the conclusion.*

Note that this fallacy is frankly context-dependent; this means that a given
argument could be fallacious in one context but not in another. That is, an argument appealing to the *Bible* to prove that Jesus was divine might be said to
beg the question if presented as proof to an audience of people not believing
the *Bible* to be the inspired word of God; but if presented to a different audience, the same argument might not beg the question.

Begging the question differs from the other fallacies we've been looking at
in an important theoretical way. All the fallacies prior to this one were defined in
such a way that arguments committing them had premisses which did not actually lead to their conclusions. In all those cases, the premisses could well have
been true without making it likely the conclusions were true. (Go back over the
list of the prior fallacies to verify this.) Those fallacies are often called *fallacies
of relevance* for that reason. But begging the question is different. There is no
reason why the premisses in a question-begging argument could not lead very
nicely and logically to the conclusion. Begging the question, then, is not a fallacy in which the truth of the premisses is irrelevant to the truth of the conclusion.
Instead, question-begging has to do with assuming too much in the premisses.

Begging the question

16.1 Each of the following arguments can fairly easily be thought of as occurring
in a context which makes the argument beg the question. For each argument, state which premiss or premisses seem to be at fault, and explain
why it is not legitimate to use these premisses without first arguing for them.

a. Joan and George watch as the tow trucks haul off the wreckage of two
cars from the highway in front of the store. Neither Joan nor George saw
the accident happen, and now they are trying to piece together what
caused it.

J: I figure the brakes on the Chevy failed.

G: What makes you say that?

J: No skid marks. There are no skid marks on the pavement
where the Chevy was. If the brakes had been working, there
would be skid marks.

b. It is immoral to intentionally take the life of another human being who
has committed no crime and who intends you no harm, unless it is necessary to take that life in order to save another life. But in the usual sort
of abortion, the doctor and the mother conspire intentionally to take a
human life, the life of the fetus, even though it is obvious that the fetus
has not committed a crime and does not intend to harm anyone. Clearly
then, it is immoral for the doctor to perform this sort of abortion.

c. Each mentally sound woman surely ought to have the right to control her own body, to decide what parts of it to keep and what parts to have removed or to remove herself. After all, we would not deny a mentally competent woman the right to decide what length of hair she wants cut off, or even the right to decide whether or not to have her appendix out. In an abortion, an unwanted part of the woman's body is removed. Clearly then, any mentally sound woman ought to have the right to decide to have an abortion.

d. I know there is a God because the *Bible* talks about God all the time, and the *Bible* is completely truthful in religious matters. How do I know the *Bible* is correct? Well, the *Bible* is the word of God and God knows everything; so either God is lying to us in the *Bible* or else the *Bible* is correct. But since God is also perfect, He wouldn't lie to us.

17

Complex questions

There's often more to asking a question than merely requesting information. In fact, some questions seem not to be primarily requests for information at all—for instance, "How are you?" uttered as a greeting. Even questions intended primarily to elicit information often do much more: if I ask, "Is the window still open?" I not only request information about the current state of the window, but I also say something which says that the window was open recently. If the window had not recently been open, my question would be out of place.

Compare these two questions: "Is the window open?" "Is the window still open?" Of course, both questions presuppose the existence of some window adequately referred to by the phrase "the window". But I am not particularly interested in that kind of presupposition—the kind that amounts to assuming that the naming phrases in the question actually have something to which they refer. However, many questions make further presuppositions, and such questions turn out to be potential fallacy sources. "Is the window still open?" is such a question, as it presupposes not only that there is a window but also that it was recently open. Here are some more questions which make the fallacy-generating kind of presupposition, along with descriptions of what they presuppose:

Questions	Presuppositions
1. Have you given up getting drunk every Saturday?	1. You used to get drunk every Saturday.
2. Was it good planning or sheer luck that won you the title?	2a. You won the title. b. The correct explanation for your winning is one of the two mentioned in the question.

3. A salesperson filling out an order blank says, "Which one of these items shall I put on your order?"

3. You are buying one of the items in question.

It would take us too far afield to give a theoretically satisfactory account of presuppositions but you can see from these examples how asking a question which makes a presupposition can be an indirect way of asserting the presupposition.

Questions which make presuppositions like the preceding ones are said to be *complex questions,* because one would need to divide them into components in order to answer them, when their presuppositions fail to be true. That is, you might need to answer question (2) by saying "I did not win the title, although I came close, but what success I did have was due to sheer luck."

There is nothing immoral, illegal, or fattening about complex questions by themselves, but they can be used in an illogical way by careless or unscrupulous arguers. (They can also be used to deceive or mislead in various ways without occurring in an argument. Political organizations wishing to bias the results of their opinion polls frequently use the complex question as a device to achieve their ends. Here's an item from a recent questionnaire mailed to prospective members by the promilitary political lobby, the American Security Council: "Should the United States balance Soviet conventional military superiority by regaining strategic military superiority over the Soviet Union?")

Since a complex question makes at least one presupposition, one can use such a question in an argument as a device for almost adding the presuppositions of the question to the argument's premises without coming right out into the open with these added "premises". Here's an example:

Have you given up getting drunk every Saturday? I certainly hope so, for your own good. This is a small town, you know, and word gets around. Folks don't mind other folks getting a little happy now and then, but they figure anyone who gets drunk regularly over an extended period must not be too reliable about anything.

This argument, presumably for the conclusion that your own good requires that you give up getting drunk every Saturday, does not explicitly state that you used to engage in that practice. But the complex question serves to introduce its presupposition into the argument as a kind of hidden "premiss".

Again, there is nothing particularly unreasonable about using a complex question to add an unspoken "premiss" to an argument. As a matter of convenience or style, using a complex question as a premiss-introducer may not be objectionable, even though the "premisses" introduced by that method are somewhat hidden from view. But sometimes, we find that if we take these hidden premisses and add them explicitly to the argument, the argument thereby becomes question-begging. This is the fallacy of complex question. Thus, *the fallacy of complex question is a special variety of begging the question, differing from ordinary question-begging only because the objectionable premiss is never actually stated outright, but only indirectly stated by means of a complex question.* The above argument about getting drunk, then, would commit this fallacy if adding the premiss that you used to get drunk every Saturday results in the argument's becoming question-begging.

Complex questions

17.1 Which of the following questions are complex? For those that are complex, state the relevant presuppositions.
   a. What time is it?
   b. Where are we?
   c. What's your name?
   d. Where will the garage be built?
   e. When did it start to rain?
   f. What is the name of that old restaurant down by the river?

17.2 Each of the following arguments employs a complex question as a device for hinting at unstated added premisses. For each, state the relevant presupposition(s) and briefly describe a context in which the argument would commit the fallacy of complex question. Explain why in that context the argument fits the description of the fallacy. Then describe a context in which the same argument would not commit the fallacy.

   a. Why did you find it necessary to flatter him to such an extent? Flattery just postpones the day when he has to face the truth about his inability to manage the office. We'll all be better off once he realizes his mistakes. So you made a mistake which will hurt us all.

   b. Shall we begin the project immediately or shall we wait a month? If we begin now, we will not be fully prepared, since the staff has not yet completed the preliminary studies. On the other hand, if we wait for a month for the staff report to be in, there will be an inadequate amount of time left to complete the project before winter. So it looks like we're in trouble.

   c. It is technologically feasible to convert most of our large oil-burning industries, such as electric generating plants, to operate instead on coal, and to do so without reintroducing dirty air, black with coal smoke, to our cities and countryside. But the conversion process is tremendously expensive. How much should we spend on this conversion? The industries affected the most will resist at first, and they will pass on their conversion costs to the consumer, thus feeding the already bright inflationary flames. But if we hold back in the face of the staggering costs, we face an even greater risk in the future. Unless we drastically reduce our dependence on foreign oil imports, we hold our whole economy open to blackmail. Unless we take bold steps now to prolong the life span of the world's oil reserves, we face the certain extinction of our way of life within a generation or two. Fertilizers, plastics, and dozens of other crucial items all require petroleum in their manufacture, and we are not likely to find quick substitutes. Thus, despite the present high costs of conversion, it is imperative that we take the long-range view and spend more now that will pay off in the next ten to twenty years.

18

Sliding down a slippery slope of argumentation

There is something wrong with the following chain of arguments, apparently invented by Eubulides, a Greek philosopher who lived over 300 years before Christ: "Anyone can see that one stone does not constitute a heap of stones. But it is also clear that by adding one stone to a collection of stones which was not a heap, one does not thereby make that collection into a heap. Thus, by adding one stone to another, obtaining a collection of two stones, one does not make a heap. Similarly, then, a collection of three stones is not a heap. Nor a collection of four. And so on. Thus, no matter how many stones are collected together, one never has a heap." Yet, although it clear that something is wrong here, it is not at all easy to say exactly where the error lies.

The argument is tricky enough that logicians have enjoyed discussing it on and off for a few centuries, and I will have to be content with a somewhat superficial account of the problem in argument chains like this one. You will perhaps agree that the trouble with the argument is that the concept of a heap is too *vague* to be used in the way it is used in the argument. Because the notion of a heap is vague there is no sharp dividing line between heaps and nonheaps. Since there is no sharp dividing line, it is easy to say that one stone can't make the difference between a heap and a nonheap. In other words, it is vagueness which makes the premisses acceptable even though their repeated use gets us into big trouble. We get into trouble through repeated use of the premisses in the chain of argument because even though one stone doesn't make *much* difference it *does* make a *little tiny bit* of difference, especially when we are dealing with collections that are getting close to being heaps. I am not saying that sentences employing the word "heap" are senseless. Most of them will be clear enough for their intended purposes. However, repeated use of a vague notion in an argument like the example will lead to trouble.

Let me try to generalize. The arguments with which we are concerned in this Section have roughly the following form:

(Premiss, placing some thing or some situation, X, into some category C.)

(Another premiss, which serves to link X with similar things or situations. This premiss will say that any thing or situation similar to X in some particular respect, R, can be classified in the same general category C.)

(Conclusion which states that some particular new thing or situation, Y, which is similar to X in respect R is in the same general category C.)

In the first stage of the sample argument about heaps category C was the category of nonheaps, and the situation X was a collection of just one stone. The similarity respect R was having just one more stone added. (Make sure you see how the heap argument fits the general pattern outlined above.)

Many nonfallacious arguments fit the general pattern above, but chains of them are sometimes fallacious. *We can say that such argument chains will be fallacious when the following three conditions are all satisfied:* 1) *Category C is described vaguely, so that its boundaries are not clear.* 2) *It is possible to think of a series of things or situations, starting with X, each of which is similar in respect R to the preceding member of the series, but the last member of the series is not in category C.* 3) *The chain repeatedly uses R as the only link between the conclusions and the original situation X.*

The fallacy described above will be called the *slippery slope fallacy.* The first premiss in these fallacious argument chains pushes the reader to the edge of the slippery slope, and then the second premiss shoves the reader over the edge, to slide on and on with no good place to stop, until finally the reader arrives at some ultimate conclusion such as the conclusion that a collection of a thousand stones is not a heap. What makes the argument deceptive is that the argument describes category C vaguely, so that there really is no one place to stop and draw the line and say we are getting outside category C.

Let's apply all this to some more realistic examples. Many of the real-life examples of slippery slope argumentation are concerned with extremely important questions of life and death, imprisonment or freedom, poverty or wealth; so our discussion is most certainly not merely about where to draw the line between heaps and nonheaps.

One of the more difficult decisions facing family members and physicians today concerns the degree to which one ought to prolong life by taking extraordinary medical measures. Although this is not a new issue, the public has become increasingly aware of it as medical technology has advanced, since this advance has made it possible to keep a body functioning through the application of various sorts of mechanical assisting devices even though there is no hope of a return to normal brain activity, no hope of the body's ever being a full-fledged person's body again.

Because of our confusion regarding such matters, there is plenty of room for fallacious argument. For example: someone who has an artificial limb, or a pacemaker for his heart, or who wears eyeglasses does not thereby become less a person. Such individuals are equally entitled to legal and moral status as are persons who do not need an artificial or mechanical aid to their bodily functioning. Now, surely if all that is true, we would not want to say that an individual who needs more than one such mechanical device to support his bodily functions thereby loses his status as a person. Someone who has *both* an artificial limb and a pacemaker is still a real person with all the same hopes and fears as a person who has only one device. In fact, it is hard to imagine that the installation of an additional mechanical assisting device on or in people's bodies suddenly destroys their humanity and leaves them without moral or legal protection from abuse. Accordingly, we see that good moral sense requires us to give just as much moral and legal protection to those individuals being assisted by many mechanical devices as we give to those who are assisted by only a few or by none. By carrying this line of argument to its logical conclusion, we arrive at the result that a body whose life functions are *all* maintained by mechanical devices is just as much a person as you or I.

Perhaps you see the similarity between the above argument and the argument chain about the heap. The above argument appears reasonable only because there can be no sharp dividing line drawn between being a full-fledged person and not being a full-fledged person. That is, because there is no natural place to draw a dividing line between individuals who are full-fledged persons entitled to moral and legal protection and those individuals, if any, who are not so entitled, the argument proceeds as though there really is *no* difference in this regard between an unassisted body and one that has *all* its vital functions performed by machines. Upon reflection, it seems to me that the lack of a natural dividing line between two types of cases *ought not* count as evidence that there is no real difference between the two types. For example, even though there is no natural dividing line between cases of red and cases of orange (because there are shades of color in between) that surely should not serve as evidence that red and orange are really the same color.

One word of caution: the conclusion of a slippery slope argument may very well be *true,* even though the argument which attempts to prove that conclusion is fallacious. It may very well be that individuals in hospitals relying on mechanical devices for bodily functions *should* be legally and morally protected. All I'm saying here is that the slippery slope type of argument cannot establish that these individuals are so entitled.

Sliding down a slippery slope of argumentation

18.1 Try to construct a fallacious slippery slope argument which has the conclusion that a human fetus is fully entitled to moral and legal protection just like normal adults are. (Begin with a premiss about how a six-year-old child is entitled to such protection, and work backwards in age.)

18.2 Construct a fallacious slippery slope argument that works in the opposite direction, with the conclusion that a six-year-old child is not a full-fledged human being entitled to moral and legal protection. (Start with a premiss about the just-conceived fetus.)

18.3 Construct a fallacious slippery slope argument for the conclusion that every student *deserves* an "A" in every course he or she takes. (Start with a premiss describing those who clearly deserve "A" 's.)

18.4 Construct a fallacious slippery slope argument for the conclusion that if the government operates a rail passenger service in this country, pretty soon it will also operate and control every major industry—that is, pretty soon we will have socialism.

18.5 Does the discussion in this Section show that all argument chains are fallacious if they are long? Explain.

18.6 Give an example of an argument chain similar to the heap argument chain, but which is not fallacious.

19

Summary of the discussion of fallacies

It is not possible to completely catalog all the conceivable logical errors which can occur in an argument, for people are quite inventive when it comes to finding ways to make mistakes. So you should understand our list of fallacies to be just a sample of some common errors in argumentation. Working through this list should help you become more aware of the sorts of problems that can come up.

At this point, some people may be bothered because in order to decide whether an argument commits certain of the fallacies (such as the fallacious appeal to authority) one needs to have background information. For example, to have background information about the reliability of a certain person to serve as an expert on nutrition, one needs to know whether that person has studied nutrition extensively. Doesn't this mean that in order to decide whether a given argument commits some of the fallacies we have been talking about, such as the fallacy of appeal to authority, one has to know the truth or falsehood of the premisses of the argument? No. Note that the background information we are dealing with in these cases does not reveal the truth or falsehood of the argument's premisses. The background information is about the context in which the argument takes place.

If you wanted to be very fussy about this, I could just make the following tricky move: take the argument which is to be evaluated, and add to its premisses enough statements to completely describe the relevant features of its context. Then you no longer need any background information about the context, since the whole context is described within the enlarged argument. Change all the fallacy descriptions to refer to enlarged arguments rather than to background information. Now evaluate just the argument, as enlarged, without worrying about whether any of the premisses are true or false. You will get the same result as you would have by not enlarging the argument and considering the context of the unenlarged argument to be background information needed in assessing the argument.

Since our list of fallacies is not complete, and there is no hope of having a complete list so long as people are inventive enough to keep coming up with new ways of arguing badly, we need a different approach if we are to have a general theory of what makes arguments "good" or "bad". That will be the aim of the remainder of this book. But before we go on, you have a chance to practice some of what you have been reading about in the following exercise set.

Summary of the discussion of fallacies

19.1 Each of the following passages may contain at least one of the fallacies we have been talking about. In each case, identify the fallacy or fallacies, or state that no fallacy occurs. Defend your answer in each case.

a. Continued nuclear weapons testing must be unsafe, for the defenders of the tests have never succeeded in presenting an argument for the safety of the tests without getting bogged down in technical disputes.

b. Gentlemen, this rent control bill is unworkable and unjust, for Mr. Hinkle and the others who have joined him in sponsoring and pushing it are all tenants and renters who have a good deal to gain if it is passed; I note there isn't a single landlord in the sponsoring group.

c. One of our most cherished principles is that every man has the right to promote and advance his ideas. I see nothing wrong, therefore, when a judge imposes regular church attendance on a person found guilty, as a condition of the criminal's being granted parole. We need more judges who will stand up for what they believe in.

d. For years the Communists in the U.S. have advocated a graduated federal income tax—the more steeply graduated, the better. Numerous strategy hints from Communist theoreticians all over the world advise the party faithful to work for such things as graduated income taxes in their respective non-Communist countries. It thus becomes obvious that all arguments in favor of such taxes must be rejected by our Congress, and we should rely on some other form of taxation to run our democratic government.

e. The Chevette is a really fine small car. I know, because one of my friends who is an excellent judge of automobiles of all kinds told me that he thinks the Chevette is about the best car in its class. I can tell that my friend is an excellent judge of automobiles because he highly recommends the Chevette which is, of course, a great small car.

f. The theory behind banning phosphates from detergents is that reducing phosphate levels in sewage discharges will slow the growth of algae and unrooted weeds in our streams and lakes. But rainfall, natural run-off, and human excrement and food wastes entering the waterways account for 75% of the phosphates present there. Only 1% of the total phosphates present in the waterways is needed to bring about excessive algae growth. Hence the theory is wrong, under present conditions, and the phosphate ban will do no good.

g. Senator Smith has come out in favor of a direct federal subsidy to help people buy health insurance from private insurance carriers. So we find our senator, who was supposed to be representing our views, coming out instead in favor of socialized medicine. Of course once medicine becomes socialized, the same arguments which were used by the likes of Smith can be equally well applied to the legal profession. One can almost hear the chorus now, chanting about the rights of everyone to quality legal care. After all lawyers have been forced into the socialistic mold, there will be no reason why all small businesses shouldn't follow suit (to provide quality commercial care for all), and if small businesses are taken out of the private economy, the large corporations must also go. So we see that Smith's course of action will ultimately lead to the government takeover of the entire economic structure, despite Smith's protestations to the contrary.

h. The instructor who teaches the erotic literature course pointed out to me yesterday that all the authors we are studying depict all healthy women as having sexual urges which are just as strong as the male sexual drive. Since my girlfriend doesn't seem to be very interested in sex, I guess she is not healthy.

i. Our economic system is based on the idea that people should contribute to the general welfare of the society through useful work in order to reap the benefits of society by being paid for that work. Hence, those employees who were unable to get to their jobs last week because of the blizzard should not be paid for the time they missed work.

j. The present train station in town is quite old and badly in need of repair, and there is not much parking space nearby for the use of passengers. Moreover, the present location of the station, far from the university campus, makes it difficult for many students to get to the station, even though students are the biggest group of potential train passengers. Since we want to promote train travel, we should not attempt to renovate the old station. Instead, a new station should be built adjacent to the campus.

k. How can we take the president's proposals on solving the country's energy problems seriously, when he has thoroughly discredited himself by choosing crooks for his closest political and personal associates?

l. Police officers lead exciting, action-packed lives, as we can see from the recreations of police life on television programs like "Adam 12", "Kojak", and "Police Woman".

m. Senator Wheatfield's arguments for farm subsidies are without merit. After all, he himself owns a farm of several thousand acres and his major campaign contributors are farmers.

n. Remember those two red-headed twins that lived down the street last year? They were sure wild! That just goes to show—redheads are always wild ones.

o. Our engineering department has been trying now for six weeks to show that the new engine design will be more durable than the old one we've used successfully for years. They have not produced a convincing case. That new engine is most likely not as good as the old one.

p. For the past ten Thursday mornings we have polled the women shopping down at the mall to determine their preferences regarding panty hose design. We found that 85 percent of these women chose the nude heel and toe design over all other types. The brand names were removed from the hose for the purposes of this test, and the women were given ample opportunity to carefully examine each pair. We employed trained survey personnel who were careful not to influence the responses. We may conclude that approximately 85 percent of all women in town prefer this style of hose.

q. Your complaints about me are worthless, since you are obviously a mean and unsympathetic person.

20

Moral considerations

It is not a part of the discipline of logic to indicate when to argue fallaciously and when not. The disciplines of rhetoric and mass psychology, not logic, address themselves to the question of what style of presentation will be most effective with a given audience. However, it is important to notice that in addition to questions about the *effectiveness* of fallacious argument, there are also serious questions about its *morality*. My purpose in the present Section is to address briefly some of the moral issues connected with presenting bad arguments. This is not part of logic, but of moral philosophy. Nevertheless, I think it is interesting and important to sketch the connnections between logic and some other disciplines.

There seem to be three different sorts of cases to consider when we ask about the morality of giving bad arguments: (1) Cases in which the person who gives the bad argument sincerely believes the conclusion of the argument to be false. (2) Cases in which he or she sincerely believes the conclusion to be true. (3) Cases in which he or she is not sure what to believe about the conclusion.

Case 1. When the author of an argument sincerely believes the conclusion of the argument is false, and presents the bad argument in an attempt to get others to believe the conclusion, we can be fairly sure that the author is aware there is something wrong with the argument. After all, if you sincerely believe the company for which you work has an inferior product and yet you are in the advertising department and you are constructing arguments to try to show your product is superior, you would be aware that there is something wrong with your arguments. In a case like this, we seem to have a clear case of lying.

What we say about cases like this, then, will depend on what we say about lying in general. There may be situations in which lying is morally justified and others in which it is not. However, it is clear that lying is not a matter which can in general be dismissed lightly, since much of our social fabric depends on truth-telling. Philosophers have often pointed out that language could not perform its informative function if people generally believed that others almost never tell the truth. I suspect that commercial advertising is approaching such a state at present; if you as an advertiser have a piece of information about your product which you honestly wish to report to the public, it may very well be nearly impossible to do so simply because nothing you say will be taken at its informative face value by the public. Each case of lying would have to be considered on its own merits, but I do think that lying is generally a more serious matter than people realize.

Case 2. In this situation—when the arguer presents a bad argument for some conclusion which he or she sincerely believes to be true—we no longer have a lie to contend with, for one cannot be telling a lie when one is saying things one sincerely believes to be true, even if one happens to be mistaken about their truth. (Actually, I'm oversimplifying here, since one can deliberately mislead others even by telling only the truth, saying things that are true, but saying them in such a way as to *suggest* to the hearers that certain *other* things are also true, when these other things are in fact not true. A full exploration of this topic would take us far astray.) In order to deal with this case, I think it important to distinguish two situations. Let Case 2a be the case in which someone argues for something they sincerely believe to be true, using an argument they believe to be bad. Case 2b then will be the situation in which someone argues for something they believe, using a bad argument, but not recognizing the argument to be bad.

Case 2a is not the same as a case of lying, because the arguer in Case 2a believes he is telling the truth. Perhaps he does not know how to argue for his beliefs effectively by using a good argument, and so turns to the knowing use of a bad argument. A lawyer is probably in this position quite often, when the lawyer believes his or her position is correct but that the jury will never agree unless some sort of deceptive fallacious argument is used.

Although fallacious argument in a case such as this may seem relatively harmless, I believe there are moral dangers here worth considering. To be sure, there is no reasonable way I could argue that it is always immoral to knowingly give fallacious arguments for things you believe in. But I can point out some problems with engaging in that sort of behavior—problems which one might not notice at first glance. When you convince somebody to believe something true by means of a bad argument, you thereby add to the stock of their true beliefs, but you probably do not add anything to that person's ability to rationally justify his or her beliefs. In fact, you probably detract from that ability by causing them to believe something on the basis of a bad argument. The moral question in such a case is whether it is justifiable to do this.

Imagine a case in which you persuade a stubborn man who needs medication to believe in the drug because a famous sports figure once took it. Perhaps no other argument would have worked to convince him. The fallacious appeal to authority works when nothing else would have worked, and the man is saved from a horrible death. Who could criticize the use of fallacious reasoning in such a case? Yet even though the man who is saved is presumably better off because of being influenced by a bad argument, he has not been encouraged to reason better in the future. In fact, he probably has been encouraged to strengthen his faith in certain fallacious appeals to authority. The next such appeal he runs into may not be so beneficial. For example, he may be conned into buying some worthless product merely because it was endorsed by the same sports figure you appealed to. Thus, although you saved his life, you helped contribute to his rational decay.

Frankly, I believe we do not take seriously enough the social problems which might be caused by the repeated use of fallacious argumentation, for I believe the continuous use of fallacious reasoning helps to create a general social climate in which careful reasoning falls into neglect. The end result would be that people are more easily duped by all sorts of political and commercial cons, and the general level of economic and political discussion would be lowered. If I am right, there are significant moral and social problems connected with the use of fallacious argumentation, even though the conclusions toward which the argumentation is directed are factually correct. It may very well be that in many circumstances the use of fallacious argument is justified, despite its dangers, but for the reasons mentioned above I doubt that it is justified as easily as people commonly believe.

Case 2b is closely related. This case is just like the one we have been discussing, except that in Case 2b the arguer does not realize the argument he or she is employing is fallacious. Of course the fact that the arguer does not realize the argument is fallacious changes none of the consequences for the hearer. Thus, from the point of view of social consequences, this case is no brighter than the last one. However, we may wish to say that the arguer in this case is not to be held morally responsible for the bad consequences of his or her actions, if any, unless we feel that circumstances suggest the arguer should have known better.

Case 3. In this case the arguer is not sure what to think about the truth of the conclusion he or she is arguing for fallaciously. I think it is clear that we do not want to accuse the arguer of *lying* in this case, but I think it is also clear that the moral burden on him or her is heavier than in Case 2. At least in Case 2 the arguer could try to justify his or her actions by saying that the fallacious argument was persuading people of the *truth*. But in Case 3 that justification is no longer open. The fallacious argument in the present case might be persuading people of something that is *false*. And since the argument is fallacious, the same problem arises here as arose before in Case 2, namely, insofar as the argument is effective it helps to promote bad reasoning rather than good reasoning.

There are some cases in which it is perfectly justified to do the sort of thing being discussed here. For example, there may be situations in which it is important to get someone to believe something even if it isn't the truth. It may be important to use fallacious reasoning to get a cancer patient to believe he will recover, even though you don't know whether he will recover. Such cases of fallacious reasoning may very well be morally completely justified. However, I have attempted in this Section to emphasize that there are also moral risks connected with fallacious argumentation, since I believe that people in general do not commonly recognize these risks. My purpose will have been satisfied if you have been encouraged to think carefully about the social consequences and the morality of fallacious argumentation.

Moral considerations

20.1 Describe a circumstance in which you believe it would be morally justifiable to argue fallaciously for a conclusion the arguer believes to be false. Describe a circumstance in which this would not be morally justifiable. Explain the difference between the cases which makes for the difference in the morality of arguing fallaciously.

20.2 Do the same thing for circumstances falling under Case 2 as described in this Section.

20.3 Are there any moral problems raised by the use of Case 2 bad arguments in addition to the one mentioned concerning rational decay?

3

Definition and
Argument Analysis

21

The role of definitions in argument

The ability to recognize fallacies can help you avoid being taken in by a bad argument. However, there are a great many other skills connected with argument evaluation. In the present Section, we will begin to develop the skill of using, recognizing, and evaluating definitions, to the extent that this skill is relevant to assessing arguments.

Definitions occur in wide variety, and serve any number of different purposes. To get some sense of the range of phenomena which are called "definitions", consider the vast differences between the way one might define "government" for a child and the way a mathematician defines a new technical term in mathematics. The mathematician, being familiar with the terminology used in mathematics, and needing a new term to use as shorthand for a complex mathematical notion, will invent a word and state that it stands for certain mathematical concepts. The definition will be labeled as a definition; it will be precisely spelled out, yielding an exact notion of what counts as satisfying the definition and what does not. The explanation for the child will differ in a number of important ways. For one thing, the explanation given the child is an attempt to tell her how a part of an already existing language is commonly used, while the mathematician is actually adding to language by creating a new meaningful word. For another thing, the explanation given the child would not be a precise list of conditions which a thing would have to satisfy in order to be a government; instead, perhaps it lists examples intended to give her the general idea. The mathematician may give some examples but the examples will not constitute the definition.

It is not my aim to give a complete account of all the different sorts of definitions. I mention the existence of a wide variety of definitions only as a warning to you not to expect all definitions to fit the same mold. In fact there is not even a precise boundary between definitions and explanations or descriptions. For instance, the explanation given the child about the government could conceivably go on at some length. Would we want to say the *whole* explanation is a *definition* of "government"? Even if it included all sorts of factual information, such as the name of the current mayor? Presumably not. Fortunately, we can make fruitful progress in our task of argument evaluation without going into a complete account of definition.

We will concentrate on two topics concerning the relation between argument evaluation and definition: arguments which employ premises that give definitions, and arguments which employ sentences that need clarification through the use of definitions. In the present Section, we will work on picking out arguments that fit into each of these types. In the next Section, the topic will be the evaluation of definitions, and in the succeeding Section, there will be a discussion of how all this helps one to evaluate an argument of either type.

Let's begin with an example of the task of picking out arguments which contain definitions as premises:

> Whenever there are "reasons" for a belief—and most of us do want reasons—then we are involved with logic, for logic, generally speaking, is primarily concerned with the relation between "reasons" and "beliefs"— more specifically with the relation between "evidence" and a "conclusion." For a belief basically is the psychological acceptance of a statement as being true or probably true—a statement is called a conclusion when it is justified by other statements, which in turn are called evidence or reasons.[1]

The ultimate conclusion of the passage is that whenever there are "reasons" (evidence) for a belief (a conclusion), then we are involved with logic. One of the premises used along the way in order to get to the ultimate conclusion is expressed by the last sentence of the passage. The last sentence also contains definitions—definitions of "belief" and "conclusion". Thus, this is an example of the sort we want, where one of the premises contains a definition.

Nothing in the argument explicitly *says* we are dealing with a definition in that last sentence. But it seems fairly obvious that the authors intend to be giving analyses of the concepts of belief and conclusion. A definition occurs whenever an author attempts to state the meaning of a particular word or phrase, or whenever an author attempts to give an analysis of a particular concept. A statement of meaning or an analysis of a concept includes only those things that the author considers essential to the meaning or the concept, and does not include things

---

[1]Manicas and Kruger, *Essentials of Logic,* (New York: American Book Co., 1968), p. 4.

that the author thinks just happen to be associated with the word or concept through some coincidence.

The above passage clearly illustrates the typical role of definition in a premiss: the authors suppose once you understand the meaning of a key word or the analysis of a key concept in the passage (in this particular case, the concept of belief or the word "believe") then you will be able to see that certain conclusions follow. The above passage also illustrates one of the typical ways in which definitions are stated in premises: we are told, ". . . a belief basically is . . .", which should give you a clue that maybe a definition follows. When authors say things like, "the concept of _____ is basically . . .", or "_____ is essentially . . .", or "_____'s are nothing other than . . .", or something similar, you should be suspicious that a definition follows.

Nevertheless, it is sometimes difficult to tell when someone intends a premiss to be a definition. You have to use clues provided by the context and wording in order to try to decide whether you are being confronted with a definition or not. And even then you may not be sure. Suppose you happened on this letter to the editor in your newspaper, for instance:

> Whenever we have been at war, there are reports of our troops killing civilians unnecessarily. The common response is horror that we have violated the rules for conducting wars. But I don't see how there can be any rules for regulating war. War is an uncivilized exercise in bloodletting, not just a game played against people wearing uniforms. Wars include a commitment from entire nations and sweep everyone into the action, directly or indirectly. It is unrealistic to suppose only uniformed people will be killed.

There seems to be more than one argument going here, but clearly the author of this letter is using the statement: "War is an uncivilized exercise in bloodletting" as an expression of a premiss. (It is because the author sees war as being uncivilized that he thinks it's impossible to have legitimate rules for the conduct of soldiers engaging the enemy in the war. Apparently the author thinks that rules of the sort he imagines are possible only in an activity which is "civilized." I suppose he is thinking of moral or legal rules, and he is supposing that both of these kinds of rules are based in social organization which can be present only in a civilized activity.) Now my question at this point is whether the premiss we are discussing is supposed to be a *definition* of "war". The question is whether we think the author is trying to give us an analysis of what he means by "war", or is he merely trying to describe a fact about war.

This sort of question will arise frequently, particularly when an author says something of the form, "X is . . . ." One often can't tell in such a case if the author is trying to *define* X or if he is just trying to say something true about X. For instance, if I say, "Today is Thursday", it is not at all likely I mean to be giving a definition of "today". But if I say "A book is a stack of sheets of paper containing visual materials and bound together in a relatively permanent fashion" I probably am trying to give a definition of "book".

The situation is complicated because people often give poor definitions and they are often satisfied to give *partial* definitions rather than complete ones. A partial definition just picks out some few essential features that go with a certain concept, rather than giving the whole list of essential features which belong to that concept. Perhaps the statement about war contained in our letter to the editor is a partial definition of "war". If so, the author would not think *all* types of uncivilized exercises in bloodletting constitute war; he would mean only that it is one essential feature of the concept of war that war is an uncivilized exercise in bloodletting.

Only a close reading of the argument in context will be likely to help you decide whether statements such as these are intended as definitions or mere true statements:

> Anything which people are willing to exchange freely for available goods counts as money. Pleasure is good; good is pleasure. The truly wealthy person is the happy person.

To sum up, we have seen that premisses sometimes state either partial or complete definitions, but that it is often difficult to tell just when this is going on. Keep in mind both the nature and variety of definitions when you are trying to decide whether a given argument contains such a premiss.

Now that we have talked a bit about recognizing definitions when we see them, we turn to the task of recognizing arguments which need definitions but don't contain them. The remainder of this Section will be devoted to sensitizing you to this type of argument.

One of the most important things I can do when talking about the need for definitions is to show when definitions are *not* needed. Many people, unskilled at argument, invariably employ the request for a definition as a tactic whenever they don't know how else to respond to a view with which they disagree. They go around saying things like: "It all depends on how we define our terms". "It's all a matter of definition", "First, you have to tell me what you mean by . . ." Of course, in any system of human communication which employs symbols the messages conveyed by the symbols will depend on the meanings of the symbols. But that does not mean that every single message remains unclear until definitions have been explicitly given for every symbol. There is no need to give definitions, and the request for a definition becomes obnoxious as a debate tactic when everyone party to a discussion knows what the relevant words mean and nothing in the arguments within the discussion requires the clarifying power of a definition. After all, the request for definitions can never be satisfied if the requester insists that *every* key term be independently defined, since each new definition will in turn contain new key terms standing in need of definition.

On the other hand, there definitely are arguments which need definitions before one can tell for certain what is being said. The author of the following passage is arguing for the legal right of women to decide for themselves when to have an abortion.

> To deny a woman the right to an abortion is to force her to carry within her body an unwanted child, and this is to deny her the right of self-determination, reducing her to the worst sort of slavery. Especially when the fetus cannot live on its own, the woman is thereby put into the degrading position of involuntary servitude, having been denied her rights as a human being.

Now I understand perfectly well what almost all of these claims mean, and I would merely be obnoxious if I said, "What do you mean by 'rights'?" But, I am genuinely puzzled about the reference to slavery. It would be helpful to know what "slavery" means in this argument, since the word seems to be used in a special sense, for a woman who is forced to carry an unwanted child has not been literally sold to a master in the way slaves are sold. Perhaps a slave in this passage is anyone who, like a real slave, has lost freedom of choice about the important things in his or her life. If so, we would want to give a definition like, "By 'slave' we mean someone who has lost the freedom to choose the sort of life she wants."

In cases like this, it is important either to get definitions from the author, or to propose definitions for the author, because the quality of the argument depends on what is actually being claimed within it. In a case where a request for definition is appropriate, the point of the request is to reveal exactly what propositions are being expressed by the author's sentences. That is, you are trying to figure out exactly what the argument actually is—what propositions make it up. This is an essential first step in evaluating the argument, for you will not be able to evaluate an argument when you don't know what the argument is. If the author is unavailable and can't supply the needed definitions, you may have to try some definitions out on his or her behalf. If you can think of more than one possible definition for some key term in the argument, each different definition will yield a new argument to be evaluated separately. Thus in the above example, if you can think of an alternative definition for "slave", you will have thought of a new version of the argument and that new version would have to be evaluated on its own.

In this Section we have been working on recognizing arguments which contain premisses giving definitions and those which contain sentences sufficiently unclear as to require clarification via definition. In the next Section we will look at ways of evaluating definitions. But first, the exercises.

The role of definitions in argument

21.1 In each of the following arguments, find a premiss or a set of premisses which express complete or partial definitions. In each case, state what terms are being defined, and whether you believe the definition is intended to be partial or complete. If you believe the definition is only partial, state why you believe this.

  a. *Alcohol is a special danger to children because it tends to cause hypoglycemia, which is a drop in blood sugar. The brain needs blood sugar to function; so if the blood sugar drops for long enough, brain damage or retardation can occur.*[1]

  b. *The summer of 1964 brought violent protests to the ghettos of America's cities, not in mobilization of effective power, but as an outpouring of unplanned revolt. The revolts in Harlem were not led by a mob, for a*

---

[1]Quote from Dr. William Atermeier in a UPI story, April 26, 1977.

*mob is an uncontrolled social force bent on irrational destruction. . . .
Even those Negroes who threw bottles and bricks from the roofs were
not in the grip of a wild abandon, but seemed deliberately to be prod-
ding the police to behave openly as the barbarians that the Negroes felt
they actually were.*[2]

c. *America has contributed to the concept of the ghetto the restriction of
persons to a special area and the limiting of their freedom of choice on
the basis of skin color. . . . Yet the ghetto is not totally isolated. The
mass media . . . penetrate . . . in continuous and inevitable communication, largely one-way, and project the values and aspirations . . . of
the larger white-dominated society. . . . If the ghetto could be contained totally, the chances of social revolt would be decreased, if not
eliminated, but it cannot be contained and the outside world intrudes.*[3]

d. *Why ban phosphates from detergents? The theory . . . is that reducing
phosphate levels in sewage discharges will slow the growth of algae
and unrooted weeds in our streams and lakes. . . . Phosphates are natural forms of an element called phosphorus. . . . Approximately 25 percent of the phosphorus entering lakes and streams comes from human
excrement and food waste. . . . Only 1 percent of the total input of
phosphorus will support excessive growth of algae when sufficient
amounts of all other nutrients are also available. . . . The only way to
reduce excessive algae growth is proper sewage treatment, which re-
moves all nutrients.*[4]

e. *It might be useful . . . to call attention to the common mistake of identifying determinism—the thesis that every event is connected with antecedent [i.e., prior] events by one or more laws—with fatalism. Fatalism
is a doctrine that is primarily concerned with the destiny of man. The
doctrine holds that man's destiny is fixed, decided upon, and recorded
in the big ledger of fate. Man's will is no match for the decrees of fate.
It is futile to take measures for his welfare, . . . for man is powerless to
escape his fate. Determinism makes no such ominous statements.*[5]

21.2 Explain the difference between giving a definition of the word "love" and
making a claim about all instances of love that is not a definition.

21.3 Some people say, "God is love". Do you think they are defining "God" or
are they doing something else? If you think they are doing something besides defining "God", what is it you think they are doing? (Consider
whether perhaps this is a partial definition. Also consider whether it is intended merely to characterize God without defining Him.)

[2]Kenneth B. Clark, "The Invisible Wall," reprinted in *Crisis*, ed. by Peter Collier (New York: Har-
court, Brace and World, 1969), pp. 52–3.
[3]Kenneth Clark, "The Invisible Wall."
[4]FMC Environmental Research Laboratories, "The Theory that Doesn't Work," Princeton, N.J.,
1974. The purpose of this pamphlet was to argue against the phosphate ban in detergents.
[5]University of California Associates, "The Freedom of the Will," reprinted in *Readings in Philosophical Analysis*, ed. by Feigl and Sellars, (New York: Appleton-Century-Crofts, 1949), p. 614.

21.4 The following arguments contain key words which seem to need definition before the arguments themselves are clear enough to be evaluated. Pick out the word or words needing definition and explain why you think these words could stand to be defined as a part of the evaluation of the argument.

a. Our own leaders refused to let the Air Force destroy Haiphong Harbor in North Viet Nam, even though 90% of the weapons used against our own Army were channelled through that port. To my mind, this makes our leaders guilty of war crimes, for which they should stand trial.

b. The local township officials are lying when they say their intent is to work closely with the people, for when I tried to obtain information about the system of government in the township I was given the bureaucratic run-around.

c. *Because a maladjusted person is not always aware of the piecemeal concepts which make up the disturbing aura of subjectivity surrounding his handicap, a main goal in speech therapy is to help him probe and review experiences, to give him insight into cause-effect relationships, and to help him reconcile the imbalance between subjectivity and objectivity.*[6]

d. *Though they are impatient, Americans are committed to the concept of democracy, and though they recognize that democratic processes are frustratingly slow, they turn toward them in solving . . . problems. Because the American culture is democratically oriented, many curriculum experiences for American children are designed to develop civic responsibility and efficiency.*[7]

e. *Does America have a political Left? . . . Our dominant political party . . . is at least nominally commited to such "socialistic" enterprises as national health insurance, . . . and stronger measures to combat corporate concentration in the economy. These programs, however, do not demonstrate that there is a political Left in America. . . . In fact, judged by their effects, rather than by what their supporters and foes hoped or feared they would accomplish, these government interventions prove not the presence of an American Left, but its absence. Most of the money the U.S. government spends on transfer payments . . . does not represent a redistribution of wealth or income.*[8]

[6]George O. Egland, *Speech and Language Problems* (Englewood Cliffs, N.J.: Prentice-Hall, 1970), p. 68.
[7]Ruth Cook and Ronald Doll, *The Elementary School Curriculum* (Boston: Allyn and Bacon, 1973), p. 107.
[8]Jeff Greenfield, "The Absent Left," *Harper's Magazine*, Vol. 255, No. 1528, (1977) p. 19.

22

The evaluation of a definition

Since many arguments either contain definitions or need to be supplemented by definitions for the sake of clarity, it's important to know how to tell the difference between a good definition and a bad one. The difference, however, is not entirely a matter of logic. In order to evaluate a definition you need to know the language containing the defined word, and knowing a language is not a matter of logic. However, it is usually conceded that meaning relationships between words and phrases are intimately involved in the assessment of argument cogency, and thus the general theory of definition belongs to logic or is at least closely related to logic. So in this Section you can expect to find general theory about the evaluation of definitions, but you should not expect to find enough information to allow you to define words or test the definitions of any given words when you are not familiar with the way those words are used in the language.

The evaluation of definitions is made more complex by the wide variety of definitions. I noted earlier that definitions serve many different purposes. It will be enough, though, if we classify these purposes as falling into one of two categories, or somewhere in between. The first category consists of *lexical definitions*—definitions which are supposed to report the actual meaning of a word or phrase in the language. Definitions of this sort often appear in a standard dictionary of the language. The second category consists of *stipulative definitions*—those definitions which give a new meaning to a word or phrase of the language. Sometimes the word or phrase itself is new, sometimes only the meaning is new. These definitions are called "stipulative" because by means of them, authors stipulate how they intend to use some particular word or phrase. Partial definitions as well as complete ones can be classified as lexical or stipulative.

In between these two categories we find other definitions which are partly lexical and partly stipulative. For instance, when a particular word is quite vague in the language an accurate lexical definition would exactly match the actual vagueness of the word. But it may suit an author's purpose to give a definition which makes the vague word somewhat more precise in meaning without completely changing the meaning. Such a definition is not strictly lexical, since it is not intended to report accurately the actual meaning of the word, but it is not strictly stipulative either, since it is not intended to provide a really new meaning, but only to alter the old meaning slightly by making it more precise. If you were writing a book on educational theory, a good candidate for a definition which was partly lexical and partly stipulative would be the vague word "education", because its vagueness allows you as author to clarify its meaning in a way that you find helpful to your theory. Here, you need not think of yourself as giving a clearer definition of the *ordinary* concept of education, but you may wish to *build on* the ordinary concept, to point out what you think are the truly important aspects of what ordinarily is called education. This kind of definition does not attempt to be completely faithful to ordinary usage, but instead makes use of ordinary usage as a key to the creative *revision* of the notion of education, a revision intended to fit into some theory you hold regarding the best way to educate. You might say, for example, that "education is nothing more than human mental improvement". By proposing this definition, you build on the common idea that education is supposed to be good for the mind, but you draw attention away from the common idea that education has something to do with certain subject matter or institutions. You are attempting by means of a definition to make a point about how broadly one should construe the notion of education. This type of definition constitutes a recommendation about how one ought to think; it is not a mere reporting of how people in fact do think, as a lexical definition would be. Yet it is not purely stipulative; the proposed definition links closely with the common meaning of the word.

The evaluation of a stipulative definition proceeds very differently from the evaluation of a lexical definition. Whereas a correct lexical definition must correspond to the actual use of the defined word or phrase, a stipulative definition is not thus restricted. I believe there are basically three tests of a good stipulative definition when that definition occurs in an argument:

1. The definition should serve some useful purpose in the argument.

2. The benefits derived from giving a new meaning to an old word should outweigh the confusion which such a procedure can cause in the reader.

3. The author should be willing to stick to using the defined terms in the way they have been defined, within the relevant section of discourse.

In order to illustrate why these tests are important, let me show you what happens when they are not satisfied. Consider this passage:

Let us agree that for our purposes "love" is any feeling of sincere appreciation which one feels toward another person. So we shall speak of love when someone helps you with the door and you feel appreciative, or when you appreciate the job your local city councilman has done for the city. There can be no doubt that people often do feel sincere appreciation for others. Thus, there is no doubt that people do often love others.

It is hard to read this passage without wondering why the author went to the trouble to give a special definition to "love" for it seems the author could have made the same points merely by saying that people often feel sincere appreciation for others. So the first test is not met.

The above passage also illustrates failure with respect to the second test. It is bound to confuse the reader when a common word like "love" is given a special new meaning which is somewhat related to the old meaning. The switch to the new meaning may even make the careless reader suppose the above argument proves a great deal more than it actually does. The last sentence only says that people often feel sincere appreciation for others—at least if the author is playing fair. However, since the word "love" is used in the sentence, the sentence may have the misleading appearance of saying something more. Our author seems only to have confused the issue by using the word "love" in an odd way.

Suppose we confront our author with these complaints and he replies that the last sentence really does say more than that people often sincerely appreciate others, for it means that people really do love others in the common sense of "love". If our author replies in that fashion, he shows that he has violated our third test of a good stipulative definition. He is not sticking to the stipulatively defined sense of "love". This is a more common problem than you might imagine. Some authors seem tempted to try to prove important things by using words in odd ways, giving stipulative definitions for them, and then supposing that the things that have been proved also apply to the words original meaning. It should be clear that in the argument given above the conclusion contained in the last sentence cannot be reached without using the special definition given in the first sentence. Thus the conclusion of the above argument makes logical sense only if we assume the word "love" is being used in its special sense throughout.

In dealing with a definition that is partly stipulative and partly lexical, we evaluate the stipulative aspects of the definition according to the standards given above for purely stipulative definitions. Accordingly, if a writer on educational theory decides to give a definition of "education" in order to make that term somewhat more precise than it otherwise is in standard English, we shall require that the definition meet the three listed tests.

Lexical definitions are evaluated differently. In checking a lexical definition for accuracy, the issue will be whether the definition accords with an established meaning of the phrase being defined. While this may sound easy, it isn't, even when you are familiar with the phrase in question. But there are some standard tests which you can use to help. These tests apply when the phrase being defined can be thought of as *referring to* or *applying to* something, even though some words or phrases cannot be thought of in that way. (I doubt that you would want to claim that the word, "because", refers or applies to anything.) Probably most of the definitions you will encounter in arguments will be definitions of phrases which can be thought of as referring or applying to something.

There are at least two tests which a good lexical definition of such an expression must pass: (1) It cannot be too *broad,* (2) nor can it be too *narrow*. A definition is too broad when it is in any way possible for there to be things which satisfy the definition, but which the original phrase being defined would not refer or apply to . In other words, a definition which is too broad applies to too many possible things. For instance, if I define "bird" as "a two-legged creature", my definition is too broad because it is possible for there to be things, such as people, which satisfy the definition (because they are two-legged) but which would not be referred to by the original word "bird". It is important that you not restrict yourself to objects which actually exist when applying this test. Even if there are no *actual* objects which satisfy the definition but which are not referred to by the original word, the definition will still be too broad if it is at all *possible* for there to be such objects. Thus, if I define "bird" as "two-legged creature with feathers", my definition is still too broad because it is possible that there could be feathered two-legged creatures that are not birds, even if there aren't any such creatures in existence this year.

A definition is too narrow when things specified in the original word or phrase can exist but do not satisfy the definition. You can see that this situation is the reverse of the problem which arises when the definition is too broad. In other words a definition which is too narrow applies to too few possible things. Again, it does not matter whether the things that do not satisfy the proposed definition actually exist or not, so long as it is in any way possible for them to exist. Thus, again my proposed definition of "bird" as "two-legged feathered creature" fails, this time because it is too narrow. Why? Because it is possible that there could be a bird without feathers. The definition is too narrow in that it fails to include all the things that could count as being birds. You will notice that my poor definition of "bird" was *both* too broad and too narrow. So it is entirely possible for one and the same definition to suffer from both problems at once. If you want an example of a definition which suffers only from the narrowness problem, here is one: Define "president of the United States" as "the man who holds the highest elective office in the executive branch of the United States government under the current Constitution". This definition is too narrow, but I don't think it is too broad. Do you see why it is too narrow? If you don't, you are probably a victim of the pervasive sexism which abounds in the U.S.

Probably a good lexical definition has to meet further tests beyond those mentioned above; however, it is quite controversial as to just what these further tests might be. I'm not going to enter into those debates here, although I can give you some idea of what the issues are. Here is a definition which is neither too broad nor too narrow, but which many people nevertheless find defective: define "Euclidean triangle" by "plane closed Euclidean figure with straight sides and interior angles summing to 180 degrees". Now it is not possible according to Euclidean geometry for anything satisfying this definition not to be a Euclidean triangle; so the definition is not too broad. Nor is it possible for anything to be a Euclidean triangle without satisfying this definition; so the definition is not too narrow. However, many people think that the definition is defective because it focusses on features which just come along automatically with the more essential or important features of triangles. People often want to say the proper definition of "triangle" would say something about the figure's having three sides or three angles. The general point is this: even though a given lexical definition is neither too broad nor too narrow, it may not mention those features which seem to most people to be the ones that are important in the concept being defined. Thus, it seems that a third test of a good lexical definition is some sort of relevance test, some sort of test which says that a good definition must define by using those features which are most relevant in the minds of people using the language. I will not worry about trying to formulate precisely such a relevance test here, because the matter is as yet too controversial to be settled in a beginning text.

When confronted with a definition that is partially stipulative and partially lexical, it is difficult to know just how to apply the broadness and narrowness tests to the lexical aspects of the definition. One needs to see to what extent the author is attempting to make his definition fit the common ordinary use of the term. One cannot expect the definition as a whole to meet the standards for a correct lexical definition; for example a definition which is intended to make a term somewhat more precise than it is in ordinary usage will normally be judged too narrow by the standards for a purely lexical definition. However, it is important to note the *extent* such a definition is too narrow or too broad to be a good lexical definition, for on occasion authors give partially lexical definitions which fail to meet the author's own purposes because the definitions are broader or narrower than intended. Let me illustrate with a passage drawn from an imaginary book on educational theory:

> The human animal from the moment of birth, and perhaps even before, is constantly learning. Impressed with the variety and ubiquity of learning behaviors, we may be tempted to refer to all of life as educational. Whatever the merits of giving in to that temptation, however, our purpose is to deal with the more formal, more structured kinds of learning situations typically found in the schools. When we speak of education, then, we shall mean to refer to activities in which one engages as a client of an organized social institution one of whose major functions is to cause learning in its clients.

The final sentence of the passage gives a definition which surely is not entirely lexical, but which one might argue is intended to be built upon one common lexical meaning of "education". Given the discussion preceding the definition in the passage, we can expect the definition to be too narrow to be a good lexical definition, for one of the author's stated purposes is to narrow down the scope of his topic. Our expectation is fulfilled; the proposed definition is indeed too narrow to be an accurate report of the meaning of "education", since for one thing it does not include the short-term educational activities of business training programs. Of course, we are not going to criticize the proposed definition on these grounds unless we find that the criteria for good stipulative definitions will not justify the narrowing of the definition in this way. One would have to see how useful this way of narrowing the meaning of "education" proves to be in the larger context of the author's work.

Even though we recognize the definition proposed above as being at least partly stipulative and thus immune from being automatically criticized on the grounds of being too narrow, it is still useful to note just how narrow the definition is, so as to be in a better position to evaluate its usefulness later. It is also useful to see if the definition is too broad to be a good lexical definition. In fact, a careful look will reveal that it is indeed too broad. The definition fails to exclude non-educational activities which one might have to engage in as a client of an educational institution. For instance, students often have to stand in line to register. This activity may be educational at first, but one rapidly reaches a point at which no more learning takes place from this activity. Yet according to the proposed definition, the activity of standing in line to register in an educational institution is always part of an education simply by virtue of the fact that the line-standing takes place in an educational institution. Once we have noticed that the definition is too broad, we have some interesting questions to ask the author, for he gave the impression that he was only interested in narrowing the definition, not in broadening it. Thus, we have learned something important when we have learned that the definition is too broad by lexical definition standards. Of course we will have to consider whether this broadening may be a legitimate stipulative move.

In sum, when dealing with a definition which is at least partly stipulative, the broadness and narrowness tests which are applicable to lexical definitions can still yield interesting results. However, one cannot say that a stipulative definition which is too broad or too narrow by lexical standards is automatically bad. Rather, by applying the broadness and narrowness tests one learns in just what ways the proposed stipulative definition differs from a good lexical definition. That is important information to retain as one evaluates the effectiveness of the stipulative aspects of the definition.

Finally a word about the evaluation of *partial* definitions is in order. A partial definition will always be too broad to be a good complete lexical definition simply because the partial definition leaves out some of the defining conditions which belong in the complete definition. When these conditions are left out, the effect is to broaden the scope of the definition. Thus, to say that by definition a bird is necessarily an animal is to give a partial definition of "bird" which is obviously too broad to be an accurate complete lexical definition. However, a partial definition that is too *narrow* is faulty. If I were to say that a bird by definition must have feathers, I would have made my partial definition too narrow, for it is possible to have featherless birds. Thus, a partial definition that is too narrow is defective unless the narrowness arises from stipulative aspects of the definition.

The evaluation of a definition

22.1 Each of the following sentences can be viewed as expressing a stipulative definition. For each, describe a possibility which shows that the proposed definition is either too broad or too narrow to be a good *lexical* definition. State whether the possibility you have come up with shows the definition to be too broad or whether it shows the definition to be too narrow by lexical standards.
  a. Any object which people value may be thought of as money.
  b. Real freedom consists of a social system in which every person develops his or her full human potential.
  c. Leadership is the ability to get other people to do things.
  d. Love is never having to say you're sorry.
  e. The real essence of being human is the ability to care about other people.

22.2 For each of the preceding stipulative definitions, describe some purpose which it is reasonable to suppose an author might have for giving the definition. Usually the best way to describe such a purpose would be to explain what conclusion the author might be driving at by setting up the definition in the given way.

22.3 For each of the above definitions, point out at least one possible interesting specific confusion which might arise in a reader as a result of the employment of the stipulative definition in an argument. For example, with respect to the first definition you could say that the definition stresses the psychological aspect of money but doesn't account for the legal aspects, thus possibly misleading readers into ignoring the role which government plays in setting up a monetary system.

22.4 A stipulative definition is not used properly in a given passage when, without warning, its author reverts to using the stipulatively defined term in its ordinary sense rather than its defined sense. You can generally detect when this has happened if you can find a phrase in the passage which contains the defined term but which quite clearly would be false or silly if the term had its defined sense in that phrase. For each of the stipulative definitions in 22.1, make up a sentence or two which would serve this purpose if it were to occur in a passage containing the definition. For example, one possible answer with respect to the first definition would be "Even though people often value art objects, these objects generally are not durable or portable enough to be used as money".

22.5 Read the following passage carefully and answer the following questions.

As we study processes taking place under a wide range of conditions, we discover that inside all of the complexity of change . . . there are relationships that remain effectively constant. Thus, objects released in mid-air under a wide range of conditions quite consistently fall to the ground. . . . We interpret this constancy as signifying that such relationships are necessary, in the sense that they could not be otherwise, because they are inherent . . . aspects of what things are. The necessary relationships between objects, events, conditions, or other things at a given time and those at later times are then termed causal laws.

At this point, however, we meet a new problem. For the necessity of a causal law is never absolute. For example, let us consider the law that an object released in mid-air will fall. This in fact is usually what happens. But if the object is a piece of paper, and if . . . there is a strong breeze blowing, it may rise. . . . Such contingencies lead to chance. *Hence, we conceive of the necessity of a law of nature as* conditional, *since it applies only to the extent that these contingencies may be neglected. Now, here it may be objected that if one took into account* everything *in the universe, then the category of contingency would disappear, and all that happens would be seen to follow necessarily and inevitably. On the other hand, there is no known causal law that really does this.*

In sum, then, we may say that the processes taking place in nature have been found to satisfy laws that are more general than those of causality. For these processes may also satisfy laws of chance. . . .[1]

a. Using your own understanding of what people mean when they speak of one thing causing another, do you think the definition of "causal law" given by this author is a good lexical definition? (Use the broadness and narrowness tests.)
b. Judging by the way the author of the passage leads up to the definition of "causal law" would you say that he intends the definition to be lexical or stipulative? Why?
c. For what ultimate conclusion is the author arguing in this passage? In what way is the definition of "causal law" important in helping him to

[1]From David Bohm, *Causality and Chance in Modern Physics* (Philadelphia: Van Nostrand, 1957), pp. 1–3.

argue for this conclusion? If the proposed definition of "causal law" were changed so that causal relationships could have exceptions, then would the author be able to describe his conclusion in the same terms?

22.6 Prove that each of the following lexical definitions is either too broad or too narrow, or both.
  a. An automobile is a vehicle with four wheels and a motor, large enough to carry at least one person.
  b. "Democracy" means that all the eligible citizens elect representatives to make the laws for the relevant governmental unit.
  c. By "human rights" we mean those areas of human life in which the government agrees not to interfere.
  d. Logic is the study of human reasoning.
  e. A teacher is a person who causes another person to learn something.
  f. A capitalist is a rich greedy person.

23

Application of the theory of definition to argument evaluation

When an argument utilizes a premiss which contains a lexical definition the truth of the premiss depends on the correctness of the definition. You can employ the broadness and narrowness tests in order to discover whether the relevant definition is correct. If it is not, then you are faced with an argument that contains a false premiss, and that means that no matter how well the argument proceeds from that point, any conclusion depending on the false premiss may itself be false. On the other hand, if the definition appears to be correct, you may go on to evaluate the other aspects of the argument using whatever techniques you can. Later in this book we will discuss some of those techniques, but you already are aware of some of the fallacies to watch for.

Similar things can be said about arguments which contain *partial* lexical definitions. To be sure you can't apply the broadness test, but you can use the narrowness test. A partial lexical definition which is too narrow is just plain mistaken, and the argument which depends on such a definition is in trouble precisely because it employs a false premiss. (The conclusion is not necessarily false, though, for it might just happen to be true, even though it was obtained through use of a false premiss.)

Arguments which employ wholly or partially stipulative definitions as premisses are more complicated. Since a stipulative definition is not supposed to be reporting the actual common meaning of a term, but instead specifies the meaning which the author wishes to give it, we cannot ever complain that the definition is false. However, as we have already seen, a stipulative definition can be good or bad, useful or useless, helpful or confusing.

Suppose you discover a stipulative definition in an argument and you think the definition is useless because it only confuses the issue and could just as well have been omitted. It seems to me that the best thing to say about the argument is just that its wording has been made unnecessarily confusing by the introduction of a stipulative definition which was not needed. Otherwise, the argument should be evaluated as it would have been had the definition never been there. You will still check for fallacies and you will still perform the other checks you will learn about in the rest of this book. The only role of the definition in all this checking is to complicate matters, for you have to keep in mind the peculiar way in which one or more of the terms in the argument are being used in order to understand what is being said in the argument.

When the stipulative definition fails to be successful because the author does not stick to the stipulation laid down in the definition, that presents special problems. In a case like that, before you can decide what to say about the argument you will have to rephrase it in whatever way seems to be closest to the author's real intentions. The problem in such a case is deciding what the author really meant. After you have decided on an interpretation of the argument, you are then ready to go on to evaluate the interpreted argument in whatever way seems appropriate.

That last point is also relevant to the evaluation of arguments which need clarification by means of definition. You will recall that some arguments which do not contain definitions nevertheless need the clarification of some key terms which a good definition can provide. If you are faced with such an argument, try to supply the needed definitions or get the author to do so, and then evaluate the original argument as clarified. In a case like this, your problem is to figure out exactly what propositions the author meant to be expressing, since the original wording was not clear. To illustrate, let me use a variation on an argument from an earlier Section:

> A woman who is forced to carry an unwanted child in pregnancy is being made a slave. No one should ever be made a slave. Thus, no woman should be forced to carry an unwanted child in pregnancy and abortions should be made readily available.

Suppose we explain how the word "slave" is being used here by saying "slave" means "a person who is not free to make some major decisions about the type of life he or she will lead." The argument then amounts to this:

> A woman who is forced to carry an unwanted child in pregnancy is prevented from being free to make some major decisions about the type of life she will lead. No one should ever be put in that position. Thus, no woman . . . .

This argument is now ready for further evaluation, utilizing whatever techniques are appropriate. Since the argument does not seem to commit any of the fallacies discussed earlier, we will have to wait until later to develop the techniques needed to evaluate it.

Application of the theory of definition to argument evaluation

23.1 The following argument employs a premiss which you worked on in Exercise 22.1.

The real essence of being human is the ability to care about other people. Unfortunately, this ability has been completely eradicated in some individuals, through the force and pressure of urban ghetto living. Once this has happened, then, we are justified in treating these individuals like the beasts they have become, for only humans are deserving of human rights.

a. If you were to treat the first sentence of this argument as an attempt at a *lexical* definition, what evaluation would you have of the argument?
b. On the other hand, suppose the first sentence expresses a stipulative, or partially stipulative, definition. What effect does this stipulation of meaning for "human" have on the meaning of "human rights" in the last premiss? It seems clear that only humans·have human rights, until one considers that "human" is being used oddly in this argument. Do you think confusion has been introduced into the argument by the employment of the definition? Can you restate in a different way exactly what point this author is making? What is your overall evaluation of this argument?

23.2 Here is an unclear argument about evaluating faculty members' work.

Some academicians judge faculty evaluation in terms of absolutes. Since evaluation techniques and procedures for faculty evaluation are less than perfect, they would throw out any advancements that could be made. This is an idealistic—and unrealistic—position.[1]

a. Presumably, the first sentence states one premiss and the first half of the second sentence states the second premiss. The first conclusion is the second half of the second sentence. In order for this argument to make logical sense, what definition of "absolute" is needed?
b. Rewrite the argument eliminating the term, "absolutes", in favor of its definition. Does the resulting argument seem to make sense?

23.3 Recently, Senator William E. Brock III has argued:

In the early days of [our policy of] containment [of the Soviet influence] the State Department stressed the revolutionary character of Soviet communism, asserting that Russian rulers sought . . . total destruction of the rival power. . . . Today, there is a tendency to identify containment as a policy of risky confrontations. This definition does not follow Webster, nor does it correspond to reality. As to the former, containment is "the policy of at-

---

[1]Richard I. Miller, *Evaluating Faculty Performance* (San Francisco: Jossey-Bass, 1972), p. 8.

*tempting to prevent the influence of an opposing nation or political
system from spreading." As to the historical past, there were actually more
incidents under containment where confrontations were evaded than
faced.*[2]

   a. Brock is here arguing that one proposed definition of "containment" is
incorrect. He offers another definition from Webster in its place. If
Brock's preferred definition is correct, what does that imply about
whether the rejected definition is too broad or too narrow?

   b. Suppose Webster's definition of "containment" is correct, and that the
U.S. pursued a policy of containment with respect to the U.S.S.R. Does
it follow that historically the U.S. did not pursue a policy of risky confrontations with the U.S.S.R.?

23.4 Later in the same article, Senator Brock discusses a label given to later
relations between the U.S.S.R. and the U.S.A., namely, "detente". He defines "detente" as follows:

*a lessening of tension without any connotation as to the depth or perma-
nency of the process.*

He concludes that it is a mistake to try to draw a sharp line between a
policy of detente and a policy of containment. Apparently, he believes that
the two types of policy could overlap with the result that one could be
pursuing both at the same time by doing the same things. Given the definitions he has proposed, is it possible that some activities could count both
as pursuit of a policy of detente and also a policy of containment? (This is
a question about the broadness or the narrowness of the definitions, because it asks whether "containment" and "detente" are defined broadly
enough to possibly overlap.)

23.5 Write a full analysis of the role of definition in the following argument.

*Real freedom exists in a social system when and only when every person
living within the system develops his or her full human potential. That is
why the claims of contemporary American politicians that the American
people are free are so absurd. All one need do to put the lie to these
claims is to visit the local skid row where the waste of potential is enormous
enough that even a casual observer cannot ignore it.*

(Hint: Think about the things you did in the previous exercises, especially
23.1)

23.6 Discuss the need for definition in order to give a full evaluation of the following argument:

*The questioning of axioms is not a simple affair of reasoning; it involves
some knowledge of ontology as well as of logic. The proper kind of edu-*

---

[2]"Detente and Containment—The Dangers of Semantics," *Journal of Social and Political Affairs,
vol. 1, no. 1, (1976): pp. 3–13.*

cation, then, must consist in the eliciting of contradictions in the matter of unconsciously held beliefs, to demonstrate elements of untenability in the implicit dominant ontology. Only when this has been done have we prepared a student for the ready reception of material furnished by the agreement between logic and fact. For to convince him that he holds contradictions is to render his present beliefs untenable and thus to put him in the way of examining others along with them.[3]

[3]James K. Feibleman, *The Two-Story World* (New York: Holt, Rinehart, & Winston, 1966), pp. 422–3.

4

Validity
and Invalidity

24

Deductive validity

So far we have discussed various techniques for revealing the structure of arguments and for recognizing some specific kinds of commonly occurring mistakes in reasoning. Starting from the most basic step in argument analysis—the separation of premisses from conclusions, we have gradually progressed to more detailed analysis of the internal structure of the arguments. All this has been aimed toward one goal, evaluation of arguments. Although not all argument evaluation depends completely on argument structure, much of it does. (Some evaluation of arguments seems to depend partly on placing the arguments into context, rather than on mere analysis of structure, as we saw in our discussion of the fallacies.)

One concept used in argument evaluation which relates entirely to the internal structure of an argument is the concept of *deductive validity*. Our focus in Chapter 4 will be on this concept and its application.

Perhaps the best way to introduce the concept of deductive validity is with examples. First, I will give two examples of simple arguments which are deductively valid. These will be followed by two examples of arguments which are not deductively valid. Your job is to pick out the relevant difference.

Deductively valid arguments

P1.   Whenever I wash my car, it always rains within twelve hours.
P2.   I'm washing my car now.
C.   It will rain within twelve hours.

P1.   All female animals are feathered.
P2.   Betsy Ross is a female animal.
C.   Betsy Ross is feathered.

Deductively invalid arguments

P1.   Whenever I wash my car, it always rains within twelve hours.
P2.   It is raining now.
C.   I must have washed my car within the last twelve hours.

P1.   Some female animals are feathered.
P2.   Betsy Ross is a female animal.
C.   Betsy Ross is feathered.

In looking at these examples, it should become apparent that it does not matter whether the premisses and conclusions are true in the real world. In fact, the first premisses of both valid arguments probably were false. Nevertheless, the arguments are deductively valid.

If the truth of the premisses or conclusions does not matter to deductive validity, what *does* matter? Perhaps you have noticed that the arguments which are deductively valid hang together logically better than the arguments which are deductively invalid. Looking at the first invalid argument, it is tempting to say that even though it is raining, that doesn't necessarily mean the arguer washed his or her car recently, since the premisses of the argument don't say that it rains *only* after the arguer washes the car. Maybe it rains at other times too. This is a problem with the argument—it has a loophole. Similarly, the second argument has a loophole as well. Its first premiss does not say *all* female animals are feathered; it only says some of them are. Thus, even if Betsy Ross is a female animal, she still might not be feathered. It is possible that its conclusion is incorrect even if its premisses are correct. Deductively valid arguments don't have this sort of loophole, and that is the crucial difference between deductively valid arguments and arguments which are not deductively valid.

This leads us to attempt a precise definition of the concept of deductive validity: an argument is deductively valid when it is completely impossible for its conclusion to be false if its premisses are all true. I think you can see why it is completely impossible for the conclusions in the two sample deductively valid arguments to be false *if* the premisses in these arguments are true. Note the "if". The definition of deductive validity does *not* say the premisses of a deductively valid argument really are true; it just says that *if* they are true, *then* the conclusion will be true also. In other words, the notion of deductive validity has to do with the *relationship* between premisses and conclusions.

Unfortunately, the definition of deductive validity proposed above lacks precision because it is not exactly clear what counts as something being "completely impossible", and people commonly use the notion of impossibility in very many different ways.

The notion of complete impossibility which we need has a special name, *logical impossibility*. But of course, to give the notion a name is not to explain it. What kind of impossibility, then, is this logical impossibility? A thing is logically impossible if it remains impossible even when the facts change, even when the principles of human psychology change, even when the laws of nature change. Thus, even though you may now say that it is completely impossible for you to like a certain person you happen to know very well, that impossibility will not be a logical impossibility, because your liking that person would not remain impossible if the principles of human psychology or your own psychological makeup were to change in such a way that you would start liking people of that sort. We might say that it is *psychologically* impossible for you to like that other person right now, but it is not *logically* impossible. Similarly, even though you might say that it is completely impossible for you to become a multibillionaire, the richest person who has ever lived, it is not logically impossible for you to become the richest person who has ever lived, since it would not remain impossible if certain facts in the world were to change in just the right sort of way. It may be impossible in some *factual* sense for you to become the richest person who has ever lived, but it is not *logically* impossible.

Some things *are* logically impossible. It is logically impossible for a triangle to have five sides, since no matter how facts change, or how the principles of human psychology change, or how the laws of nature might change, all triangles still have three sides, since by definition a triangle is a three-sided figure. It is also logically impossible for your mother to fail to be herself. No changes in the facts, no changes in human psychology, no changes in the laws of nature would make it possible for your mother to be different from herself. She might have had a different name from the one she actually has, or she might have been a bit heavier, or lighter in weight, but she still would have been whoever she is, not someone different from the person she is.

You should be getting the idea that most of the things we normally refer to as being impossible are not *logically* impossible. Most of the things we commonly call impossible are really only psychologically impossible, or physically impossible, or impossible given the current facts. These things would all become possible if human psychological principles, or the laws of nature, or the facts were to be different. These things are thus logically possible.

Given the notion of logical impossibility, one can give a final definition of deductive validity: *an argument is deductively valid when and only when it is logically impossible for the conclusion of the argument to be false if the premisses of the argument are all true.* If you have some trouble understanding that definition, here is an equivalent way of phrasing it: *an argument is deductively invalid when and only when it is logically possible for the premisses of the argument to all be true even if the conclusion of the argument is false.* You should now verify that these definitions fit the examples of deductively valid and deductively invalid arguments already presented.

Unfortunately, the word "valid" is used in English to refer to many different things. For instance, we often hear people say "You have a valid point", or

"That's a valid observation". These uses of the word "valid" may have little or nothing to do with argument, and should be carefully distinguished from my use of the word "valid" which is always just short for "deductively valid", and is always used in reference to arguments that satisfy the definitions given in the last paragraph. (This use of the word is fairly standard in logic.)

The concept of validity is extremely useful. For one thing, it is a concept which can be applied to an argument whether or not anyone *knows* if the premisses of the argument are true. Remember that a valid argument merely has the feature that *if* its premisses are all true, then its conclusion will be true also. We don't need to know whether the premisses are in fact true in order to test an argument for validity. This is a big advantage, since we are often not in a position to know enough facts to tell whether certain premisses are true, even we want to evaluate arguments employing these premisses. One concept we can use in such a case is the concept of validity, since we don't need to worry at all about the actual truth or falsehood of the premisses or the conclusion in order to find out if the argument is valid. We merely ask ourselves whether the premisses of the argument relate to the conclusion in such a way that *if* the premisses did happen to be true, the conclusion would have to be true also. If the answer is "yes", then the argument is valid.

Since an argument can be valid even when its premisses are not true, you cannot determine whether an argument is valid merely by knowing whether its premisses are true or false. We can find examples of valid arguments and of invalid arguments in each of the following categories:

(A)	P1.	true	(B)	P1.	true	(C)	P1.	false	(D)	P1.	false
	P2.	true		P2.	false		P2.	false		P2.	false
	P3.	true		P3.	false		P3.	false		P3.	false
	C.	true		C.	true		C.	true		C.	false

But the definition of validity rules out *one particular arrangement:*

(E)	P1.	true
	P2.	true
	P3.	true
	C.	false

Here are some examples of each of the arrangements:

(Arrangement A)

P1. All cats are mammals.

P2. All mammals are animals big enough to be seen with the naked eye.

P3. All animals big enough to be seen with the naked eye contain cells.

C. All cats contain cells.   (This argument is valid.)

P1. Some ships are painted gray.
P2. Some ships are made of wood.
P3. Some ships are made of metal.
C.   Some sailors are men.   (This argument is invalid.)

Arrangement B

P1. Some men are arrogant.

P2. If a man is arrogant, he can always make friends easily.

P3. Anyone who can always make friends easily will become wealthy by the age of 30.

C.   Some men will become wealthy by the   (This argument is valid.)
     age of 30.

P1. Some men are arrogant.

P2. If a man is arrogant, he can always make friends easily.

P3. Anyone who can always make friends easily will become wealthy by the age of 30.

C.   Some men are tall.   (This argument is invalid.)

Arrangement C

P1. Everyone legally eligible to vote for the president in the last election likes the president.

P2. Everyone who likes the president is a Christian.

P3. Everyone who is a Christian was subject to the election laws of the U.S. at the time of the last election.

C.   Everyone legally eligible to vote for the (This argument is valid.)
     president in the last election was subject to the election laws of the U.S.

P1. (Same premisses as those directly above.)
P2.
P3.

C.   Everyone who voted in the last election   (This argument is invalid.)
     was over forty years old.

Arrangement D

P1.   All birds are mammals.
P2.   All mammals can fly.
P3.   All mammals have fur.
_____
C.   All birds have fur and can fly.   (This argument is valid.)

P1.   (Same premisses as those directly
      above.)
P2.
P3.
_____
C.   All birds are blue.   (This argument is invalid.)

Arrangement E

P1.   Some birds can fly.
P2.   Some animals that fly are mammals.   (Since this argument fits
P3.   Some mammals weigh over one ton.   arrangement E, it must be
                                               invalid.)
C.   Some birds weigh over one ton.

It should now be clear that it is not appropriate to speak of arguments as being either true or false. Calling an argument "true" is like calling an automobile genetically sound. Automobiles don't have genes; arguments are neither true nor false. Arguments are *valid* or *invalid* while *premisses* and *conclusions* (which are all propositions) are true or false. So don't say things like "This argument is true", or "This premiss is valid".

Maybe you would like to have an easy way to refer to arguments that not only are valid but also have premisses which are actually true. All right. The label will be "sound". *A sound argument is an argument which is valid AND which has all true premisses.* For the most part, logic doesn't reveal whether an argument is sound; logic can only show whether the argument is valid.

(One little sidelight: even a deductively sound argument might be fallacious. It might commit the fallacy of begging the question, for example, if one of its premisses is not an acceptable premiss to use within the argument's context of discussion, even though that premiss happens to be true.)

There is one special case in which the application of our concept of validity causes special problems. That is the peculiar case in which the premisses of the argument *can't* all be true—when it is logically impossible for them all to be true. Here's an example:

P1.   Jones is the best person for the job, and ought to be selected rather
      than Smith.

P2.   The best person for the job is a woman with courage and tact.

P3.   Jones is a man of courage, but he lacks tact, while Smith is a woman
      of both courage and tact.
_____
C.   Smith should be selected.

In order to determine whether this argument is valid, we are to ask whether it is logically possible for its conclusion to be false if its premisses are all true. But how can we tell in this case, since it is not possible to conceive of all the premisses' being true? The standard interpretation of the definition of validity will startle you in this case, for on the standard interpretation, we must say that this argument is *valid,* because its premisses *cannot* all be true. In other words, we must say it is *not* logically possible for the conclusion of this argument to be false if its premisses are all true, precisely because it is not logically possible for all its premisses to be true!

If you find this point confusing, you have considerable company. A number of logicians have been working on revisions of the notion of validity to modify this result. However, it is simplest to just accept the standard definition of validity, and swallow this result along with it. This result can be made to sound more reasonable by saying it really means that once we allow people to adopt premisses which cannot possibly be true, we might as well allow them to conclude anything they want, for all rational controls are gone once impossible premisses are employed. Accordingly, with the standard interpretation of deductive validity *any* argument with impossible premisses automatically becomes valid, no matter what its conclusion may be. (Of course such an argument cannot be *sound,* since its premisses cannot all be true; so perhaps it's not so bad to say that it's valid.) Fortunately, most of the arguments which interest us do not have impossible premisses.

Another peculiar, but less troublesome, consequence of our definition of validity comes to view when we consider arguments containing conclusions which cannot possibly be false. For example, the conclusion, "A boy is a boy", or the conclusion, "If it rains, it rains". Obviously, if a proposition cannot ever be false, it cannot be false when the premisses of the argument are true. Since it cannot be false when the premisses are true, the argument is automatically valid, no matter what the premisses may be. Thus,

Today is Thursday.
A boy is a boy.

is valid! (Test it against the definition of validity.) Again, this may bother you a bit, since the premiss in such an argument need not have anything to do with the conclusion. The conclusion stands on its own, so to speak. But that is exactly why the argument is sensibly called "valid"—the conclusion is strong enough to stand on its own. If you argue like this all the time, you won't be able to conclude anything except things that must be true, no matter what. Thus, you will never be able to start with true premisses and end up with a false conclusion arguing this way. That is why these strange arguments are called "valid".

Not all good reasonable arguments are deductively valid or sound. We will discuss *invalid* arguments that are reasonable later on, in Chapter 10. But for now we will concentrate on deductive validity. It turns out that the notion of validity is very useful. You might think of the valid argument as a kind of standard or measuring stick against which other arguments can be measured.

The concept of validity ties in directly with the concept of logic with which this book began. Recall that logic studies those aspects of propositions which relate to what makes the truth of one proposition good evidence for the truth of another proposition. In a valid argument, the propositions that make up the premisses relate to the conclusion in such a way that the truth of the premisses would provide absolutely conclusive grounds for the truth of the conclusion. There would not be any possible way in which the conclusion of the argument could be false if the premisses of the argument were true. Thus, in studying what makes an argument valid, we are studying part of what is involved in making one proposition good evidence for another proposition.

To oversimplify matters just a bit, we could break up the total process of argument evaluation into two parts. One part has very little directly to do with logic, where we ask if the premisses of the argument are true. The other part is the heart of logic, where we ask whether the premisses would provide good grounds for accepting the conclusion if the premisses happen to be true. In order to carry out the first part of this two-part process, one needs to have the appropriate sort of background information to decide whether the premisses are in fact true. In order to carry out the second part of the evaluation process, one does not need to know whether the premisses are true.

Deductive validity

24.1 Which ones of the following statements are correct?
   a. An argument can be valid even though it has false premisses.
   b. An argument can be valid even though it has a false conclusion.
   c. All valid arguments have true premisses.
   d. No valid arguments have true premisses.
   e. If the premisses in a valid argument are true, then the conclusion of the argument is true also.
   f. In an valid argument, some of the premisses might be true and some might be false.
   g. According to the definition of "valid argument", if the premisses of a valid argument are false, then the conclusion of the argument will have to be false also.
   h. If an argument is sound, some of its premisses might be false.
   i. The conclusion of a sound argument is true.
   j. An argument could be sound even though it is not valid.
   k. If a valid argument has a false conclusion, then it has at least one false premiss.

24.2 Use the definitions of "valid argument" and "invalid argument" to help you decide which of the following arguments are valid and which are invalid.

   a. All submarines are painted gray, and that ship over there is painted gray; so it must be a submarine.

   b. Some people who like catfish also like trout, but no one who likes catfish likes halibut. Trudy likes catfish and trout. Thus, she does not like halibut.

   c. The men have been trapped in the mine shaft now for ten days without food or water. We can conclude they are all dead.

d. It rains only if there are clouds overhead, and it is raining now. Thus, there are clouds overhead.

e. No one enjoys doing busywork. Sally's job as receptionist requires her to do busywork most of the time. So naturally she is unhappy with her job.

f. The universe exhibits a very complex pattern of organization; hence, there must be a God who designed the whole thing.

g. The universe contains so much evil and suffering! So it is absurd to believe in God.

h. On the average, blacks score lower than whites on the standard written intelligence tests administered throughout the nation in our school system. Therefore, the unfortunate fact is that blacks are less intelligent than whites on the average.

i. A very small number of Americans are U.S. senators—in fact, only one hundred Americans hold that position. However, there are considerably more women in the U.S. than there are men, for a variety of reasons, one of which is that women tend to live longer than men. So, there must be at least a few women who have served in the Senate.

25

Proof of invalidity by means of counterexamples

Now that we know what a valid argument must be like, we can discuss techniques for proving that an argument is invalid. The first such technique is one you can use all the time. I refer to this technique as the method of counterexample construction.

A counterexample is a logically possible story which describes one way in which the premisses of a given argument could all be true even though the conclusion of the argument is false. A moment's thought will show you, I hope, that if such a story can be told, the given argument is invalid, since a valid argument cannot possibly have all true premisses when it also has a false conclusion. The story involved in a counterexample need not be true or believable but: (1) *All the premisses of the given argument must be true in the story.* (2) *All the parts of the story must hang together to form a coherent whole which contains no explicit or hidden logical impossibilities.* (3) *The conclusion of the given argument must be false in the story.* Any story meeting these requirements will be a counterexample and will prove the argument it deals with to be invalid.

Example 1

Argument to be proved invalid:
All submarines are painted gray. That ship over there is painted gray. Therefore it is a submarine.

Counterexample:
Suppose that in fact all submarines are painted gray since all ships of any sort are painted gray. Suppose also that that ship over there is not a submarine, but a destroyer. Since it is a ship, it is naturally painted gray like all other ships.

Check for yourself to see that the story told in this counterexample satisfies all three tests for being a good counterexample. Note also that the story is not a true story, since in the story all ships are painted gray, while in the real world, not all ships are gray.

Example 2

Argument to be proved invalid:
If you study hard, you will get an A. But you don't study hard. Therefore, you will not get an A.

Counterexample:
Suppose that everyone who studies hard always gets an A. Of course that means that you, too, will get an A if you study hard. However, suppose also that you are lazy and do not wish to study at all, and that as a result you in fact fail to study very much. Nevertheless, you are also a skilled cheater and you find ways to cheat on all the tests, giving you an A average. Your instructor does not catch you, and you end up getting an A.

This is a perfectly fine counterexample. It satisfies all three tests for a counterexample, even if it is not a true story. (Verify this for yourself.) Naturally, there are other stories which would do the job just as well. I prefer stories with extra details thrown in to make the story come alive. That helps to make the counterexample convincing.

Sometimes people get the mistaken idea that a counterexample is an *argument* for a conclusion opposite that of the original argument. Someone who had that idea about counterexamples would think that the counterexample given above is an *argument* that you will get an A. This is a mistake because the counterexample is not an argument at all. It is a *story* in which you get an A. It doesn't matter at all whether I even believe my own story, for the point of the story is not to argue that you will get an A. Rather, the point of the story is just to show that the original argument is invalid.

Even wild unbelievable stories count as perfectly good counterexamples. For instance, a crazy science fiction story will be a fine counterexample if it meets the three tests of a good counterexample. If you use a weird story as a counterexample note that it is often hard to tell whether the story is genuinely logically possible. I often get the feeling when reading science fiction about time machines, for example, that the story is logically incoherent and nonsensical.

Example 3

Argument to be proved invalid:
> If you study hard, you will get an A. But you don't study hard. Therefore, you will not get an A.

Attempt at a counterexample:
> You are the sort of person who never seems to be able to get an A, no matter how hard you study. You have finally come to realize this fact about yourself, so you don't bother to study hard. However, in this particular course you get lucky and get an A anyway.

This attempted counterexample is a failure, because it does not meet the three tests for a counterexample. Why? The first part of this supposed counterexample seems to say that you never get an A. But the last part says that you do get an A in this course. If that is what is going on in the story, the story contains a logical impossibility. On the other hand, perhaps you will say that I am being too hard on this story because the first sentence of the story probably just means that you *hardly ever* get an A. All right. Then the logical impossibility I have been fussing over disappears. But there is still trouble, since now the counterexample does not say if you study hard you will get an A. It is absolutely essential to a good counterexample that *all* the premisses of the argument under attack be true in the story. However the attempted counterexample above does not seem to make the first premiss of our argument true. Hence, it fails.

What happens if the original argument is *valid*, and I try to construct a counterexample to prove it invalid? I will fail. It is impossible to construct a counterexample which meets all three tests of a good counterexample when the argument under attack is valid.

Example 4

Argument to be attacked:
> If you study hard you will get an A. You do study hard. Thus, you will get an A.

Attempt at a counterexample:
> Suppose that you are able to learn well enough when you study hard that you will get an A if you study hard. Suppose also that you do study hard, but you don't get an A since the tests were poorly designed and you didn't do well on them.

This supposed counterexample flops. It contains a hidden logical impossibility, and thus does not meet all three tests for a good counterexample. The logical impossibility lies in the internal conflict between saying you will get an A if you study hard and saying that in this course you will not get an A even though you did study hard. The argument given to attack in this example is *valid*, and for that reason *no* good counterexample can be constructed for it.

Proof of invalidity by means of counterexamples

25.1 Construct counterexamples for each of the following arguments that are invalid. If you think the argument is valid, merely say that you think it is, and thus no good counterexample will be possible.

a. If Angela Davis was guilty of conspiring to aid the convicts in their escape, she ran from the police when the escape attempt occurred because she knew she would be implicated. She in fact did run from the police when the escape occurred. Hence it is obvious that she was guilty of the conspiracy.

b. I agree that if our senator is dishonest, he should be defeated in the next election, but fortunately he is not dishonest. Therefore, he should not be defeated in the next election.

c. The men have been trapped in the mine shaft now for ten days without food or water. Consequently, they are all dead.

d. All human beings are mortal. Jones is a human being. Therefore, Jones is mortal.

e. No industrialists are socialists. All socialists are radicals. Hence, no industrialists are radicals.

f. All sheep are warm-blooded. This fuzzy white animal is warm-blooded. Thus, it is a sheep.

g. All mammals are warm-blooded. This animal is warm-blooded. Hence, it is a mammal.

h. Some women are poor drivers. Some poor drivers get into accidents. Therefore some women get into accidents.

i. The only punishments which are justified are those which prevent further infractions of just rules and regulations. Whenever someone is punished secretly, the punishment does not prevent further infractions of any rules whatsoever. Hence, it is never justifiable to punish someone secretly, even when the rules and regulations which the person violated are just.

j. The only punishments which are justified are those which prevent further infractions of just rules and regulations. Jones is guilty of violating a just rule. Hence, if he is being punished it is justifiable.

k. The only punishments which are justified are those which prevent further infractions of just rules and regulations. Jones is being executed as punishment for violating a just rule. Hence, Jones is justly punished.

l. Most people who use marijuana are under the age of thirty. Most people under the age of thirty have incomes below the national average. Thus, most people who use marijuana have incomes below the national average.

26

Proof of invalidity by constructing absurd logically analogous arguments

For relatively simple short arguments, there is another effective and sometimes amusing technique for proving invalidity, illustrated by an example drawn from the previous Section. Let's use the argument from Exercise 25.1 (I), to illustrate the new technique. Now instead of constructing a counterexample, I will construct a *new* argument which is similar in structure to the argument of 25.1 (I):

> Most people who use the New York City subway system daily are Americans under the age of sixty. Most Americans under the age of sixty live outside the New York City area. Thus, most people who use the New York City subway system daily live outside the New York City area.

Several features of this new argument are important here: (1) This is an argument, not a story; hence it is not a counterexample. (2) This argument has the same structure as the argument in 25.1 (I), although the subject being discussed has changed. (3) The premises of this argument are known to be true, while its conclusion is known to be false. (4) Since its premises are known to be true and its conclusion is known to be false, the argument I constructed above must be invalid.

The above combination of four features explains why my new argument about the subway riders show the old argument (in 25.1 (I)) about the marijuana users to be invalid. How so? Basically, I'm saying here that *if* the marijuana argument were valid, then my subway argument would also be valid, because my subway argument is put together exactly like the marijuana argument. But my subway argument is known not to be valid as explained above. Hence, the marijuana argument can't be valid either.

In general terms, this technique for proving invalidity works as follows: in order to prove a given argument invalid, you (cleverly) construct a new argument about any subject whatsoever, which is put together in exactly the same way as the given argument you want to attack. But you carefully choose the premises and the conclusion of your *new* argument in such a way that the premises are generally known to your audience to be true, while its conclusion is generally known to be false. You then say that *if* the *given* argument were valid, your *new* argument would also be valid, but since your new argument is obviously *not* valid, the given argument cannot be valid either.

Here's another example: in order to prove the argument (from Example 3, Section 24)

> If you study hard, you will get an A. But you don't study hard. Therefore, you will not get an A.

invalid, using the technique we are now discussing, you could use the following new argument:

If it becomes overcast, the sun will be hidden from view. But it does not become overcast. Therefore, the sun will not be hidden from view.

The new argument clearly has a structure identical to the old argument, and yet under the proper circumstances—namely, on a clear night—the premisses are known to be true when its conclusion is obviously false. Hence, this new argument is invalid, and we may conclude the old argument is also invalid. Of course, my new argument is only one of many new arguments which could have been used for this purpose.

Theoretically, *any* invalid argument could be proved invalid by this technique. However, in actual practice it is simply too difficult to come up with an appropriate new argument when the given invalid argument is long or complicated. Nevertheless, for proving relatively short simple arguments invalid, this technique can be very effective.

Of course this technique does depend on one somewhat hazy notion—the notion of exact structural similarity between two arguments. No logician knows how to explain this notion completely, and you will have to wait until later in this book to see how some kinds of logical structure can be diagrammed quite precisely; so at this point you will have to rely on common sense to tell you when two arguments are put together in exactly the same way in all relevant respects. For simple arguments, this may not be too much to ask.

Certain key words provide some help in the matter of preserving structure, for they are structure indicators, and as such should appear in analogous positions in both the original argument and the one you have created to attack it. Some of these words occurred in the example—"if", "but", "therefore", and "not". Other such words are "all", "every", "some", "a few", "most", "many", "or", "and", "nevertheless", "unless", "provided that", "until", "except for", "because", "when", "since", "for the purpose of", and so on. Perhaps you can see that each of these words (or phrases) could remain unchanged while you alter the subject matter of an argument considerably. Check to see how the key words in the example stayed put when the new argument was constructed.

In order to have a standard label by which we may refer to the new arguments one constructs when using the technique discussed in this Section, I will call these new arguments "absurd logically analogous arguments", because the new arguments are absurd and they are also logically analogous to the given argument which you are attempting to prove invalid.

When you actually use this technique for proving a given argument invalid, you need to say something like this, pointing toward the given argument: "That argument is clearly invalid, because to argue that way is just like arguing . . . ." Then you trot out your absurd logically analogous argument.

Proof of invalidity by constructing absurd logically analogous arguments

26.1 Produce two absurd logically analogous arguments for each of the following.

   a. All submarines are painted gray, and that ship over there is painted gray; so it must be a submarine.

   b. If it is raining, there must be clouds overhead. There are clouds overhead now. So it must be raining.

   c. The universe exhibits a very complex pattern of organization; therefore, there must be a God who designed the whole thing.

   d. On the average, blacks score lower than whites on the standard written intelligence tests administered throughout the nation in our school systems. Therefore, the unfortunate fact is that blacks are less intelligent than whites on the average.

   e. Some students are brilliant thinkers. Some brilliant thinkers are obnoxious. Thus, some students are obnoxious.

26.2 Here is an invalid argument:

   Avon cosmetics must be the best available, because they are the most popular.

   Below are four attempts to construct an absurd logically analogous argument, only one of which is a fully successful attempt, meeting all four criteria for absurd logically analogous arguments mentioned in this section. Which attempt is successful? In what respects are the others unsuccessful?

   a. Dying must be good for you, because so many people do it.

   b. Henry must be the best candidate for the presidency because he got the most votes.

   c. McDonald's hamburgers must be the best hamburgers available because they are the most popular.

   d. Green must be the best available color for leaves, since it is the most common color for leaves.

26.3 In what respects are counterexamples different from absurd logically analogous arguments, with respect to proving the invalidity of a given argument?

26.4 Prove the following arguments invalid by using the techniques of this Section, or by producing counterexamples to each argument. (Some are easier to handle by one technique rather than the other.)

a. Until such time as a fetus is capable of supporting life on its own—outside its mother's body—that fetus owes its existence to its mother. It therefore can have no rights which conflict with the mother's rights.

b. *Of all the figures one could recite to show the success of free enterprise, none is more impressive than this. One half of all the goods produced in the past 10,000 years have been produced in the United States in the past 200.*[1]

c. (Ad for a toilet bowl cleaner): You see, once household germs are in your toilet, they're in your home. So, once they're out of your toilet, they're out of your home.

d. Perhaps there was a time when college was only for the elite, but that time has passed! College graduates by the thousands are working for the minimum wage. Employers sure don't see students as elite.

[1]From an undated letter written in the late 1970's, soliciting funds for the Center for the Defense of Free Enterprise.

PART TWO

Logical Systems

5

Propositional Logic Without Logical Symbols

27

English connectors

Now that you have seen the importance of argument structure, you will be interested in learning techniques for more systematically diagraming the logical structures of various kinds of arguments. Logicians continue to devote considerable effort toward developing and refining such techniques. However, the most modern methods that will be discussed here have been known in one form or another for some eighty years or more, thanks to the pioneering work of the German mathematician and philosopher, Gottlob Frege and his successors.[1] The maturation of these techniques and their uses in systems of logical argument construction has been, in my opinion, one of the greatest intellectual achievements of the late nineteenth and early twentieth centuries.

As a first step we will focus on some seemingly simple and apparently unimportant features of the ways in which certain complex propositions may be

---

[1]See Frege's *The Basic Laws of Arithmetic*, Montgomery Furth, trans. and ed., (Berkeley, Ca.: University of California Press, 1964). The development of techniques like these began with the ancient Greek philosopher, Aristotle, and has continued off and on through the twenty-three centuries following his death.

expressed in the English language. It will turn out that these features are neither simple nor unimportant. Recall from English grammar that a *clause* is a sentence or a piece of a sentence which contains a subject and a predicate (and which can therefore function by itself to express a proposition). We seek certain key words which can connect two clauses, or which operate on single clauses to form new clauses. For example, "the store" is not a clause, but "the store was open on Thursday" is a clause. Another clause, "the store was closed on Friday", can be connected to the first clause by the word "but" to yield the sentence: "The store was open on Thursday but the store was not open on Friday". "But" is one of the key words that can connect two clauses. "Not" is also a key word of the type we seek; it can operate on a single clause to form a new clause, as it does in "The store was not open on Thursday".

For the sake of emphasis, key words used to connect clauses or operate on them will often be written in capital letters when I write premisses or conclusions. However, for the sake of clarity, I will always try to move clauses and words around in a sentence to make it clear which clauses are being operated on by the capitalized key words. Accordingly, a capitalized connecting type of key word will be placed *in between* the clauses it connects. Thus, I will not write

EVEN THOUGH the store was open on Thursday it was closed on Friday.

Instead,

The store was closed on Friday EVEN THOUGH it was open on Thursday

will be preferred, since the capitalized "EVEN THOUGH" now occurs *between* the clauses it connects without introducing any change in the proposition expressed by this sentence. Key words which operate on only one clause will be written *directly in front of* the affected clause, and parentheses will be used to show how far the key word's effect reaches. Thus, the sentence

The store was NOT open on Friday

will be written

NOT (the store was open on Friday)

and this sentence will be called the *negation* of the clause in parentheses. Again,

Perhaps he'll come soon

will be written

PERHAPS (he'll come soon).

All these key words will be called *connectors,* although some of them, such as "not", do not really "connect" anything since they apply to only one clause rather than two. There are hundreds of connectors in the English language, but we will talk about only a few.

Some connectors which consist of two or more words naturally occur split up within the sentence. For example, the connector "if . . . then":

IF a processing center has work in process at the end of a period, THEN completed units alone will not accurately measure total output.[1]

I will allow such connectors to be written in such a way provided that at least part of the connector appears between the clauses connected. In the preceding example "then" meets this requirement. How would you handle

If a processing center has work in process at the end of a period, completed units alone will not accurately measure total output

given that the word "then" is missing here? One possibility is just to add the word "then" in the appropriate place, as above. Another correct method is to rearrange the clauses:

Completed units alone will not accurately measure total output IF a processing center has work in process at the end of a period.

You may have noticed that there is another connector in the above sentence— the word "not". Capitalizing and correctly moving this connector yields either

NOT (completed units alone will accurately measure total output) IF a processing center has work in process at the end of a period

or

IF a processing center has work in process at the end of a period, THEN NOT (completed units alone will accurately measure total output).

Study these carefully for structure.

Suppose we have to work instead with the slightly more complex sentence

If a processing center has work in process at the beginning or the end of a period, counting completed units alone will not accurately measure total output for the period.

In this sentence we have the word "or" which, as you can tell with just a bit of thought, is often a connector. But as written in this sentence "or" does not connect two clauses. ("The end" is not a clause.) Nevertheless, "or" counts as a connector in this sentence because the sentence can be *expanded* as follows without changing the total complex proposition expressed by the original sentence:

IF a processing center has work in process at the beginning of a period OR a processing center has work in process at the end of a period, THEN NOT (counting completed units alone will accurately measure total output for the period).

Note that in this expanded version of the sentence, "or" *does* connect two clauses. The reason for expanding a sentence like that is to reveal important

---

[1]Garrison, *Managerial Accounting.*

structure in the propositions expressed by English sentences. The amount of structure you need to reveal is determined by how you wish to use the revealed structure. Possible uses will become clear presently.

Sometimes when we are interested in the structure of a sentence rather than its subject matter I will substitute single capital letters for the clauses in the sentence, generating a *diagram* of the sentence. A sentence diagram of the last sample sentence above is

IF A OR B, THEN NOT C.

If you examine such diagrams you will see that some kind of grouping markers are sometimes needed to avoid ambiguity. For example,

IF A THEN B OR C

is ambiguous in much the same way that the arithmetic expression "2 + 3 × 5" is ambiguous. If we interpret the latter expression as "(2 + 3) × 5" it equals 25, but if we interpret it instead as "2 + (3 × 5)" it equals 17. Turning to the sentence diagram, you can see that it might be read as

IF A THEN (B OR C)

or as

(IF A THEN B) OR C.

These readings are different. Thus, "If it rains today, I'll stay home or go for a walk" becomes

IF it rains today, THEN (I'll stay home OR I'll go for a walk).

The general rule for adding parentheses is simple: when there are three or more clauses in the sentence the safe thing is to group the clauses in the most natural way by adding parentheses.

The grouping of clauses by parentheses must be done with care to avoid changing the meaning of the sentence. For example, suppose someone asked you why you were not going to John's party, and you replied

If I go, I'll be bored, and I hate John.

Grouping this sentence incorrectly would yield the result

IF I go THEN (I'll be bored AND I hate John).

This result is incorrect because it makes your dislike of John depend on your going to the party. Presumably, what you meant to say is better represented by

(IF I go THEN I'll be bored) AND I hate John.

Many key words which sometimes function as connectors do not *always* do so. Hence, caution must be exercised in identification of connectors. For example, there are cases in which "and" is clearly not a connector, when it cannot be made to connect whole clauses even with rewriting. The following sentences contain such uses of "and":

Judy and John were able to get back together again.
Two and two equals four.

If the "and's" occurring in these sentences were connectors, it would be possible to rewrite these sentences as follows, without changing the meaning:

Judy was able to get back together AND John was able to get back together.
Two equals four AND two equals four.

It's obvious that the original sentences do not express the same propositions that these new sentences express. The crucial test which will help you to decide whether something that looks like a connector is being used as a connector in a given sentence is this: does the word in that sentence, after needed rewriting, apply to whole clauses? If not, then it is not a connector.

Let me summarize the techniques discussed in this Section:

1. We will put into capital letters some words or phrases which connect clauses or apply to clauses to form larger clauses. All such words or phrases are called connectors.
2. If necessary, sentences should be rewritten so as to put at least part of each connector between any two clauses which that connector joins, and so as to put connectors applying to only one clause out in front of that clause. Such rewriting should not make the sentence express a proposition different from the proposition it expressed prior to rewriting.
3. Sentences should be rewritten so phrases are expanded into whole clauses when expansion is required to show which clauses are being operated on by connectors.
4. Parentheses should be added around each clause that is preceded by a connector applying only to that clause. (See the treatment of "not".)
5. Parentheses should be added whenever there is ambiguous grouping of clauses otherwise.
6. Generate sentence diagrams by replacing all the different clauses in a rewritten sentence with different single capital letters.

These rather informal techniques can reveal an amazing amount of important logical structure. At this point we are dealing with individual sentences rather than whole arguments, but soon we will be in a position to utilize these techniques to diagram whole arguments, and this in turn will give us a way to better appreciate and evaluate arguments on many different subjects. But in order to accomplish these ultimate aims, you must first practice the analysis of individual sentences.

To help you practice, the following examples are offered for your examination, adapted from Gilder's article on sex discrimination.

(1) Today, the burdens of childbearing no longer prevent women from performing the provider role, and if day care becomes widely available it will be possible for a matriarchal social pattern to emerge.

(1) can be rewritten as (1a):

(1a) NOT (today the burdens of childbearing continue to prevent women from performing the provider role) AND IF day care becomes widely available THEN it will be possible for a matriarchal social pattern to emerge.

Or, in more detail:

(1b)   NOT (today the burdens . . . provider role) AND IF (day care . . . available) THEN POSSIBLY (a matriarchal social pattern will emerge).

(1a) may be diagrammed as

(1a diag)   NOT A AND IF B THEN C.

(1b) may be diagrammed as

(1b diag)   NOT A AND IF B THEN POSSIBLY C.

Parentheses may be added, if you like, although in this case they are not strictly needed to prevent ambiguity:

(1a diag)   NOT A AND (IF B THEN C).
(1b diag)   NOT A AND (IF B THEN POSSIBLY C).

The next sentence is harder.

(2)   These practices are seen as oppressive by some; but they make possible a society in which women can love and respect men.

It is tempting to rewrite (2) as

(2a)   These practices are seen as oppressive by some; BUT (they make possible a society in which women can love men AND they make possible a society in which women can respect men).

However, I believe (2a) is not quite accurate, since (2a) does *not* seem to say a single society in which women can do *both* things is made possible by the practices in question, while the original sentence (2) *does* say such a single society is made possible. Hence, I reject (2a) in favor of

(2b)   These practices are seen as oppressive by some; BUT they make possible a society in which women can love and respect men.

which has the simple diagram

(2b diag)   A BUT B.

English connectors

27.1 Rewrite each of the following sentences and put their connectors into capital letters, following the rewriting procedures listed in this Section.
a. Kyle and JoAnn both ran for election.
b. I don't like movies.

    c. I'll go with you only if you promise to be quiet.

    d. Unless we get some action on this soon, we'll have to give up.

    e. The patient will die unless we operate on him.

    f. If the county is allowed to keep the money, there will not be enough to run the town.

    g. Unless some other finances are made available, if the county is allowed to keep the money, there will not be enough left to run the town.

    h. You will be covered for the full amount provided that you are healthy at the time of application.

    i. We'll make money this year if the new line does well, but only if the new line does well.

    j. This step should be taken without full consideration of the social and economic consequences.

    k. We were able to complete the work on time because we employed extra help, despite the fact that it was expensive to do so.

    l. We couldn't get to him, because a person can get up there only if the trail is open.

27.2 In which of the following sentences can "and" or "or" be construed as connectors of the sort we have been working with?

    a. John or Joe will have to do it.

    b. Amtrak and the airlines both serve Denver.

    c. The students were asked to fill out a questionnaire which assessed their scholastic aptitude and their drug use habits.

    d. Unemployed women may perform valuable work in creating and maintaining families.[1]

    e. The unemployed male can contribute little to the society and will often disrupt it.

    f. Nothing is so important to the sexual constitution as the creation and maintenance of families.

27.3 Rewrite the following passage separating premises from conclusions, labeling each with "P" or "C", and laying out each separate argument with premisses written on top of a horizontal line. *Within* each premiss or conclusion, capitalize each connector, insert parentheses when needed, and do whatever rewriting is necessary.

Man's natural instinct moves him to live in civil society, for he cannot, if dwelling apart, provide himself with the necessary requirements of life, nor procure the means of developing his mental and moral faculties. . . . Civil authority, therefore, comes from nature, and the end [i.e., the goal] of such authority is the preservation of order in the State. Consequently, if for the preservation of order capital punishment is necessary, the right to

---

[1]d., e., and f. are from Gilder, *Suicide of the Sexes.*

inflict capital punishment must be comprised in the civil authority. For nature never gives the end without the means. Capital punishment then is not legalized murder.[2]

28

Reducing the number of connectors

Here's a list of some important connectors that come to mind:

and	because
but	provided that
not	on the condition that
neither . . . nor	since
or	for (when it means "since")
both . . . and	for the reason that
either . . . or	moreover
if . . . then	except when
only if	which means that
if and only if	although
unless	before
after	with the exception that

(Most of these words or phrases can be used in various ways, but when I included them in the above list I was thinking of the ways they can be used to apply to whole clauses for the purpose of forming larger clauses.)

A complete list of such words and phrases would be quite long, but a great many items on the list are not really needed. As far as logic is concerned, we don't care about literary elegance or style. In fact, from the point of view of elementary logic, we need not be concerned with the difference between two connectors unless that difference has something to do with the conditions under which sentences containing these connectors are true or false. That is why I claim that a great many items on my list, and on the complete list which I haven't given you, are unnecessary. They are redundant because there are no significant differences between them relating to truth or falsehood.

The words "and" and "but" serve as a good example of a redundant pair. Why? Compare a sentence containing "and" with one containing "but":

(1)   Greg was at the party AND he ate too much.

(2)   Greg was at the party BUT he ate too much.

[2]John J. Ford, "Capital Punishment—a Defense," in *Ethics and Metaethics,* ed. by R. Abelson (New York: St. Martin's Press, 1963), pp. 20–1.

It seems that sentence (1) claims both that Greg was at the party and also that he ate too much. So does sentence (2). However, the difference between (1) and (2) amounts to the introduction in sentence (2) of the suggestion that Greg's eating too much is surprising, or unacceptable, or not completely in harmony with his being at the party. In a sentence of the form 'A BUT B' the word "but" suggests that there is something about the 'B' part of the sentence which stands in contrast in some special way to the 'A' part of the sentence. This feature of the word "but" makes "but" taste a little different from "and", but it does not affect the fact that both 'A AND B' sentences and 'A BUT B' sentences essentially just assert both A and B are true.

Pushing this point a bit further, I would claim that the slight difference in emphasis which "but" gives a sentence does not in any way affect whether that sentence is true or false. This means when I rewrite a sentence containing a "but" I may very well choose to use "AND" to represent the connector. (I render "but" with "AND" rather than rendering "and" as "BUT" because "AND" does not have the extra connotations which "BUT" seems to carry.) "AND" will be one of the select group of *preferred connectors* I will choose to use later.

It would be nice to have a complete list of preferred connectors together with a table showing how each is equivalent to various other connectors. Unfortunately, no one knows how to provide such a list at present. Nevertheless, we can work with a partial list and some useful common equivalences. This turns out to be useful enough to be worth doing.

"Neither . . . nor" will serve as a good example of a connector which can be treated as equivalent to something else. In dealing with "neither . . . nor" we have to be concerned about sentences like "Neither rain nor snow will stop the U.S. Mails". I think you can see that this sentence is equivalent to the sentence

Rain will not stop the U.S. Mails and snow will not stop the U.S. Mails.

It is then easy to see how "neither . . . nor" finally may be analyzed:

NOT (rain will stop the U.S. Mails) AND NOT (snow will stop the U.S. Mails).

This example illustrates that sentences which read roughly as "Neither . . . nor . . . " amount to

NOT A AND NOT B.

(There are other ways to eliminate "neither . . . nor", but for now we won't worry about them.)

Another good example is provided by "if and only if". "If and only if" occurs in sentences like "You can have an ice cream cone if and only if you are good." Which amounts to

(IF you are good, THEN you can have an ice cream cone) AND
(you can have an ice cream cone ONLY IF you are good).

This analysis isn't half bad. It gets rid of "if and only if" in favor of "IF . . . THEN", "AND", and "ONLY IF".

We don't need to stop here though, since we can get rid of "ONLY IF" in favor of "IF . . . THEN", thus making the analysis more elegant. This turns out to be an important move, for it shows the power of "IF . . . THEN". So, we now focus on the following sentence:

You can have an ice cream cone ONLY IF you are good.

It is tempting to say that this sentence is equivalent to

IF you are good, THEN you can have an ice cream cone.

However, that just can't be right. If it were right, then our original IF AND ONLY IF sentence would amount to

(IF you are good, THEN you can have an ice cream cone) AND
(IF you are good, THEN you can have an ice cream cone).

That's silly. So, look again at the "ONLY IF" part of the sentence. It does *not* say that being good will get you an ice cream cone. What it says instead is that you can't have a cone unless you are good. Thus, it says that if you get a cone, you must have been good. There we have it. The sentence should be rewritten as follows:

IF you can have an ice cream cone, THEN (that means) you are good.

In other words,

A ONLY IF B

should be rewritten as

IF A THEN B

not as

IF B THEN A

You must keep the 'A' and 'B' in the proper order when you change from "ONLY IF" to "IF . . . THEN". We have thus learned how to handle "ONLY IF", even when it is *not* part of an "IF AND ONLY IF" sentence.

Notice how I inserted the words "that means" after "THEN" above. This is not strictly necessary, but it helps a great deal to make the sentence sound more natural, and it avoids confusion which otherwise is quite likely to arise. Compare

IF you can have an ice cream cone, THEN you are good.

to

IF you can have an ice cream cone, THEN that means you are good.

The first one sounds wrong, while the second one sounds clearer, and maybe almost right. The point is that "IF . . . THEN" carries all sorts of connotations, some of which we don't want in this case. We can express ourselves more clearly by inserting the phrase "that means". In fact, you may insert that phrase in as many of the "IF . . . THEN" sentences as you wish, if it helps to make them sound right.

Applying all this to our original IF AND ONLY IF sentence, we get the following complete and elegant analysis:

(IF you are good, THEN you can have an ice cream cone) AND
(IF you can have an ice cream cone, THEN that means you are good).

In general, one can often replace

A IF AND ONLY IF B

with

(IF B THEN A) AND (IF A THEN B)

which of course is equivalent to

(IF A THEN B) AND (IF B THEN A).

If you have problems making these replacements for "IF AND ONLY IF" sound right insert "that means" as I did earlier.

This sort of careful thinking about what various types of sentences really mean may make the difference between your understanding and not understanding a contract you enter into some day, or the difference between understanding and not understanding some technical writing you have to deal with. We are not just playing with words. We are dissecting sentence structure to get important clues about the messages conveyed by language. Let me try to drive this point home by quoting a sentence from my automobile insurance policy:

This provision applies *only when* you have two or more autos insured in your name.

Clearly, this says

This provision applies ONLY IF you have two or more autos insured in your name.

But does that mean

IF you have two or more autos insured in your name THEN this provision applies?

No, for I may have to pay extra for this provision, or I may have to satisfy other requirements. Rather, the sentence does say

IF this provision applies, THEN (that means) you have two or more autos insured in your name.

Although we found a way to replace it, "IF AND ONLY IF" and its equivalent phrases occur frequently enough to make it convenient to keep "IF AND ONLY IF" on the list of preferred connectors. We will not keep "ONLY IF" on the list though, since it may be eliminated in favor of "IF . . . THEN". To our present list we can add "OR" in order to obtain a useful set of connectors:

AND   NOT   IF . . . THEN   IF AND ONLY IF   OR

This is the final list of *preferred connectors* in English.

You should be aware that even these preferred connectors are ambiguous in various important ways. For example, "AND" is ambiguous, in that it occasionally means "AND THEN". Moreover, this type of ambiguity is important for logic, because it affects the situations in which sentences are true or false. Although it is not likely you would have noticed the particular ambiguity I just discussed, there are some other ambiguities hiding in our list of key words which might complicate doing the exercises in this Section. It seems only fair to warn you. For now, try to take the attitude that you will forge ahead, ignoring some of the finer points that involve ambiguity of these words and phrases.

It would be nice to try to find ways to translate all the other key words and phrases into combinations of the members of the preferred list. I don't think it can be done. That means the preferred list is incomplete. But an amazingly large amount of work can be done with just this small list, and we will do our best to get by with it. Ultimately, that will mean that some of the connectors will have to be *approximated* by items on the preferred list, since no combination of preferred connectors will be equivalent to them.

Reducing the number of connectors

28.1 Each of the sentences below contains an italicized key word or phrase from the list given at the *beginning* of this Section. You are to rewrite each sentence in such a way that the italicized word or phrase is eliminated and some other key word or phrase drawn from the list at the *beginning* of this Section replaces it. However, this rewriting is to be done in such a way that the rewritten sentences express the same propositions as the original sentences.
   a. Children need the demonstration of love and concern from their parents, *for* otherwise they will grow up feeling insecure.
   b. Peterson's prices are lower; *moreover* he is a very reliable supplier.
   c. *Since* you left early, you missed all the fun.
   d. It rained *before* the floods came.
   e. The property is yours, *on the condition that* you continue using it for farming purposes.
   f. This experiment must be rerun, *for the reason that* the data has been lost from the first run.
   g. Your insurance policy covers this sort of loss *except when* the loss is due to your own carelessness.

28.2 Rewrite each of the following sentences using only *preferred* connectors in place of the italicized connectors. Make sure that the rewritten sentences express the same propositions as the original sentences. Use capitalization and parentheses as needed to bring out the structure of the sentences.
   a. My lease runs for ten years, *but* I *don't* believe I'll stay that long.
   b. You'll get rich doing that kind of work, *provided that* you are able to keep it up for a while.
   c. The patient will die *unless* we operate.
   d. The patient will die *if* we operate.
   e. I wanted to go shopping tonight; *however,* the stores are all closed.
   f. The barometer will fall *only if* it will rain.
   g. *If* George doesn't show up pretty soon, I'm going to leave, *but* I know he's going to be unhappy with me in any case.

h. Our team will win *if and only if* Jones plays, *unless* a replacement can be found for her.
i. *Neither* Jim *nor* Holly has any idea whether the train is coming.
j. *Although* one might expect more of him, he will *not* perform up to the required level.

28.3 Assume that your answers to 28.2 are correct, and that they provide an accurate guide to the way one can usually translate connectors not on the preferred list into ones that are on the preferred list. Use these assumptions to help you provide sentence *diagrams* containing only preferred connectors equivalent to the following sentence diagrams. (You may find it useful to keep this list.)
a. A BUT B
b. A BUT (NOT B)
c. A PROVIDED THAT B
d. A UNLESS B
e. A IF B
f. A ONLY IF B
g. A IF AND ONLY IF B, UNLESS C
h. ALTHOUGH A, B
i. A; HOWEVER B
j. (IF (NOT A), B) BUT C

28.4 Can you find some connectors on the list at the beginning of the Section which cannot be replaced by any combination of the preferred connectors?

29

Using connectors to diagram argument structure

Many arguments can be at least partially analyzed by looking for connectors, and those arguments will be emphasized in this chapter. Here is an example:

Logic is really just common sense, and it will come easily to someone who can grasp the basic ideas which lie behind it. But if logic is just common sense, anyone is capable of learning at least some of it. Hence, everyone is able to learn at least some logic.

First, separate the premises from the conclusion, and in each capitalize all the important connectors, converting to preferred connectors whenever possible.

P1. Logic is really just common sense AND it will come easily to someone who can grasp the basic ideas which lie behind it.

P2. IF logic is just common sense; THEN anyone is capable of learning at least some of it.

C1. Everyone is able to learn at least some logic.

We are just using the techniques from the last section here several times. Note that "Anyone is capable of learning at least some logic" is equivalent to "Everyone is able to learn at least some logic". So, a diagram for this argument could look like this:

P1. A AND B
P2. IF A, THEN C

C1. C   (Diagram I)

Before reading further, compare the diagram with the argument above it, to see the fit.

There is nothing magical about going in alphabetical order with the single capital letters used in a diagram. A perfectly fine diagram for the above argument could use 'Q', 'H', and 'M' instead of 'A', 'B', and 'C'. It is, however, important to keep the structure of the diagram intact. It would not do, for example, to change the "AND" to an "OR", or to delete the "IF . . . THEN". Nor would it do to insert identical single capital letters where they are not justified, as happens here:

P1'. Q AND P
P2'. IF Q, THEN P

C1'. P   (Diagram II)

This last diagram is not correct, because it requires the conclusion of the argument to equal the part of the first premiss which follows the "AND". Note that when the same proposition is expressed twice in the same argument, the same capital letter can be used repeatedly to mark the positions of the expressions. This occurs in P2 and C1 in Diagram I where the letter 'C' appears twice. It is often important to mark repeated expressions of the same proposition in that way, since the internal connections within the argument are likely to depend on such repetitions.

Such diagrams make it easier to grasp some aspects of the structure of arguments. Moreover, once one has an argument diagram, one can generate other arguments from it, simply by plugging a new set of propositions into the slots indicated by the single capital letters. For example, I can generate a new argument from Diagram I as follows:

P. Profit-making and legal protection are necessary in order for a firm to stay in business.

P. IF profit-making is necessary in order for a firm to stay in business THEN profit-making serves a useful social function.

C. Profit-making serves a useful social function.

In fact, there is nothing about the diagram of P1 which requires A and B be related to each other; so even the following dismal argument would fit Diagram I:

P.   Today is Wednesday AND I'm tired.
P.   IF today is Wednesday, THEN it's raining.
C.   It's raining

Note, however, that whenever the same single capital letter appears more than once in a diagram, that means that any argument which fits the diagram *must* have expressions for the *same* proposition appearing in each place where that single letter appears.

One little technicality about argument and sentence diagrams sometimes causes a bit of trouble. Even though *different* single capital letters appear in a diagram, it is *not* necessary that different propositions be expressed in these places. For example, I can substitute the very same expression in *all* the slots in Diagram II getting something crazy like this:

P.   Today is Wednesday AND today is Wednesday.
P.   IF today is Wednesday, THEN today is Wednesday.
C.   Today is Wednesday.

This argument makes no point. Still, it does fit the diagram.

Given the technicality that allows us to plug in the same proposition for two different single capital letters in a diagram, we can use the following diagram for the original argument about logic:

P1.   Q AND P
P2.   IF R, THEN S
C.   G

We can get the original argument from this diagram by substituting the appropriate expressions into the slots marked out by the single capital letters. However, this diagram is not as useful as Diagram I because this diagram hides some of the repetitions that existed in the original argument. Generally speaking, we will always try to use the diagram which reveals any repeated occurrences of propositions.

___

Using connectors to diagram argument structure

29.1 In the argument diagrams below, substitute the following expressions for the letters as indicated, and express the resulting arguments in *ordinary smooth* English.
   A: Laura let him have it.
   B: His parents will be surprised.
   C: His father will be angry.
   D: His mother will be amused.
   E: He will be embarrassed.

a. IF A, THEN C
   A
   _____
   C

b. C IF AND ONLY IF (D AND E)
   IF (NOT A), THEN (NOT E)
   C
   _____
   A

c. (C OR D) AND B
   IF B, THEN (NOT C)
   D

29.2 Construct diagrams for each of the following arguments, making these as
detailed as possible (don't leave out important connectives). Use preferred
connectors whenever possible. (This may require some rewriting.)

a. The answer to the fifteenth problem is right only if the book contains a
misprint. But the book surely does not have any misprints. So, the answer to the fifteenth problem is not right.

b. Every man has a right to promote and advance his ideas; therefore,
judges and other public officials are justified in using their official positions to forward their religious views.

c. The unemployed male can contribute little to the society and will often
disrupt it, while the unemployed woman may perform valuable work in
creating and maintaining families. Hence, it is more important to prevent
unemployment among males than to prevent it among females.[1]

d. Since the role of the male as principal provider is a crucial prop for the
family, the society must support it one way or the other. (Treat this sentence as an expression of an argument.)

e. Plainly, the only students who do well in elementary logic courses are
those who work hard at learning the material. After all, those who come
to understand logic either have a natural ability for puzzle-solving and
abstract reasoning, or they work hard at mastering logic techniques. But
those who have the natural ability, which makes logic easy for them,
quickly become overconfident and bored in elementary logic classes,
and bored students don't perform well. So those to whom logic comes
easiest don't do well in elementary logic courses. (This is a short argument chain, rather than just one argument.)

f. *Man's natural instinct moves him to live in civil society, for he cannot,
if dwelling apart, provide himself with the necessary requirements of
life, nor procure the means of developing his mental and moral faculties. . . . Civil authority, therefore, comes from nature, and the end of
such authority is the preservation of order in the State. Consequently, if
for the preservation of order capital punishment is necessary, the right*

---

[1]c. and d. are derived from Gilder, *Suicide of the Sexes.*

to inflict capital punishment must be comprised in civil authority. For nature never gives the end without the means. Capital punishment then is not legalized murder.[2]

30

Truth functions and truth tables

Suppose we were evaluating an argument which had the following diagram:

IF A THEN B
A

B

A little thought should reveal to you that any such argument must be *valid*, since the truth of such premises as would fit the diagram will guarantee the truth of the conclusion. In this case the diagram tells us a crucial fact about the quality of the argument.

The above case is fairly simple. Many argument structures are considerably more complex. So we need a systematic way to tell us when an argument's particular structure is sufficient to guarantee its validity. Since validity depends on what is possible when the premises are all true, then what we really want is a systematic way of keeping track of all the possible things that can happen when an argument's premises are all true, somehow using the argument's diagram. The purpose of this Section is to help you develop one such system. This development will be the subject of the next several Sections, so you will have to wait until Section 33 to see it completed.

I will begin by discussing the relation between the *parts* of a sentence and the truth or falsity of the *whole* sentence. Although this means we will now be turning once again to the analysis of individual sentences as opposed to the analysis of whole arguments, you should keep in mind that we will eventually apply all this to the determination of the validity of arguments.

First, it will be handy to introduce a bit of special terminology. I will often refer to the *truth value* of a proposition or of a sentence. By "truth value" I shall mean merely the truth or the falsity which that proposition or sentence happens to possess. So if a given proposition is true, then it has truth as its truth value. If the proposition is false, it has falsity as its truth value.

It is clear that in many cases the truth value of a compound sentence is determined by the truth value of the clauses it contains. For example, the sentence "The book does not contain a misprint", can be rewritten as "NOT (the book contains a misprint)", and it is clear that this compound's truth value is

---

[2]Ford, "Capital Punishment." See exercise 27.3.

totally determined by the truth value of the clause "The book contains a misprint". If the clause is true, then the sentence is false, while if the clause is false, then the sentence is true. Again, the truth value of "Bob is a man AND John is a man" depends solely on the truth values of its clauses. If "Bob is a man" is true and if "John is a man" is also true, then the compound is true, but if either one or both of the clauses is false, then the compound is also false.

When the truth value of a compound is solely dependent on the truth values of the clauses appearing within it, there is a good way to summarize the relationships between the truth value of the compound and the truth values of the clauses, using a device called a *truth table*. Here are truth tables which summarize the results for the two compounds we have just been talking about:

Table I

The book contains a misprint.	NOT (the book contains a misprint).
T	F
F	T

Table II

Bob is a man.	John is a man.	Bob is a man AND John is a man.
T	T	T
T	F	F
F	T	F
F	F	F

Table I says that when the first sentence is true, then the second sentence is false, and when the first sentence is false, then the second sentence is true. The double vertical line is used to separate the compound from the clauses. Since there is only one clause to the left of the double line in table I, there are only two possible truth value combinations to consider. On the other hand, table II contains two independent clauses to the left, and says that when the two clauses are both true, then the compound on the right is true also, but when either one or both of the clauses on the left is false, then the compound on the right is false. There are *two* sentences appearing to the left of the vertical double line, since in this case the compound contains two clauses rather than one. Because there are two clauses there will be *four* possible combinations of truth values to consider.

When setting up truth tables, it is good practice to follow a standard procedure for generating all the possible combinations of truth values for the clauses on the left. There is an easy standard way to set up such tables, which follows. Study the following partial tables until you see the pattern and can duplicate it or expand it to deal with four clauses.

Clause	Compound
T	
F	

Clause	Clause || Compound
T	T	
T	F	
F	T	
F	F	

Clause	Clause	Clause || Compound
T	T	T	
T	T	F	
T	F	T	
T	F	F	
F	T	T	
F	T	F	
F	F	T	
F	F	F	

Every time one more clause is added on the left, the table doubles in size. This means that for *n* clauses, the table will have 2 to the *n*th lines. (One clause needs two lines; two need four; three need eight; four need sixteen.)

So far, we have been able to construct good truth tables to represent the connections between the truth values of clauses and the truth values of their compounds. Unfortunately, in many cases one cannot tell what truth value the compound will have even if one knows what truth value each clause within it has. In such cases the right-hand part of the table cannot be completed with "T"'s and "F"'s. For example, consider the sentence, "She loved him because he was charming". (Assume that this sentence merely expresses an explanation of her love rather than an argument to prove her love.) It is not too difficult to decide the truth value of this compound under most circumstances. When it is false that she loved him, or when it is false that he was charming, then it cannot be true that she loved him because he was charming. Accordingly, we can confidently say that in these circumstances we need an "F" under the compound in our truth table.

But what happens to the compound when it is true that she loved him, and it is also true that he was charming? Is it then *true* that she loved him BECAUSE he was charming? Maybe. Is it then *false* that she loved him BECAUSE he was charming? Maybe. We don't have enough information to know. She might have loved him for his money. Or his physique. Or his brains. Or whatever. In other words, it is not possible to determine the truth value of the compound in this case from the truth values of the clauses. I will indicate this by putting a question mark in the appropriate place in the table (see Table III).

Table III

She loved him.	He was charming. || She loved him BECAUSE he was charming.
T	T	?
T	F	F
F	T	F
F	F	F

The question mark appearing in table III indicates that the compound sentence might be true or might be false when the two clauses are both true. Thus, in this case, the truth value of the compound sentence is not a function of the truth values of the clauses making it up.

When doing truth tables, use columns to the left of the double vertical line for the sentences which serve as the building blocks for the compound sentences to the right of the double line. Whenever possible, clauses which are entered in this part of the table should not contain connectors.

Clauses lacking connectors are often called *atomic sentences* because, like the atoms in nature which combine to form compound molecules, they go together to form compound sentences with connectors acting as the glue that holds them together. The double vertical line in the table indicates that all the truth values entered to its right are obtained *solely* from the truth values entered to its left. Later, we will sometimes write tables with simple *compounds* to the left of the double line and more complex compounds to the right. But the double line will *always* indicate that the truth values to its right are determined solely from those to its left.

If the information to the left of the double line is insufficient, enter a "?" in the appropriate place to the right of the double line.

It may help you to think about truth tables if you conceive of each line of a truth table as representing a kind of world determined by the truth values on the left.

For instance, in table III the first line represents the kind of world in which she loved him and he was charming, while the second line represents the kind of world in which she loved him but he was not charming. To figure out what truth value to enter in each line to the right of the double vertical line, consider each kind of world in turn. Ask yourself whether the sentence at the top of the column would always be true in the kind of world represented by that line of the table. If the answer is "yes", then you enter a "T"; if the answer is "no", you ask whether the sentence at the top of the column would always be false in that kind of world. If the answer is "yes", you enter an "F". If the answer is "no" again, you enter a "?".

When it is possible to complete a truth table for a compound sentence on the right without having to enter any question marks in the table, the truth value of the compound sentence is completely determined by the truth values of whatever sentences appear to the left of the double line in the table. *When the truth value of one proposition is completely determined by the truth values of some other propositions, then the first proposition is said to be a truth function of the others*. Thus, "The book does not contain a misprint" is a truth function of "The book does contain a misprint". Also, "Bob and John are men" is a truth function of "Bob is a man" and "John is a man". But "She loved him because he was charming" is *not* a truth function of "She loved him" and "He was charming".

I would like to remind you that we are not dealing with whole arguments yet. We are making tables which represent single compound sentences, and we are not as yet identifying any of these sentences as expressing premises or conclusions. In particular, the clauses on the left of a table are not premises for the compound on the right.

Also, in this Section I have often been talking as though our tables are tables for sentences, rather than propositions. In fact, it is more accurate to view our tables as tables for *propositions*, since the sentences which we write as column headings may be ambiguous. When there is an ambiguous sentence in a table, you first have to decide what proposition is being expressed by that sentence before you can attempt to fill in the table. For example, the first line of the following table cannot be filled in until you decide exactly what the compound sentence means!

They got married.	They had a baby.	They got married and had a baby.
T	T	

If the compound expresses the proposition that they got married and *then* had a baby, as it probably does, you would have to enter "?" in the blank above, since the information on the left does not tell you in what order the events occurred. On the other hand if the compound only means that both things happened, you would enter a "T" in the blank.

Truth functions and truth tables

30.1 Break each of the following sentences into atomic sentences and determine whether the resulting compounds are truth functions of the atomic sentences by attempting to complete a truth table for each compound.
   a. The driver either fell asleep or he was drunk.
   b. I took the checkbook and the sack with me.
   c. Possibly, you were coming in too fast.
   d. I didn't make it on time because I missed the bus.
   e. Since I was late, I got fired.
   f. The explosion happened before the fire broke out.
   g. The contract was signed after the court declared the company bankrupt.
   h. I had hoped for a win but we only got a tie.
   i. If it's raining, then it must be overcast.
   j. We weren't able to find the eggs or the butter, but we brought the beer.
   k. We were tired and hungry because we had not eaten for five hours and we had been working hard.

31

Truth tables for common connectors

It is useful to be somewhat familiar with the way the truth values of compound sentences do or do not depend on the truth values of their clauses. The purpose of this Section is to help you complete truth tables for commonly encountered compounds.

Let's take "NOT" first, since it's easiest. "NOT" always applies to a whole clause, although that clause may itself be either simple or compound. In "NOT (I'm coming)" the clause is simple, while "NOT (John is coming OR Fred is coming)" illustrates the case in which the clause is compound. In either situation, the function of "NOT" is the same: 'NOT X' amounts to saying that the proposition expressed by 'X' is not true, no matter whether 'X' is compound or simple.[1] I can construct the following standard truth table to summarize how "NOT" works:

X	NOT X
T	F
F	T

This table tells you how to complete a truth table for *any* sentence 'NOT X', once you have computed the value of the 'X' part of the sentence whether 'X' is simple or compound.

Let's take "AND" next. Recall that not all occurrences of "and" may be rewritten properly as "AND", because "and" is not always a connector. Those nonconnector occurrences of "and" do not count as "AND". One standard way in which "AND" works can be represented in the following table:

X	Y	X AND Y
T	T	T
T	F	F
F	T	F
F	F	F

(Check this by going over the above table line by line.) However, "AND" occasionally means "and then", as it probably does in "They got married AND they had a baby". In that case, we would get the following table:

---

[1]There are some sophisticated problems lurking here. One might wonder, for example, whether denying 'A' is true amounts to exactly the same thing as asserting that 'A' is false. This would be a difficulty if we allowed clauses to be substituted for 'A' which were neither true nor false, for then we would wish to deny 'A' is true without thereby claiming 'A' is false. Such issues are beyond the scope of this text. I assume throughout that all propositions are either true or false and none are both. I also make various other simplifying assumptions, such as that asserting 'A' amounts to the same thing as asserting that 'A' is true.

They got married.	They had a baby.	They got married AND they had a baby.
T	T	?
T	F	F
F	T	F
F	F	F

(See the end of Section 30 for an explanation.) Thus, "AND" is ambiguous, sometimes requiring the table

X	Y	X AND Y
T	T	?
T	F	F
F	T	F
F	F	F

One moral to this sad tale is simple: no purely mechanical method for splitting up sentences containing "and" has been presented. Each case must be treated with care and sensitivity to the meaning of the original sentence. Unless there is a particular reason to choose the second table for "AND", I will always use the first table. Note that the two tables differ only in one place.

The next connective with which I will deal is "OR". There are several difficulties about the meaning of "OR" which all arise because "OR" is ambiguous. The ambiguity can be illustrated by comparing the following three sentences:

(1) The person I saw was either a college student or a person of college age.

(2) Either he should lose a week's pay or he should have to do extra duty for a month.

(3) Either we operate or the patient dies.

The most natural interpretations of these sentences will result in each receiving a differently structured truth table. Looking first at (1), I would say that the person who utters this sentence would not be likely to think he had made a mistake if it turned out that the person he saw was in fact *both* a student *and* a person of college age. Thus, the following table would be reasonable for (1):

The person I saw was a college student.	The person I saw was of college age.	The person I saw was a college student OR the person I saw was of college age.
T	T	T
T	F	T
F	T	T
F	F	F

Note the single occurrence of an "F" in the last column, for the world in which the person seen was neither a college student nor of college age.

On the other hand, it most likely would seem wrong to the utterer of (2) if that sentence received a similar table, because (2) is probably intended to lay out two alternative punishments. If the utterer of (2) were told that in fact the person he was talking about should have to do extra duty *as well* as losing a week's pay, I would think the utterer of (2) would find that situation incompatible with his statement. (2) is probably intended to *rule out* both penalties' being imposed. That is, (2) is probably intended in such a way that if it really is true that both penalties should be imposed, (2) is incorrect. That yields the following truth table as the most likely one for (2):

He should lose a week's pay.	He should have to do extra duty for a month.	He should lose a week's pay OR he should have to do extra duty for a month.
T	T	F
T	F	T
F	T	T
F	F	F

Note that this table has two "F"'s in its last column, making it different from the table for (1). The topmost "F" represents the case where both penalties in fact should be imposed, making the statement (2) false.

Sentence (3), on the other hand, seems different again. Unlike sentence (2), sentence (3) is not proved false when both its component sentences are true—when it is true that we will operate and it is also true that the patient will die—becuse sentence (3) does not assert that the operation is bound to be a success. Thus, we cannot put an "F" in the top place in the last column of the table, as we did for (2). On the other hand, we surely cannot put a "T" there, either, because the mere truth of both components does not tell us that the only chance for the patient is our operating on him. That is, we are stuck with "?" for the top place in the last column. Similarly, in the next line of the table, where it is true that we will operate but false that the patient will die, we lack sufficient information to tell if the operation was necessary, as sentence (3) claims, and once again are thus stuck with "?" in this slot in the last column. In the third line of the table, where it is false that we will operate but true that the patient will die, we once again don't have enough information to tell whether our operating was the only chance the patient had, as sentence (3) claims. Once again, then, we must enter a "?" in the last column. However, in the fourth line of the table, where we consider the world in which the patient will not die and we will not operate, it is clear that (3) is false, since (3) claimed the patient would die if we did not operate. We therefore obtain this table for (3):

We will operate	The patient will die.	We will operate OR the patient will die.
T	T	?
T	F	?
F	T	?
F	F	F

Obviously, this table differs from the previous tables for "OR".

The three common tables for "OR", then, look like this:

X	Y	X OR Y	X	Y	X OR Y	X	Y	X OR Y
T	T	T	T	T	F	T	T	?
T	F	T	T	F	T	T	F	?
F	T	T	F	T	T	F	T	?
F	F	F	F	F	F	F	F	F

The first is known as the inclusive "OR", the second as the exclusive "OR", because the first includes the possibility that both 'X' and 'Y' are true, while the second excludes that possibility. Whenever possible, it is safest to interpret English "or" as inclusive unless there is good reason from the context or the sentence containing it to treat it otherwise. Sometimes, the inclusive "or" is written as "and/or"—a cumbersome device. Sometimes, the exclusive "or" is indicated by adding the words, "but not both" to the end of the sentence containing it.

We come now to "IF . . . THEN". A great deal has been written about this connector, and there is continuing controversy about the construction of the truth tables for various ways in which it is used. (I will pretend that this controversy does not exist, and proceed on the assumption that everything I say about this connector is plainly correct.) Let's consider some typical examples:

(5) IF the sun shines tomorrow, THEN it will get warmer.

(6) IF that line of tires sells well, THEN we'll have to get more of them in.

I think (5) makes a prediction about what will happen to the temperature if the sun should in fact happen to shine tomorrow. It does not tell us for sure what will happen if the sun does *not* shine tomorrow, since there might be other things which could make it warmer. I think (5) also says that if the sun shines tomorrow then it will get warmer *because* of the sun's shining. If you agree that (5) says all this, then you and I will both have some trouble filling out the truth table for (5). Let's try. We'll get the following result:

The sun shines tomorrow.	It will get warmer.	IF the sun shines tomorrow, THEN it will get warmer.
T	T	?
T	F	F
F	T	?
F	F	?

The first line of the table gets a "?" because we are not told in the left-hand columns whether the temperature's rising happens *because* of the sun's shining. Thus, if (5) really does claim that the warmth occurs *because* of the sun, then we have to put a "?" in the first line.

The second line of the table clearly needs an "F" in its last column, since in this line we are considering the kind of world in which the sun shines but the temperature does not go up. This situation is in clear violation of the prediction made by (5).

In the third and fourth lines, we are considering what to say when the sun does not shine tomorrow. Now we have already agreed that (5) does not tell us what will happen if the sun fails to shine tomorrow. And we have agreed that (5) claims there is a connection between the sun's shining and the temperature's rising. I think it is then clear that in the last two lines we have inadequate information given in the left-hand columns for deciding whether (5) is true or not.

Similar problems will arise in connection with (6). I believe (6) says that the tire's selling well would be a reason for getting more of them in. But now look at the truth table:

That line of tires sells well.	We'll have to get more of that line of tires in.	IF that line of tires sells well, THEN we'll have to get more of them in.
T	T	?
T	F	F
F	T	?
F	F	?

I put "?"'s in the last column in three places, because in these three lines it seemed to me that we had inadequate information for telling whether the tire's selling is a reason for having to get more. (Even in the first line, where we are told that the tires do sell well, and that we have to get more of them in, we still do not know that there is any connection between the two things.) The only clear line seems to be the second one, in which the tires sell well but it remains false that we have to get more of them.

All of this suggests that the usual truth table for "IF . . . THEN" looks like this:

A	B	IF A THEN B
T	T	?
T	F	F
F	T	?
T	F	?

Some sentences involving "IF . . . THEN" seem to have different tables. But *every* such table will have to have the "F" which is included in the table above. The main thing that is important for our purposes turns out to be the one solid "F" which occurs in the table for standard "IF . . . THEN" sentences.

We are left with just one preferred connector yet to deal with, namely, "IF AND ONLY IF". Suppose someone says something of the form, 'X IF AND ONLY IF Y'. It is not hard to see that what has been said is equivalent to

(7)   (IF Y THEN X) AND (X ONLY IF Y).

Moreover, in previous Sections we have seen that often 'X ONLY IF Y' means the same as 'IF X THEN Y'. We can then rewrite (7) as

(8)   (IF Y THEN X) AND (IF X THEN Y).

(8) then gives us one reasonable rewriting of 'X IF AND ONLY IF Y'. This should help us come up with a truth table for "IF AND ONLY IF". When we say 'X IF AND ONLY IF Y' we are really saying "IF X THEN Y' and 'IF Y THEN X'. Look at the typical truth table for these two sentence diagrams:

X	Y	IF X THEN Y	IF Y THEN X
T	T	?	?
T	F	F	?
F	T	?	F
F	F	?	?

(Notice that the definite "F" occurs in the *third* line in the last column because of the order in which the letters 'Y' and 'X' occur in it.) Putting together these results into one table for (8), we get this:

X	Y	(IF X THEN Y) AND (IF Y THEN X)
T	T	?
T	F	F
F	T	F
F	F	?

This is the table which I believe is typical of English "IF AND ONLY IF" sentences.

We have now gone through each of the preferred connectives and for each one we have come up with one or more typical truth tables which show the effect of the connective on the truth value of the compound sentence formed through application of the connective.

In order that we might be able to use truth tables in analyzing a wider variety of arguments, it's useful to see how a few more connectors work. I will somewhat arbitrarily choose four additional connectors to discuss, since there are too many to allow discussion of them all.

"Because" actually functions in two different ways. Sometimes it serves as an indicator that an argument is being given; in this role it separates premises from the conclusion, and is not a connector. For example, one might say "You ought to go along with me tonight because you promised you would". If we were to write out this argument to indicate its structure, we would write

You promised you would go along with me tonight.
You ought to go along with me tonight.

As you can see, there is no connector appearing in this argument.

On the other hand, "because" sometimes does not so much seem to indicate that premises are about to be given as it does to indicate that an explanation is about to be given. Here are some examples: (a) "It's raining because a warm front is passing through." (b) "His disposition has mellowed because he has found satisfaction in life and his financial affairs are straightening out." I believe it is fairly easy to see how each of these sentences could be thought not to offer any *arguments* at all, but only *explanations* of the events they report. If so, we would not write

A warm front is passing through.
It's raining.

Instead, we would write

(9) It's raining BECAUSE a warm front is passing through.

We already have seen what the truth table for a sentence of the form 'X BECAUSE Y' looks like (in Table III, Section 30):

X	Y	X BECAUSE Y
T	T	?
T	F	F
F	T	F
F	F	F

The same remarks and truth table apply to "since" or "for". Next, let's take a brief look at "unless". Consider the sentence

(10)   The patient will die unless we operate.

One way to handle this sentence is to rewrite it as follows:

(11)   IF NOT (we operate) THEN the patient will die.

On the other hand, I think (10) probably does not also mean

(12)   IF we do operate, THEN NOT (the patient will die).

(12) says the operation is guaranteed to save his life, while (10) does not seem to me to imply that. Working with (11) then, and using the usual technique for an "IF . . . THEN" truth table, we get the following table:

We operate.	The patient will die.	IF NOT (we operate) THEN the patient will die.
T	T	?
T	F	?
F	T	?
F	F	F

(The "F" occurs in the bottom line of this table rather than in the second line, because of the "NOT" in the column heading. Think about it. The last line of the table concerns the situation in which we do not operate and yet the patient lives.)

This suggests handling "unless" by treating 'X UNLESS Y' as equivalent to 'IF NOT Y THEN X'. You may choose this strategy if you like. An alternative strategy treats 'X UNLESS Y' as equivalent to 'X OR Y'. This is an easier strategy to remember. In order to justify it, compare the preceding table to the table for the third sense of "OR" earlier in this Section. Since these tables are the same, there is no difference between "UNLESS" and "OR" (in its third sense) for our purposes.

Given this information about typical truth tables for some connectors, we have gathered enough tools to handle an interesting variety of sentences as they occur in the roles of premiss and conclusion in arguments. I leave it to you to work out the tables for other common connectors.

Truth tables for common connectors

31.1 Write out typical truth tables for each of the following diagrams. Try to do them without referring back to the material in this Section.
   a. A AND B
   b. A OR B (In all its senses.)
   c. A, IF B
   d. A ONLY IF B
   e. A IF AND ONLY IF B
   f. A OR (NOT B)
   g. A IF AND ONLY IF (NOT B)
   h. A BECAUSE (NOT B)
   i. A SINCE B
   j. A UNLESS (NOT B)
   k. IF A THEN (B UNLESS C)

31.2 Write out truth tables for each of the following sentences.
   a. If there is a train coming, you'd better get off the tracks.
   b. He went to London or to Istanbul (and maybe both).
   c. They got married and had a baby.
   d. You'll pass if and only if you take the final exam.
   e. You'll pass if and only if you don't flunk the final.
   f. Jones will get rich if he works hard.
   g. I'll get an "A" on the test or I'll be very unhappy.
   h. We'll have to put Hank in the starting lineup or we won't win.

31.3 Construct truth tables for each of the following sentences. (You might have to figure out how to rearrange the parts in order to make them fit the typical patterns.)
   a. Unless you give my doll back, I'll cry.
   b. Only if you're brave do you deserve to win the crown.
   c. Because I'm doing well in my piano lessons, I'm going to continue.
   d. Since this is the first opportunity we've had to discuss business in a serious way, I think we should try to find some common ground on which to start.
   e. Business has dropped off and profits are slumping because the economy is undergoing a retrenchment. (Break this sentence into three parts.)
   f. The books have not come in because there is a truck strike, and we don't expect another shipment for five weeks.

32

Truth tables for arguments

It is possible to prove certain arguments valid by using truth tables, but thus far, you have only seen how to construct truth tables for one compound sentence at a time. In order to deal with arguments it will be necessary to construct tables that treat more than one compound sentence, because arguments generally contain several such sentences and we want the table to reveal the relationships between them. Accordingly, the first step in learning how to prove an argument valid using truth tables is to learn how to construct a truth table for a whole argument.

Suppose you went on a picnic and you were supposed to bring the eggs. You took both a sack and a box to the picnic, and when it came time to eat, someone asked you where the eggs were. You replied that they were either in the sack or in the box (and maybe in both). Whoever it was that asked you then looked in the box and found no eggs. You could have replied with a little argument:

P1.   The sack has eggs in it OR the box has eggs in it.
P2.   NOT (the box has eggs in it).
C.   The sack has eggs in it.

Since you can see that this whole argument is built from just two atomic sentences, the truth table for it can be set up using just these two sentences to the left of the double line. All the rest of the premises or conclusions in the argument should be placed to the right of the double line in separate columns. The resulting table looks like this:

The sack has eggs in it	The box has eggs in it.	The sack has eggs in it OR the box has eggs in it.	NOT (the box has eggs in it).
T	T	T	F
T	F	T	T
F	T	T	F
F	F	F	T

In general, when constructing a truth table for an argument, put all the *atomic* sentences occurring *anywhere* within the argument to the left of the double line and put all the *compound* sentences which occur as premises or conclusions to the right of the double line. Note that the double line does not necessarily separate premises from the conclusion. In order to complete the columns of "T"'s and "F"'s in the *right*-hand portion of the table, you use whatever information from the *left*-hand portion of the table is relevant. In the above table, I used the information from the second column in the table to compute the values that went into the far right-hand column. (The first column of the table was irrelevant to that computation.)

By placing all the propositions from the argument in one truth table, we can study some of their relationships. By carefully checking the table, we can see if any given combination of sentences in the table is true or false. For instance, we can see that there is only one kind of world in which both of the compound sentences are true, namely, the kind of world in which the sack has eggs in it but the box does not (as represented by the second line of the table).

Sometimes the truth table for an argument will contain a "?" instead of "T" or "F". Don't let that bother you. Just enter the "?" whenever it is appropriate. This point is illustrated in the next and final example for this Section.

P1. I didn't make it on time because I missed the bus.
P2. I missed the bus because it was early.
_____
C. I didn't make it on time because the bus was early.

The table for this argument follows.

You should examine this table carefully to make sure you can tell where each of the entries comes from. Note that when it is true that I made it on time, then it is false that I didn't make it on time, and that is the reason why the first four "F"'s occur in column four. This table, like the one we did before, reveals some interesting relationships between the various propositions in the table. For instance, is there any chance at all that the three compound propositions could all be true? Only one such chance. What kind of world would allow that to possibly happen? (Look at the fifth line of the table.) Why are there no lines in which all the compounds are listed as definitely being true? (Think about the relationships between the parts of the compounds as revealed by the table, and think about why the question marks appear where they do.)

I made it on time.	I missed the bus.	The bus was early	(premiss) NOT (I made it on time) BECAUSE I missed the bus.	(premiss) I missed the bus BECAUSE it was early.	(conclusion) NOT (I made it on time) BECAUSE the bus was early.
T	T	T	F	?	F
T	T	F	F	F	F
T	F	T	F	F	F
T	F	F	F	F	F
F	T	T	?	?	?
F	T	F	?	F	F
F	F	T	F	F	?
F	F	F	F	F	F

Truth tables for arguments

32.1 Construct a truth table for the following argument.

The driver fell asleep, or he was drunk. But he did not fall asleep. So, he was drunk.

32.2 Are there any worlds in which both the premisses and the conclusion in the above argument are true? If so, in what line of the truth table are they represented? Describe these worlds in English; that is, say whether the driver fell asleep or did not fall asleep in them, and say whether or not he was drunk. Are there any worlds in which the premisses are true and the conclusion is false? If so, describe them.

32.3 Construct a truth table for the following argument.

The explosion happened before the fire broke out. The roof collapsed after the fire broke out. Thus, the roof collapsed after the explosion happened.

32.4 Are there any worlds in which there is at least a chance that both the premisses and the conclusion in the preceding argument are true? Are there any worlds in which you know that all these propositions are definitely false? Pick out the worlds, if there are any, and describe them in each case.

32.5 Construct a truth table for the following argument. (Treat its second sentence as an explanation rather than as a brief argument by itself.)

I had hoped for a win but we only got a tie. Since we only got a tie, we won't be able to go on to the tournament. So the season is over for us.

32.6 Are there any lines in your truth table for this argument in which the premisses of the argument are all true? If there are, pick them out and list them. Are there any lines in which the premisses are all false? List these, if any. Do you think this argument is valid? Why, or why not? Does your truth table show you any possibilities where the premisses might all be true when the conclusion of the argument is false?

33

Using truth tables to prove validity

We can *prove* the egg argument in the last section *valid* by examining its table! Here is the table, once again, except that this time I have marked the premisses and the conclusion in the column headings.

(conclusion) The sack has eggs in it.	The box has eggs in it.	(premiss) The sack has eggs in it OR the box has eggs in it.	(premiss) NOT (the box has eggs in it).
T	T	T	F
T	F	T	T
F	T	T	F
F	F	F	T

How does this table prove the argument valid? Recall that the table displays at least all the logically possible kinds of worlds, line by line, and for each kind of world the table tells us what happens to the truth values of the compound sentences. Since the table reveals at least all the logical possibilities, the table shows whether it is logically impossible for the premisses all to be true in the same kind of world in which the conclusion is false. Of course, if that arrangement of truth values is not possible, the argument is valid. (See Section 24.)

We look, then, for a line in the table where the premisses are all true and the conclusion is false. Scanning the two premiss columns above reveals that both premisses are true in only one kind of world, namely, the kind represented by the second line of the table. However, we then note that in that kind of world, the conclusion is *true.* Therefore, we may assert that the argument has been proved valid, for *whenever* the premisses are all true, so also is the conclusion.

Let's apply these ideas to tables which contain question marks. The table for the following simple valid argument will do:

P1.   If you have a high cholestrol level in your blood, you run a greater than normal risk of heart trouble.

P2.   You do in fact have such an elevated cholestrol level.

C.   Your risk of heart trouble is abnormally high.

(P2) You have a high cholesterol level in your blood.	(C) You run a greater than normal risk of heart trouble.	(P1) IF you have a . . . in your blood, THEN you run . . . of heart trouble.
T	T	?
T	F	F
F	T	?
F	F	?

A quick look at this table shows that there are no lines in which the premisses are both definitely true; hence there are of course no lines in which the premisses are definitely true when the conclusion is false. (When looking at a table be careful to note which columns represent the premisses—in the above case, it's the first and last columns.)

Can we say, then, that the argument is proved valid by the table? No, not yet, for we need to take the question marks into account. Since we are interested to know whether there is *any* possibility at all that the premisses are all true when the conclusion is false, we must consider if this could happen if some of those question marks represented "T'"s or some of them represented "F'"s.

Since in this particular table, all the question marks are under a *premiss* rather than under a conclusion, we don't care if they turn into "F'"s; after all, what we are worried about is what happens to the conclusion when the premisses are all *true.* So, in the above table, where the question marks are under a premiss, the troublesome case is not when the question marks are thought of as "F'"s, but rather when they are considered as "T'"s. So consider what happens when some or all of the "?'"s are considered as "T'"s. When we consider the top "?" as a "T", we have a line in which both premisses are true. Fortunately, in this line the conclusion is also true; so this line does not raise questions about the argument's validity. The remaining two question marks don't cause any trouble for validity either, since in their lines *P2* is false; this means that in neither of these lines can both P1 and P2 be true, even if P2's question marks are turned into "T'"s.

We have now considered *all* the lines, and we have found that in no case can all the premisses be true when the conclusion could be false. Thus, the table shows this argument is valid.

Next, consider an argument which is clearly invalid:

P1. I got drunk.
P2. I was fired.

C.   I was fired because I got drunk.

The table for this argument corresponds to our judgment that this argument is invalid:

(P1) I got drunk.	(P2) I was fired.	(C) I was fired BECAUSE I got drunk.
T	T	?
T	F	F
F	T	F
F	F	F

The first line of this table indicates that the premises might be true when the conclusion is false (because the "?" could represent an "F").

However, the unfortunate fact is that one normally cannot just look at the entries in a truth table and declare an argument invalid. A truth table by itself does not usually show that an argument is invalid, even when there is a line in the table in which all the argument's premises are marked true and in which its conclusion is marked false. Hence, the above table does *not* definitely prove the argument invalid; it only indicates the argument *might* be invalid. A truth table *can* prove some *valid* arguments are valid, but a truth table by itself normally *cannot* prove any argument invalid. This limitation of truth tables has nothing to do with the question marks in the tables, but rather comes from a more fundamental logical feature of how the tables are constructed.

Why? Validity depends on what is logically possible. Truth tables list at least every logical possibility, but sometimes they list more. Some of the worlds represented in some tables are logically impossible, and some ways of changing question marks into "T"'s and "F"'s in our minds when we are looking for the possibility of true premises and false conclusions do not make any sense. When this happens, the table can be misleading, for the table will make it look as though there is a possibility that true premises are accompanied by a false conclusion when no such possibility exists. Hence, in order to actually demonstrate that an argument is invalid we need to do more than merely examine the entries in the truth table for that argument. We need, in fact, to examine the relationships between the propositions at the top of the table. This in effect was done by the techniques for proving invalidity discussed in Sections 25 and 26. Truth tables can, however, be a clue to where the problems with an invalid argument lie. See Section 40.

You may object that in the example last discussed, the table accurately revealed the invalidity of the argument. *That* table was accurate; but not all tables are. For instance, the simple argument, "I arrived at work today at 9 A.M.; therefore, I arrived before noon", clearly must be valid. Yet its table looks like this:

(P1) I arrived at work today at 9 A.M.	(C) I arrived at work today before noon.	
T	T	
T	F	
F	T	
F	F	

The second line of this table *seems* to say it is possible for P1 to be true when C is false; so this line *seems* so say the argument is invalid. But this line of the truth table does *not* represent a genuine logical possibility, for it is simply not logically possible for it to be true that I arrived at 9 A.M. and false that I arrived before noon.

I can only tell what is genuinely possible and what is not by bringing in background information about the logical relations between propositions—background information not presented by the table itself. That is why the table *by itself* can not prove any argument invalid.[1]

Let me summarize what we have found. First, when a truth table for an argument contains no line in which all the premisses could be true while the conclusion could be false, we can be confident the argument has been proved valid. This is true because the truth tables represent *at least* every logical possibility. On the other hand, when the table *appears* to say the premisses could be true when the conclusion could be false, we must conclude that the table fails to prove the argument valid. The argument might be valid, but the table fails to prove it.

It will be handy to have a label for arguments which cannot be proved valid through the use of any truth table. I'll call such arguments *T-invalid,* meaning that the most detailed truth tables one can construct for them still make them look invalid. Some T-invalid arguments, such as the last argument used as an example above, are really *valid,* while others are not.

An argument will be T-invalid when the most detailed truth table for that argument contains at least one line in which it *appears* possible that the argument's premisses are *all* true when the argument's conclusion is false. I will call such a line a *bad line.* Here are examples of different sorts of bad lines:

Example 1:

(conclusion)	. . .	(premiss)	(premiss)	(premiss)
F		T	T	T

(In example 1 you see the conclusion does not necessarily appear to the right.)

Example 2:

(premiss)	. . .	. . .	(conclusion)	(premiss)
T			?	T

---

[1]There is one special case in which the truth table *can* by itself prove an argument invalid, a case which I will ignore because it occurs rarely. It occurs when all the premisses are marked "T" in all lines of the table, and the conclusion is marked "F" in all the lines of the table. Such an argument must be invalid.

In example 2, we have a bad line because the "?" could indicate "F".

Example 3:

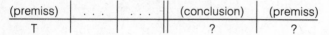

(premiss)	. . .	. . .	(conclusion)	(premiss)
T			?	?

In example 3, we have a bad line because the first "?" could indicate "F" and the second "?" could indicate "T".

Here is a precise definition of "bad line": *a line in a truth table is a bad line if and only if in that line all the premisses are marked with a "T" or a "?", and in that line the conclusion is marked with an "F" or a "?".* When a bad line occurs in the most detailed truth table for an argument, then (and only then) that argument is T-invalid. If no bad lines occur in a truth table for a given argument, then that argument is valid.

One final technical point. Some arguments' truth tables contain no lines at all in which all the premisses could be true. When this happens, the table automatically has no bad lines in it. (Why?) When this happens, the argument in the table is *valid.* This result seems odd to most people, because this result means that if I use the following propositions as premisses, I will not have any bad lines in my table:

P1.   Today is Wednesday.
P2.   Today is not Wednesday.
C.

It doesn't matter what conclusion I draw from these premisses. My argument will still be valid (since its table will contain no bad lines)! This consequence was discussed earlier where we considered what happens when the premisses are logically incompatible—when the premisses cannot all be true together.

Using truth tables to prove validity

33.1 Determine whether or not the truth tables for each of the following arguments contain bad lines.

a. We didn't get the full amount promised in the contract. But either we got the full amount promised, or the contract was violated. So, the contract was violated.

b. Neither Maude nor Philip wants to go to the party. So Philip does not want to go.

c. I don't expect her to show up, but I'm still hoping that she will. Either I'm hoping that she will show up or I'm despondent (but not both). So, I'm not despondent (yet).

d. It will either rain or it will snow (but not both). It will snow. Thus, it will not rain.

e. It will rain and snow. Thus, it will rain.

f. It will rain and snow, but it won't sleet. Thus, it will neither sleet nor rain.

g. The sun rises every day only in the morning. The sun is rising now and the birds are singing. So, it must be morning.

h. I love him because he is charming. Thus, he is charming and I love him.

i. He is charming because he is polite and intelligent. Therefore, he is polite.

j. I was late to work today and I got fired. Thus, I got fired because I was late to work today.

33.2 On the basis of the truth tables constructed in response to 33.1, answer the following questions for each argument whenever the table gives you enough information to do so. When the table does not yield enough information say so.
a. Is the argument T-invalid?
b. Is the argument valid?

34

Some common valid patterns of arguments

You now know enough about truth tables, validity, and argument diagraming to put together an interesting and clear theory about their connections.

It is reasonably obvious that the following two arguments fit the same argument diagram, and that both are valid *because* they fit that diagram:

(I) P1. If gasoline prices rise above $3 per gallon, many people will be forced to curtail their driving habits considerably.
P2. Gasoline prices will indeed rise above $3 per gallon.

C. Many people will be forced to curtail their driving habits considerably.

(II) P1. If Sarah loves John, she will marry him.
P2. Sarah does in fact love John.

C. She will marry him.

The diagram, of course, is this:

IF A THEN B
A
_____
B

We can use truth tables to show that *all* arguments which fit this diagram are valid; the only restriction on our proof of this fact is that the proof assumes "IF A THEN B" has a *typical* truth table. (That is, the proof might not work for any arguments which fit the diagram but which have nontypical truth tables.)

A	B	IF A THEN B
T	T	?
T	F	F
F	T	?
F	F	?

A quick check of the table will reveal (1) that the premises and the conclusion in the above argument diagram are all present, and (2) there are no bad lines in the table. We have given a proof that the preceding argument diagram typically has only valid arguments fitting it. I will say this means the diagram itself is *valid*. This diagram is one simple pattern of argument found very often in everyday life in one guise or another. It has been known for many centuries and bears the name *modus ponens*.

Given our proof of this diagram's validity, you can show that an argument in English is valid just by pointing out that it fits the diagram, provided there is nothing about the argument which would give it a special truth table.

Many such valid patterns of argument can be readily constructed. It is a good idea to learn to recognize as many of these as possible, and to see how variations on them are also possible. I will give a final example, leaving you to work on several others as an exercise.

The diagram

A UNLESS B
NOT A
———————
B

can be proved valid by reference to the following tables:

A	(con) B	(prem) A UNLESS B	(prem) NOT A
T	T	?	F
T	F	?	F
F	T	?	T
F	F	F	T

A	(con) B	(prem) A UNLESS B	(prem) NOT A
T	T	F	F
T	F	?	F
F	T	?	T
F	F	F	T

The table on the top includes the typical table for "UNLESS"; the one on the bottom contains an alternate treatment of "UNLESS" which is sometimes appropriate when 'A UNLESS B' means '(IF B THEN NOT A) AND (IF NOT A THEN B)'. (A sentence such as "I'll love you unless you turn against me" is probably best treated as the second kind of "UNLESS".)

Verify for yourself that neither table above contains a bad line, and thus that the diagram is valid under its typical interpretations.

It should not be too difficult, then, to see that the following diagrams are also valid:

(NOT A) UNLESS B	A UNLESS (NOT B)
A	NOT A
B	NOT B

The diagram on the left was obtained from the original diagram by substituting 'NOT A' for 'A' throughout—and changing 'NOT (NOT A)' to 'A'. The diagram on the right was obtained by substituting 'NOT B' for 'B' in the original. Of course these diagrams have their own tables, but as you might imagine, the substitutions which generated these new diagrams do not affect their validity. In fact, these new diagrams are just variations on the old diagram, substituting something more complex for one of the letters in the old diagram. Such substitutions will never change a valid diagram into an invalid one, since the valid diagram is proved valid by the truth tables no matter how complex a proposition 'A' and 'B' stand for in the tables. (However, if you wish, you may check up on me by constructing new truth tables for each of our new argument diagrams.)

Here are a few more commonly encountered valid diagrams. Look them over carefully. Try to see why each would be valid.

(1) IF A THEN B
NOT B
———————
NOT A
(Called modus tollens)

(2) A OR B
NOT A
———————
B
(Called disjunctive syllogism)

(3) A OR B
IF A THEN C
IF B THEN D
———————
C OR D
(Called constructive dilemma)

(This diagram is valid for both the inclusive and the exclusive "OR" if "OR" is interpreted the same way in both the premiss and the conclusion. It is not valid if the third sense of "OR" is used for both premiss and conclusion.)

(4) A BECAUSE B
IF B THEN C
———————
C

(5) A BEFORE B
IF B THEN C
———————
C

When an argument diagram contains a bad line in its typical truth table, we may call it "T-invalid", meaning that truth tables cannot prove it valid. Many such T-invalid diagrams are not ultimately valid. That means there are some arguments fitting the diagram which are invalid arguments, but it does not mean that all arguments fitting the diagram are invalid. The diagram

(6)   A OR B
      IF B THEN C
      ―――――――
      C

is T-invalid for all interpretations of "OR". (Check this for yourself.) In fact, this diagram is invalid, as can be seen by considering an invalid argument in English fitting the diagram, like this:

Today I'll study for my German test or I'll work on my lab.
If I work on my lab, I'll need more graph paper.
―――――――――――――――――――――――――――――――
Therefore, I'll need more graph paper.

On the other hand, although the diagram is truly invalid and not merely just T-invalid, it still has valid arguments fitting it, like this one:

The train is due in half an hour, or at least within an hour.
If it's due within an hour, we'll be late getting to the station to meet it.
―――――――――――――――――――――――――――――――――――――
We're going to be late getting to the station to meet the train.

Some T-invalid argument diagrams are nevertheless valid, which means that all arguments fitting these diagrams and possessing typical truth tables, are valid. Here are two examples of such diagrams, which of course cannot be *proved* valid using our techniques.

(7)   IF A THEN B   (8)   IF A THEN B
      IF B THEN C   ―――――――――――
      ―――――――――   IF NOT B THEN NOT A
      IF A THEN C

Verify that these diagrams seem valid by thinking about the relations between their parts. The reason we cannot prove these diagrams valid is that their validity does not depend on truth tables alone.

In sum, the situation with respect to argument diagrams parallels that for arguments themselves: we can prove some diagrams valid through truth tables, and we cannot prove others valid even though they really are.

―――――――――――――――――――――

Some common valid patterns of arguments

34.1 Prove each of the argument diagrams (1) through (5) valid by constructing truth tables for them. In some cases, you will need to construct more than one table for a given diagram to cover all the typical interpretations of the connectors it contains.

34.2 Give an example of an argument in ordinary English which fits each of the diagrams you proved valid above.

34.3 Construct three valid argument diagrams which are valid variations on diagram (1). Prove one of them valid with a truth table.

34.4 Construct a new valid argument diagram which is not just a variation of one of the above and prove it valid by a truth table.

34.5 Many times when people *say* "if . . . then", in ordinary life contexts, they *mean* "if and only if". Arguments in which this happens should be diagramed using "IF AND ONLY IF" rather than "IF . . . THEN". This helps to ensure that those premisses and conclusions which actually do fit "IF . . . THEN" diagrams, rather than "IF AND ONLY IF" diagrams, have a typical "IF . . . THEN" truth table. On this assumption, test the following diagram for validity, using the typical "IF . . . THEN" tables:

IF A THEN B
B
_____
A

Now test this one, using the typical "IF AND ONLY IF" table:

A IF AND ONLY IF B
B
_____
A

Make up arguments which fit each of these.

34.6 Prove the following diagram valid on one meaning of "OR" and T-invalid on another typical meaning of "OR".

A OR B
A
_____
NOT B

What do your results show about the validity of English arguments fitting this diagram? Give an example of an English argument fitting the diagram for each interpretation of the "OR" in the diagram.

34.7 Prove the following diagrams cannot be shown valid using our techniques.

a. A AND B
_____
A BECAUSE B

b. IF A THEN B
IF B THEN C
_____
C

c. IF A THEN B
NOT A
_____
NOT B

d. A IF AND ONLY IF B
_____
A

Are any of these diagrams really valid? Prove the same thing for diagrams (7) and (8) from the Section. Are these diagrams valid?

34.8 Make up two English arguments which fit the diagram of 34.7c, one of which is *valid* and one which is invalid. (You might not be able to *prove* the valid one valid, but try to find one that is fairly obviously valid.)

34.9 Construct a diagram for each of the following arguments in as much detail as possible using any connector needed and then prove the arguments valid by proving the diagrams are valid. You will sometimes need to exercise judgment in choosing to use the same capital letter in your diagrams for two or more sentences from a given argument which are worded differently but which seem to express roughly the same proposition. Try to capture the spirit and inner connections in these arguments when you diagram them. This can be done in each case with fairly simple diagrams, employing no more than three single capital letters.

a. Unless Smith can garner enough votes to block Jones' nomination on the first few ballots, Jones will be the next Republican Presidential nominee. But Smith has angered enough of the Republican faithful to weaken her cause sufficiently to prevent her getting those crucial votes. So, it looks as though Jones will be the next Republican standard-bearer.

(Hint: Treat the second sentence as being equivalent to "Because Smith has angered enough of the Republican faithful to weaken her cause, she will not be able to get those crucial votes".)

b. Most college students are missing out on the opportunity of a lifetime to broaden their knowledge and become more interesting, useful people, since the popular attitude among students these days is expressed in the slogan, "If it isn't going to make you more marketable when you graduate, avoid it." However, our society's tremendous investment of dollars in higher education makes sense only if most students do not blow their opportunity to become broader and more useful citizens. So our current practice of pouring money into the higher educational enterprise has now become senseless.

34.10 Convert each of the following common valid argument diagrams to equivalent ones which contain only preferred connectors. Prove the resulting diagrams valid by constructing typical truth tables for them.

a. A UNLESS B	b. A PROVIDED THAT B	c. A BEFORE B
NOT B	B	IF B THEN C
A	A	C

6
Symbolic Propositional Logic—Symbolization and Truth Tables

35

Specific symbols for the preferred connectives

As we have already seen in detail, there are considerable variations in meaning among the uses of simple English words like "and" and "or". Sometimes, for instance, the word "and" cannot even be called a connective, as in the case of "two and two equals four". Even when "and" is a connective, it can have more than one truth table.

Although it is possible to try to keep track of all the variations in meaning associated with each connector, it becomes cumbersome and time-consuming, especially when dealing with lengthy or otherwise complex arguments. For this reason, logicians developed some special symbols which can be used in place of the preferred connectives. However, the real advantage of having a special symbol comes only if the special symbol has a more precise definition than the word or phrase it replaces. Accordingly, each of the special symbols has just one truth table instead of several, and the truth table for each symbol contains no question marks. This makes the special symbols much easier to use.

But there is a price to be paid for the aid and comfort given by the special symbols. Part of the cost consists of the effort required to translate English sentences into sentences containing the special symbols. The rest of the cost arises because many of these translations cannot be carried out with complete accuracy, for there is no way that a special symbol with no "?" 's in its truth table can equal exactly a connective such as "IF . . . THEN" which does have "?" 's in its table. If the two were truly equal, they would have the same tables.

I say all this by way of both promise and warning. The promise is that the special symbols can provide a degree of precision and clarity in argument diagraming that English words cannot. The warning is that the special symbols do not exactly equal the English connectives they are intended to replace. First, we will explore the promise. Later we will take more account of the warning.

To begin, you must become acquainted with the special symbols. Here is a list:

Preferred connective in English	Symbol which will replace the preferred connective
AND	&
OR	∨
NOT	~
IF . . . THEN	⊃
IF AND ONLY IF	≡

(The symbols I'm using come largely from the pioneering work of Bertrand Russell and Alfred North Whitehead done near the beginning of this century.[1] The only difference is that I use the "&" instead of the Whitehead-Russell "·" for "AND", because I believe the "·" is too easily confused with a period used at the end of a sentence.)

The precise definition of each of the special symbols is given by means of a truth table. It is absolutely essential that you learn these definitions backwards, forwards, upside down, and inside out.

The negation sign, the "~", always works like this:

X	~X
T	F
F	T

That table constitutes the complete definition of "~". It shows that "~" always reverses the truth value of the proposition to which it is applied. In order to understand this definition, however, it is necessary to realize that the proposition which is expressed by the sentence that is substituted for 'X' in this table could be extremely complex. 'X' might stand for 'A & B', in which case '~X' would stand for '~(A & B)' where the parentheses indicate that the whole of 'A & B' is affected by "~". In this case the truth value of '~(A & B)' is always the opposite of the value of 'A & B', as the above definition indicates.

Our other symbols will each connect two propositional expressions to form a compound expression. The function of each of these is defined as follows:

X	Y	X & Y	X ∨ Y	X ⊃ Y	X ≡ Y
T	T	T	T	T	T
T	F	F	T	F	F
F	T	F	T	T	F
F	F	F	F	T	T

In the remainder of this Section I will discuss, and attempt to justify (in part), these definitions.

The definition of "&" (called an ampersand) is just what you ought to expect given our previous examination of "AND". Essentially, 'X & Y' asserts that both

[1]*Principia Mathematica* (London: Cambridge University Press, 1910).

'X' and 'Y' are true; it asserts nothing more. Thus, when both 'X' and 'Y' are true, 'X & Y' is correct, or true; otherwise it's false.

The definition of "$\vee$", called "the wedge", picks out one of the typical tables associated with "OR", and makes that the definition of "$\vee$". You will recognize this table as the *inclusive* sense of "OR". Note that while 'X & Y' is true in only one case, 'X $\vee$ Y' is false in only one case.

In order to get the definition of "$\supset$" called a "horseshoe", we take the table typical of "IF . . . THEN", containing "?" 's, and we convert all the "?" 's to "T" 's. We know that one of the characteristics desired for all our special symbols is that no "?" 's appear in their tables; so in defining a special symbol to take the place of "IF . . . THEN", something had to be done about the "?" 's which typically appear in its tables. It happens to work out better for what we will do later on if "T" 's are used for all the "?" 's in setting up the definition of "$\supset$". You can look on 'X $\supset$ Y' as a truth-functional *approximation* for the nontruth-functional 'IF X THEN Y'. Because 'X $\supset$ Y' is only an approximation for its English counterpart it will cause trouble on some occasions, but for now we will ignore that problem.

It is fairly easy to remember the definition of "$\supset$" by noting the single case in which a sentence of the form 'X $\supset$ Y' is false. In a sentence of the form 'IF X THEN Y' or of the form 'X $\supset$ Y', 'X' is called the *antecedent* and 'Y' is called the *consequent*. The whole sentence is called a *conditional*.

Finally, in order to get a definition for "$\equiv$", called "triple bar", we do the same thing to "IF AND ONLY IF" that we did to "IF. . . THEN". That is, we take the typical table for 'X IF AND ONLY IF Y', which contains some "?" 's, and we convert all the "?" 's to "T" 's. The result is simple: 'X $\equiv$ Y' essentially says 'X' has the same truth value as 'Y'. Perhaps the symbol, "$\equiv$", will suggest the idea of equality to you. Thus, when 'X' and 'Y' do in fact have the same truth value, 'X $\equiv$ Y' is true; otherwise, 'X $\equiv$ Y' is false.

In all cases, the 'X' and the 'Y' involved in these definitions can be extremely complex or extremely simple sentences. Moreover, several of the special symbols can appear in one sentence, just as long as it is clear what they hook together or apply to. So the following are all acceptable forms:

	English rough equivalent
A $\supset$ ~B	IF A THEN (NOT B)
~(A $\supset$ B)	NOT (IF A THEN B)
A $\vee$ (B & ~ C)	A OR (B AND (NOT C))
(A & B) $\supset$ (C $\vee$ D)	IF (A AND B) THEN (C OR D)
~~A	NOT (NOT A)
~(A & ~A)	NOT (A AND (NOT A))

But the following are not acceptable forms:

	Attempt at an English equivalent
A $\supset$ $\equiv$ B	IF A THEN IF AND ONLY IF B
$\supset$ A	IF . . . THEN A
~$\supset$A	NOT (IF . . . THEN A)
& (A $\vee$ B)	AND (A OR B)

The following rules may help you to avoid writing nonsensical formulas:

1) The negation sign always applies to one sentence, which may be complex if desired. The sign will apply to the smallest sentence following the sign unless parentheses dictate otherwise.

2) The other signs must always connect exactly two sentences, each of which may be complex if desired.

' ⊃ A' violates the second rule, and is not a well-formed sentence form as a result. Since ' ⊃ A' is not well-formed, '~ ⊃ A' will violate the first rule, because in ' ~ A' the negation sign is not applied to a well-formed sentence. Some people are bothered by ' ~ ~ A' in this connection, but ' ~ ~A' is perfectly fine. The first negation sign applies to one sentence, namely, the sentence with the form, ' ~ A'. The second negation sign applies to 'A' itself. You could think of '~ ~A' like this: '~ (~ A)'. You can add the parentheses if you want to, but they are unnecessary.

When you want one of the symbols to apply to something complex, indicate this by using parentheses. Thus, if you want to negate 'A ⊃ B', don't write '~A ⊃ B'. Instead, write '~(A ⊃ B)'. Or, if you want "&" to conjoin 'A ∨ B' to 'C', don't write 'A ∨ B & C'. Instead, write '(A ∨ B) & C'.

Special symbols for the preferred connectors

35.1 Fill in the blank columns in the following table:

A	B	A ⊃ B	B ⊃ A	A ≡ B	B ≡ A	A & B	B & A	A ∨ B	B ∨ A
T	T								
T	F								
F	T								
F	F								

35.2 What is the one line in which 'A & B' is true? What is the one line in which 'A ∨ B' is false? What is the one line in which 'A ⊃ B' is false? In which lines is 'A ≡ B' false? (Remember the answers to these questions.)

35.3 Fill in the blanks in the following table:

A	B	~(A & B)	~(A ∨ B)	~(A ⊃ B)	~(A ≡ B)
T	T				
T	F				
F	T				
F	F				

35.4 What is the truth value of 'A ⊃ B' when 'A' is false? Do you need to know the truth value of 'B' to answer? What is the truth value of 'A & B' when 'A' is false? What is the truth value of 'A ⊃ B' when 'B' is true? What is the truth value of 'A ∨ B' when 'B' is true?

35.5 Which ones of the following are sensibly put together?

a. ~(~A & B)
b. A ∨ & B
c. A ∨ ~~B
d. (A ⊃ B) & C
e. A ⊃ B & C
f. ~~~A
g. ~(A & ~B) ⊃ C
h. ≡ (A ∨ B)
i. ~≡ (A & B)
j. & A

k. & (A B)
l. & (A ∨ B)
m. (A ⊃ B) ⊃ C
n. A ⊃ (B ⊃ C)
o. A ⊃ B ⊃ C
p. & ∨
q. A B
r. ~A & B
s. ~A & ~B
t. ~(A ∨ (B & ~C))

36

Computing the truth value of a complex formula

In order to work with sentences or sentence diagrams containing more than one special symbol, it is useful to have a method for computing the truth value of complex formulas. The strategy for computing the value of a complicated formula is first to compute the values of the smaller parts of the formula and then to use these values in combination to obtain the value for the whole formula.

Example 1
Formula: ~(A ≡ ~B)
Step 1.
Set up the table's standard left-hand columns and write the compound formula in a spread-out fashion at the top of a right-hand column of the table.

A	B	~ (A ≡ ~ B)
T	T	
T	F	
F	T	
F	F	

Step 2.
Copy the values of the single capital letters in a column directly under each occurrence of these capital letters in the right-hand table column. (This step can be omitted once you get better at doing the tables.)

A	B	~ (A ≡ ~ B)
T	T	T T
T	F	T F
F	T	F T
F	F	F F

Step 3.
Find the smallest units in the compound formula and compute their values, using the "T" 's and "F" 's from Step 2. Write these values under the symbol which creates the unit. (In this example, the next smallest unit is '~B'. The symbol which creates the unit is "~".) Cross off any "T" 's or "F" 's used in this computation.

A	B	~	(A	≡	~	B)
T	T		T		F	~~T~~
T	F		T		T	~~F~~
F	T		F		F	~~T~~
F	F		F		T	~~F~~

↑

These values represent
the values of '~B'.

Step 4.
Find the next smallest unit in the larger formula and compute its value. (In this example, the next smallest unit is created by the "≡", and consists of 'A ≡ ~B'. Since we already have computed the values of '~B', we can use these, together with the definition of "≡", to get the values for 'A ≡ ~B'. Think of '~B' as the unit which follows the triple bar, and of 'A' as the unit which precedes the triple bar.) Cross off the "T" 's and "F" 's used in this computation.

A	B	~	(A	≡	~	B)
T	T		~~T~~	F	~~F~~	~~T~~
T	F		~~T~~	T	~~T~~	~~F~~
F	T		~~F~~	T	~~F~~	~~T~~
F	F		~~F~~	F	~~T~~	~~F~~

↑   ↑

Use these two columns
to get the middle column.

Step 5.
Continue as above, working from smaller units to bigger ones, crossing off used "T" 's and "F" 's as you go, until the entire formula is evaluated. The values for the whole formula are the ones which appear under it and are not crossed off.

A	B	~	(A	≡	~	B)
T	T	T	~~T~~	~~F~~	~~F~~	~~T~~
T	F	F	~~T~~	~~T~~	~~T~~	~~F~~
F	T	F	~~F~~	~~T~~	~~F~~	~~T~~
F	F	T	~~F~~	~~F~~	~~T~~	~~F~~

↑   ↑

These entries were obtained from these.

Example 2
Formula: (A $\lor$ B) $\equiv$ ~(B & ~A)
Steps 1 through 3 (omitting step 2).

A	B	(A $\lor$ B)	$\equiv$	~	(B	&	~A)
T	T	T					F
T	F	T					F
F	T	T					T
F	F	F					T

Step 4.

A	B	(A $\lor$ B)	$\equiv$	~	(B	&	~A)
T	T	T				F	F̷
T	F	T				F	F̷
F	T	T				T	T̷
F	F	F				F	T̷

(In general, do not cross off the entries appearing
to the left of the double vertical line in a table.)

Step 5.

A	B	(A $\lor$ B)	$\equiv$	~	(B	&	~A)
T	T	T̷	T	T̷		F̷	F̷
T	F	T̷	T	T̷		F̷	F̷
F	T	T̷	F	F̷		T̷	T̷
F	F	F̷	F	T̷		F̷	T̷

Do this computation first.

Do this one last.

Computing the truth value of a complex formula

36.1 Construct truth tables for each of the following formulas.
  a. ~(A $\lor$ ~B)
  b. A $\supset$ ~B
  c. ~A $\supset$ B
  d. (A & B) $\lor$ C
  e. ~(A & (B $\lor$ C))
  f. A $\equiv$ ~(B $\lor$ D)
  g. (A $\lor$ B) $\supset$ (C & ~D)

37

Symbolization of English sentences

Any compound sentence in the English language which is a truth function of its component sentences can be rewritten without important distortion with our new symbols. In order to accomplish the translation from ordinary English into sentences containing the symbols, one merely needs to find some symbol or combination of symbols which will yield a truth table identical to the table for the sentence in ordinary English. It is always possible to construct some combination of symbols and sentences to do this job when the original sentence is a truth function of its components.

Accordingly, to use the special symbols in rewriting

Everyone who was there had a good time but they all left early

we need to find some symbol or combination of symbols which will yield the same truth table as the following one:

Everyone who was there had a good time.	Everyone who was there left early.	Everyone who was there had a good time, BUT they all left early.
T	T	T
T	F	F
F	T	F
F	F	F

As you already know, "BUT" works like "AND", which suggests use of "&" to replace "BUT". This may be verified by actually constructing the following truth table, using the definition of "&" as a guide for determining the values in the right-hand column.

Everyone who was there had a good time.	Everyone who was there left early.	Everyone who was there had a good time & they left early.
T	T	T
T	F	F
F	T	F
F	F	F

We can readily see that this table matches the table for the original English sentence, line for line, entry for entry. Thus, "&" is a good replacement for the "but" which appeared in the original sentence. The rule to follow then, is: *whenever the English sentence is a truth function of its components, choose some symbol or combination of symbols to use in the rewriting so that the resulting truth table will exactly match the table for the original sentence.*

Example 1

English sentence: Either politics or inefficiency has caused wastefulness in this project.

Step 1.

Isolate the smallest component sentences.

> Politics has caused wastefulness in this project.
> Inefficiency has caused wastefulness in this project.

Step 2.

Identify the connectors, replacing them when possible by preferred connectives.

> Either . . . or
> (Replaceable by the preferred connective, "OR".)

Step 3.

Build the original sentence up out of its smallest component sentences and the connectors isolated above.

> Politics has caused wastefulness in this project OR inefficiency has caused wastefulness in this project.

Step 4.

Think about or construct the truth table for the sentence obtained in Step 3.

> In this example, we have to decide whether to treat "OR" exclusively or inclusively. Suppose we treat it inclusively. Then there is really no need to construct an actual table, since we are familiar with what the table would look like.

Step 5.

Find a symbol or set of symbols to replace each of the key words or connectives in the sentence, taking care to use these symbols in such a way that the resulting truth table matches the table for the sentence obtained in Step 4. In this example, it's easy; use the "$\lor$" for "OR". The result looks like this:

> Politics has caused wastefulness in this project $\lor$ inefficiency has caused wastefulness in this project.

Example 2

English sentence: (Same as in Example 1.)
Steps 1 through 3. (Same as in Example 1.)
Step 4.

Suppose we decide to treat the "OR" in the exclusive sense. The table will look like this:

Politics has caused. . .	Inefficiency has caused. . .	Politics has caused. . . OR inefficiency has caused. . .
T	T	F
T	F	T
F	T	T
F	F	F

Step 5.
The "$\vee$" will no longer work, since its use gives the wrong result in the last column of the table. (It gives a "T" in the top slot rather than an "F".) Find something that will work. A survey of the definitions of the connectives reveals that none match the table in Step 4. So we must use a *combination* of symbols to get the desired result. There are different ones that will do the job. For instance, you might notice that the table we want is just the reverse of the table for 'A $\equiv$ B'. We could reverse the entries in the table for 'A $\equiv$ B' by placing "$\sim$" in front of 'A $\equiv$ B', like this: '$\sim$(A $\equiv$ B)'. That would give us the table we want:

Politics has caused. . .	Inefficiency has caused. . .	$\sim$(Politics has caused. . . $\equiv$ inefficiency has caused. . .)
T	T	F ~~T~~
T	F	T ~~F~~
F	T	T ~~F~~
F	F	F ~~T~~

The above procedure showed that an English sentence of the form 'A OR B' where the "OR" is meant *exclusively,* rather than *inclusively,* can be thought of as being of the form '$\sim$(A $\equiv$ B)'. But, looking back over the steps which led to this result, you should be struck by the fact that I came up with '$\sim$(A $\equiv$ B)' out of thin air, so to speak. Nevertheless, it is perfectly correct to use '$\sim$(A $\equiv$ B)' as a replacement for 'A OR B' when the "OR" is taken exclusively. What makes it correct is the simple fact that '$\sim$(A $\equiv$ B)' has the same truth table as 'A OR B' when "OR" is taken exclusively.

Alternate Step 5.
If one didn't happen to notice that the table obtained in Step 4 can be duplicated by using "$\equiv$" and "$\sim$", one could try rephrasing the English sentence using English to try to bring out the meaning of the exclusive "OR". That might provide a clue. In this particular example, we could accomplish this by adding "but not both" to the sentence, resulting in a clearer way of saying the same thing the original sentence was saying:

Either politics or inefficiency, but not both, has caused wastefulness in this project.

Having rewritten the English sentence to bring out the hidden meaning of the exclusive "or", we go back to Step 2, where we think of rewriting the sentence in terms of the preferred connectors again.

Steps 2 and 3, repeated.
Rewriting the English sentence in terms of the preferred connectors yields

(Politics has caused wastefulness in this project OR inefficiency has caused wastefulness in this project) AND (NOT (politics has caused wastefulness in this project AND inefficiency has caused wastefulness in this project)).

There is no need to do Step 4 again, unless you want to double check to make sure that the sentence we've just constructed really does have the same truth table as the original. I'll leave that to you.

.

Step 5, again.
We can now use the symbols which replace each preferred connector occurring in Step 4, to see if we get the right result. Just replace each preferred connective by the symbol which normally replaces it, and then construct a truth table to check the result. That means we have to do a table for the sentence

> (Politics has caused wastefulness in this project $\lor$ inefficiency has caused wastefulness in this project) & ~(politics has caused wastefulness in this project & inefficiency has caused wastefulness in this project).

In the future, instead of writing out such long sentences in truth tables, I will construct truth tables for the diagrams of the sentences. The relevant table in the present example looks like this:

A	B	(A $\lor$ B)	&	~	(A & B)
T	T	T	F	F	T
T	F	T	T	T	F
F	T	T	T	T	F
F	F	F	F	T	F

Inspection of this table reveals that it gives exactly the result we want. This process of rewriting English sentences (or sentences of any natural language such as Spanish, German, Russian) so as to use the special symbols of logic is called "symbolizing" the sentences.

Steps similar to those followed in the above examples may be taken with *any* sentence that is a truth function of its component sentences in order to ultimately arrive at a detailed diagram of the original sentence. However, as we saw in Example 2, it is possible that one English sentence, taken as the expression of just one proposition, can nevertheless have two (or more) perfectly adequate symbolizations. Remember, the goal is always the same: *find a symbolic formula which has the same entries in its truth table as the original English sentence has in its truth table.* This merely means that a completely acceptable symbolic version of an English sentence will be true under exactly the same circumstances in which the English sentence itself is true.

Unfortunately, even sentences constructed with the help of our preferred connectors in English are often not truth functions of their components. In order to have a useful symbolization technique it will be necessary to consider how best to symbolize such sentences. Obviously, we cannot produce a symbolic sentence diagram which has exactly the same truth table as a nontruth-functional English sentence, since all symbolic sentence diagrams have truth tables which contain no "?" 's.

One rule for producing the best symbolization of a nontruth-functional English sentence is this: *in converting the English sentence into a symbolic diagram, it is acceptable to produce a diagram which has either a "T" or an "F" in its truth table where the English sentence has a "?", but it is not acceptable to produce a diagram which has a "T" where the English sentence has an "F" or a diagram which has an "F" where the English sentence has a "T".*

According to this rule, it would not be acceptable to rewrite "I was fired because I was late" as "I was late $\supset$ I was fired" because this attempted rewrite does not have the proper table, as you can see by comparing the last two columns of the following table.

I was fired.	I was late.	I was fired because I was late.	I was fired ⊃ I was late.	I was late ⊃ I was fired.
T	T	?	T	T
T	F	F	F	T
F	T	F	T	F
F	F	F	T	T

The horseshoe changes an "F" under "because" to a "T". That is unacceptable. The idea behind the rule for symbolizing is that the symbolic version of a sentence should not differ from the original sentence more than it absolutely has to.

What then is the correct symbolization of our English sentence? One way to symbolize the sentence without violating our rule is to change "because" to "&". Note the resulting table:

I was fired.	I was late.	I was fired & I was late.
T	T	T
T	F	F
F	T	F
F	F	F

The last column differs from the entries under the "because" sentence only in one place, namely, where the "?" occurred in the original table.

You can see that we may use the horseshoe to replace many "IF. . .THEN" 's, following our rule. This is verified by comparison of the typical "IF. . .THEN" table with the "⊃" table:

X	Y	IF X THEN Y	X ⊃ Y
T	T	?	T
T	F	F	F
F	T	?	T
F	F	?	T

The use of "⊃" does not convert anything to "T" except for "?" 's; hence, the use of "⊃" is acceptable by our rule as a replacement for the typical "IF. . .THEN".

You may be asking yourself, "How do I know whether to convert "?" 's to "T" 's rather than to "F" 's when symbolizing a sentence?" I haven't given any rule yet to answer that question. The only rule I will give at this point is deliberately evasive: generally, it's better if the "?" 's are changed to "T" 's rather than "F" 's.

I realize that isn't much of rule. The problem is that better rules are hard to come by and difficult to explain. In Section 41 I'll discuss this problem more fully.

So far, we have seen that "&" is a decent approximation for the typical "because", and "⊃" can often replace "IF . . . THEN". It should come as no big surprise that "≡" usually replaces "IF AND ONLY IF" by the same reasoning. But how about "unless"? The typical "unless" table probably looks like this:

X	Y	X UNLESS Y
T	T	?
T	F	?
F	T	?
F	F	F

This table helps justify the practice of treating all of the following as being equally adequate approximations for 'X UNLESS Y':

$$X \vee Y \qquad {\sim}X \supset Y \qquad {\sim}Y \supset X$$

Proof:

X	Y	X ∨ Y	~ X ⊃ Y	~ Y ⊃ X
T	T	T	F̶ T	F̶ T
T	F	T	F̶ T	T̶ T
F	T	T	T̶ T	F̶ T
F	F	F	T̶ F	T̶ F

(These symbolizations are acceptable because they all contain an "F" in the bottom line, where the table for "unless" contains an "F".) Since all three symbolizations are equivalent, there's no reason to say one is better than the others. I suggest you choose the one which is easiest for you to remember. "Only if" has often been treated as "IF . . . THEN" earlier in this book. The typical truth table for 'X ONLY IF Y' looks like this:

X	Y	X ONLY IF Y
T	T	?
T	F	F
F	T	?
F	F	?

This table helps justify using the horseshoe to replace "ONLY IF", since the table for 'X ⊃ Y' (which should be familiar to you by now) contains an "F" in the appropriate place. (Careful: 'Y ⊃ X' is wrong, yielding a "T" in the second line.)

The following list summarizes the usual symbols used to replace the connectors we have discussed.

Connectors	Symbolization
X AND Y	X & Y
X OR (inclusive) Y	X ∨ Y
NOT X	~ X
IF X THEN Y	X ⊃ Y
X IF AND ONLY IF Y	X ≡ Y
X BECAUSE Y	X & Y
X UNLESS Y	X ∨ Y, or ~X ⊃ Y, or ~Y ⊃ X
X ONLY IF Y	X ⊃ Y (Watch out here for the order of the parts.)
X OR (exclusive) Y	(X ∨ Y) & ~(X & Y) or ~(X ≡ Y)

Symbolization of English sentences

37.1 Symbolize each of the following sentences as well as possible. Use single capital letters instead of writing out the atomic sentences. Whenever there could be any question about which atomic sentence should be substituted for your single capital letters, provide a translation key which lists the appropriate substitutions.

Sample.
>He'll marry her only if she's rich.

Answer: A ⊃ B
>A: He'll marry her.
>B: She's rich.

a. I'd like to go with you but I have work to do.
b. Unless you're still eating, I'll take your plate.
c. We need the wood or the coal.
d. If we get the wood, we don't need the coal.
e. Neither IX Corp. nor CDT Co. was able to bid on the job.
f. I can't come because I have work to do.
g. Everyone loved the dinner but not everyone liked the dessert.
h. If the election were held tomorrow, the Republicans would win unless the voter turnout were low.
i. Although she displays the symptoms of measles, I don't believe she has measles at all.
j. Justice requires that she be imprisoned even though she is not capable of being reformed.
k. This could be a fruit or a berry, but not a root.
l. I can only help you if you'll cooperate.
m. Even though he's saddened by the loss, he's been able to show her sympathy.
n. The land is yours, on the condition that you use it for farming purposes.

38

Using truth tables to prove validity for symbolized arguments

Now that you are an expert symbolizer, you will be able to use truth tables much more efficiently to see what's going on in certain simple arguments. Our goal is to develop a technique for using truth tables to show which symbolic argument diagrams are valid. This in turn will show that all the arguments which fit those valid diagrams are valid.

Let me begin by asking you to figure out what conclusions could be added to the following sets of premisses so as to produce valid diagrams.

(1) A ⊃ B   (2) A ∨ B
<u>   A   </u>   <u>   ~A   </u>

Before reading on and finding out what conclusions I had in mind, try on your own.

The conclusions I had in mind were 'B' for both arguments. 'B' will complete both argument diagrams to make them valid.

Let's prove the diagrams valid by using truth tables. The following tables do the trick:

For 1

| (prem) | (concl) | (prem) |
A	B	A ⊃ B
T	T	T
T	F	F
F	T	T
F	F	T

No bad lines.

For 2

| | (concl) | (prem) | (prem) |
A	B	A ∨ B	~A
T	T	T	F
T	F	T	F
F	T	T	T
F	F	F	T

No bad lines.

Think for a moment about what these tables show. They show that no matter what propositions might be expressed by 'A' or 'B', it *is impossible to have an invalid argument which exactly fits one of these diagrams since no matter what propositions you might be talking about, the table covers every possibility, and there never is the possibility of true premisses occurring together with a false conclusion in any of these tables.*

Here is an argument diagram which is not valid:

A ∨ B
<u>B   </u>
A

Test it with a table:

| (concl) | (prem) | (prem) |
A	B	A ∨ B
T	T	T
T	F	T
F	T	T ←bad line
F	F	F

The presence of the bad line indicates the *possibility* that the premisses of an argument fitting this diagram are true at the same time the conclusion is false. We can say that generally this is not a valid way of constructing an argument. Now that our tables no longer contain any question marks, all T-invalid argument diagrams are actually invalid. That is, the presence of a bad line now indicates a genuine possibility that the argument fitting the diagram is invalid. However, I must immediately add an important caution: even though a bad line indicates the diagram is not valid; it does *not* indicate that *every* argument fitting that diagram is invalid. (Recall what we saw earlier in Section 33.)

To illustrate, the above diagram generates a bad line, but there are both valid and invalid arguments fitting it:

Jones is a male or Jones is a bachelor.

Jones is a bachelor.

Jones is a male.   (valid)

That fellow is sick or high on drugs.
In fact, he's high on drugs.

He's sick.   (invalid)

Remember the moral: *if a diagram is valid, all arguments which fit it exactly are valid. If the diagram is not valid, some arguments which fit it are invalid.* In most cases, some *valid* arguments will fit an *invalid* diagram. A bad line for a *symbolic* argument diagram definitely indicates the diagram is invalid, while a bad line for a *nonsymbolic* diagram containing question marks does not necessarily indicate invalidity of the diagram.

We can obviously apply this moral to the evaluation of specific arguments in English. Of course, the strategy will be to diagram an argument symbolically in as much detail as possible, and then test the diagram with a truth table. There is only one hitch: when we deal with arguments expressed in English, our symbols often only approximate the connectives occurring in the argument. This becomes a very serious problem when dealing with some arguments in English, but we cannot tackle that here. In order to avoid dealing with it, when I present an argument which is to be evaluated in this Section I will merely approximate the argument as well as I can using our symbolizing techniques, and I will assume that the approximation is good enough to be useful. Strictly speaking, though, whenever an *approximation* is used, that introduces the chance of error in the evaluation of the argument, and a thorough treatment of the argument would demand that it be investigated whether the approximation is good enough.

Here are some examples of the application of our symbolizing techniques:

Example 1
Argument to be evaluated:

Students will become confident of being able to find a job after graduation only if inflation is arrested and the economic situation stabilizes. However, if the economic situation stabilizes, food prices will hold steady. Thus, if food prices hold steady, students will become confident of being able to find a job.

Step 1.
Symbolize the argument as accurately as possible, providing a translation key:

Translation Key:
A: Students will become confident of being able to find a job after graduation.

B: Inflation is arrested.
C: The economic situation stabilizes.
D: Food prices will hold steady.

Approximate symbolization:

A ⊃ (B & C)
C ⊃ D
―――――――
D ⊃ A

This symbolization does a very good job of bringing out the approximate structure of the argument, even though the causal connections which are probably
implied by the premisses are left out of the symbolic version of the argument.

Step 2.
Test the diagram for validity.

A	B	C	D	(prem) A ⊃ (B & C)		(prem) C ⊃ D	(concl) D ⊃ A	
T	T	T	T	T	T	T	T	
T	T	T	F	T	T	F	T	
T	T	F	T	F	F	T	T	
T	T	F	F	F	F	T	T	
T	F	T	T	F	F	T	T	
T	F	T	F	F	F	F	T	
T	F	F	T	F	F	T	T	
T	F	F	F	F	F	T	T	
F	T	T	T	T	T	T	F	←
F	T	T	F	T	T	F	T	
F	T	F	T	T	F	T	F	←
F	T	F	F	T	F	T	T	
F	F	T	T	T	F	T	F	←
F	F	T	F	T	F	F	T	
F	F	F	T	T	F	T	F	←
F	F	F	F	T	F	T	T	

There are four bad lines, marked by the arrows, which indicates that the diagram is not valid. Thus, the original argument looks as though it might be valid
or it might be invalid. Our truth table test is inconclusive.

The table shows us the problem with the original argument. For example,
the first bad line suggests the possibility that 'A' is false while 'B', 'C', and 'D'
are all true. On the basis of the table alone, without consideration of the relations
between the propositions making up the argument, one can only say that the
argument's diagram is invalid and thus the argument itself *might* be invalid.

Example 2
Argument to be evaluated:

A woman has a general moral right to have an abortion at her own request
only if a fetus is not a person. But the fetus is a person unless there is some
point in its development at which it changes from nonperson to person.
Since there is no such point, a woman does not have a general moral right
to have an abortion at her own request.

Step 1.
Translation key:
> A: A woman has a general moral right to have an abortion at her own request.
> B: The fetus is a person.
> C: There is some point in its development at which a fetus changes from nonperson to person.

Approximate symbolization:

A ⊃ ~B
B ∨ C
~C
_____
~A

Step 2.

A	B	C	(prem) A ⊃ ~B	(prem) B ∨ C	(prem) ~C	(concl) ~A	
T	T	T	F	T	F	F	
T	T	F	F	T	T	F	
T	F	T	T	T	F	F	Check for
T	F	F	T	F	T	F	bad lines.
F	T	T	T	T	F	T	
F	T	F	T	T	T	T	
F	F	T	T	T	F	T	
F	F	F	T	F	T	T	

There are no bad lines in this table; hence, the diagram is valid. That means that any argument which *exactly* fits the diagram is valid. Our original argument does not exactly fit the diagram since we had to use approximation techniques when symbolizing it. However, the diagram does give a good idea of the structure of the original argument and we might assume the original argument is close enough to the diagram that it proves the original argument valid.

That means women do not have a general moral right to have abortions at their own request, unless one of the premises of the argument is wrong! In other words, we have here a valid argument, but perhaps not a sound argument, about an important moral issue.

Try to see by looking back at the diagram and at the argument how the diagram captures the important structure of the argument. Many people have told me that it was not until they looked at argument diagrams that they began to see how arguments in English are really put together.

Using truth tables to prove validity for symbolized arguments

38.1 Test each of the following argument diagrams to see whether they are valid. Try to *think* each one through before doing the test to see if you can tell in advance how the test will come out.

a. A ⊃ B   b. A ⊃ B   c. A & B   d. A ∨ B   e. A ⊃ B
<u>~B</u>   <u>B ⊃ C</u>   <u>A ⊃ C</u>   <u>~A</u>   <u>~A ∨ B</u>
~A   A ⊃ C   C & B   ~B

38.2 Symbolize each of the following arguments as accurately as possible. Assume that the resulting diagrams actually represent the structure of these
arguments—ignore the fact that you have to use approximation techniques. On that assumption, which of the arguments come out looking
valid? For which arguments is the truth table test inconclusive?

a. You should buy a new stereo only if the old one can't be repaired. But,
as you just found out, the old one can be repaired. So, you shouldn't
buy a new one.

b. If we are going to avoid a severe shortage of fuel in the next few years,
we will have to adopt strong conservation measures now. But if we
adopt such measures, many people will be very unhappy. Thus, we are
going to avoid a severe shortage of fuel in the next few years only if
many people are going to be unhappy.

c. I'm going to go out tonight with John or with Fred. Since John is so
corny, I'll not go with him. But if I go with Fred, I'll be bored. So, I guess
I'll be bored.

d. If God can do anything God desires, God can prevent evil. But there is
evil in the world only if God cannot prevent it. As we all know, there is
evil in the world. It follows that God cannot do anything God desires.

e. If Angela Davis was guilty of conspiring to aid the convicts in their escape, she ran from the police when the escape attempt occurred, because she knew she would be implicated. She in fact did run from the
police when the escape occurred. Hence it is obvious that she was
guilty of the conspiracy.

f. I agree that if our Senator is dishonest, he should be defeated in the
next election, but fortunately he is in fact not dishonest. Therefore, he
should not be defeated in the next election.

g. The men have been trapped in the mine shaft now for ten days without
food or water. Consequently, they are all dead.

h. All human beings are mortal. Jones is a human being. Therefore, Jones
is mortal.

39

Shortcut methods in truth tables

It doesn't take much observation to see that truth tables are tedious tools when the sentences they represent are lengthy or complex. Finding a bad line in a table which is sixty-four lines long can take quite a while, and the chance for error is greater than one would like. For these obvious reasons it would be nice to have a shorter method. It also turns out that shortcut truth tables reveal more clearly what is going on in an argument.

The best way to see how the shortcut method works is to look at its operation in an example:

(A & B) ⊃ C
~C & A

~B

Set up the table in the usual way:

A	B	C	(prem) (A & B) ⊃ C	(prem) ~C & A	(concl) ~B
T	T	T			F
T	T	F			F
T	F	T			T
T	F	F			T
F	T	T			F
F	T	F			F
F	F	T			T
F	F	F			T

After setting up the table, I chose an easy-to-handle column to begin with. In this example, I thought the last column would be a good choice. It wouldn't be necessary to do the whole column; I could skip around the table, but this might cause some confusion. The verdict regarding validity will not be affected by the choice of starting place.

You are looking for a bad line. The idea of the shortcut method is to find any bad lines quickly. Already, on the basis of the part of the table completed above, some lines of the table can be eliminated from consideration, since they can't possibly be bad lines. *Draw a circle around all the truth values in the table which show that the line in which they occur cannot be bad.* In the table above, we can draw some circles since in a bad line the conclusion is false. Hence I would draw circles under all "T" 's which occur under the conclusion. The result looks like this:

A	B	C	(prem) (A & B) ⊃ C	(prem) ~C & A	(concl) ~B
T	T	T			F
T	T	F			F
T	F	T			(T)
T	F	F			(T)
F	T	T			F
F	T	F			F
F	F	T			(T)
F	F	F			(T)

Once a line has a circled value in it, there is no longer any reason to do any more computing of values for that line in any other columns.

Next, pick a column that is fairly easy to work with and enter some more values in the table, leaving all lines with circled values in them blank. For instance, you might choose to work with the second column in the middle of the table. That will be a good choice because the main connective in this column is "&", which is easy to work with, and because the "&" will yield mostly "F" 's as entries in this column. Every "F" can be circled, since this is a premiss column and in a *premiss* column a bad line would have to have a "T". The result will look like this:

A	B	C	(prem) (A & B) ⊃ C	(prem) ~C & A	(concl) ~B
T	T	T		(F)	F
T	T	F		T	F
T	F	T			(T)
T	F	F			(T)
F	T	T		(F)	F
F	T	F		(F)	F
F	F	T			(T)
F	F	F			(T)

You are thus left with only one line that could possibly be a bad line, namely, the second one. Complete that line, and thus the table, as follows:

A	B	C	(prem) (A & B) ⊃ C	(prem) ~C & A	(concl) ~B
T	T	T		(F)	F
T	T	F	(F)	T	F
T	F	T			(T)
T	F	F			(T)
F	T	T		(F)	F
F	T	F		(F)	F
F	F	T			(T)
F	F	F			(T)

All the lines of this table have circles in them. That means there are no bad lines in this table, and that the argument diagram is valid.

As you seek a bad line, you are seeking a kind of world in which it would be possible for the premisses to be true and the conclusion false. The way in which line after line is progressively ruled out as circles are entered perhaps indicates to you how the premisses interact with the conclusion to prevent certain kinds of worlds from counting against the argument. That is, one premiss may rule out the first four worlds, leaving the second four. Then another premiss may rule out all but one of those worlds. And so on.

Here is a list of helpful hints:

(1)   When and if you find a bad line, you can quit, if you're satisfied with just one bad line. One bad line means the symbolic diagram is not valid. However, it sometimes pays to have more than one bad line to work with, as we shall see in the next Section when we do counterexamples again.

(2)   When trying to decide what column to work on next in using the short-cut method, pick a premiss column that will have many "F" 's in it, if possible, and pick a conclusion column only when it will have many "T" 's in it. This cuts down on future work, since all the premiss "F" 's get circled and all the conclusion "T" 's get circled.

Shortcut methods in truth tables

39.1 Do shortcut truth tables for the following diagrams, and indicate for each where the bad lines occur, if any.

a. A
$$\frac{\sim B \supset \sim A}{B}$$

b. $\sim\sim R$
$$\frac{(P \ \& \ Q) \lor (\sim P \ \& \ \sim Q)}{\sim P \lor \sim R}$$
$$\sim Q$$

c. $P \equiv (R \lor Q)$
$$\frac{P \supset \sim P}{\sim R}$$

d. $(P \ \& \ Q) \lor R$
$R \supset S$
$$\frac{(P \ \& \ Q) \supset V}{S \lor V}$$

e. P & Q
$$\frac{(P \lor R) \equiv S}{S}$$

f. P
$$\frac{\sim P}{D}$$

39.2 Do shortcut truth tables for the symbolic approximation diagrams of the following arguments. Draw arrows pointing to any bad lines you discover.

a. The state will increase its financial support for our university if and only if the priorities within the thinking of the legislature are shifted in favor of higher education. But if such a shift were to occur, the people who benefit from other state projects would complain bitterly. If the state does not increase its financial support for the university, tuition will have to be raised. So, tuition will have to be raised.

b. If a man is to have a role to play in society, that role must be determined by nature or by society. However, if his role is determined by nature, that role will be the role of the selfish hunter on the make. Hence, either society determines a role for man or men will play the role of selfish hunter always on the make.[1]

[1]Based on Gilder, *Suicide of the Sexes*.

40

Counterexamples, again

We can use the bad lines in the truth tables of invalid diagrams to help us construct counterexamples for all the invalid arguments which fit the diagram. The bad lines serve as clues to what might be wrong with these arguments. Even when we are using approximation techniques we can still try to use the bad lines as clues.

A relatively simple theory lies behind the use of bad lines as guides to counterexamples. Each bad line represents what happens in a particular sort of world. If the world or worlds in which this happens are genuinely logically possible worlds, involving no hidden logical impossibilities, the argument with which we are dealing is truly invalid, provided that it fits the diagram exactly. Sometimes, however, a bad line does not represent a genuine logical possibility, because there are hidden logical relationships which make the world described by the bad line an impossibility. When we construct a counterexample, we are showing in detail how there is a genuine logical possibility for the premises of the argument to be true and the conclusion of the argument to be false. Therefore, any bad line which does represent a genuine logical possibility will also represent the basic outline of a world whose description will be a counterexample.

Things sometimes go wrong when we are dealing with diagrams containing approximations, but even in these cases, things often go right. At least in complicated arguments doing the symbolization and finding the bad line can go a long way toward showing you what might be wrong with a given argument. The bad line can be used as an aide, but you will still have to be careful that your counterexample passes all three tests of a good counterexample, especially when you are dealing with approximations.

Example
Argument to be evaluated:

> If the corn crop this year were the largest ever harvested, the price per bushel would hold steady or drop. But the crop is not the largest. Moreover, if the price per bushel of corn neither holds steady nor drops, the price for beef will go up. Thus, the price for beef will go up.

Symbolic argument diagram:

$A \supset (B \lor C)$
$\sim A$
$(\sim B \,\&\, \sim C) \supset D$
D

This diagram employs approximation techniques. Truth table for the diagram, revealing bad lines:

A	B	C	D (concl)	A ⊃ (B ∨ C) (prem)	~A (prem)	(~B & ~C) ⊃ D (prem)	
T	T	T	T	T　t	F	f　f　f　T	
T	T	T	F	T　t	F	f　f　f　T	
T	T	F	T	T　t	F	f　f　t　T	
T	T	F	F	T　t	F	f　f　t　T	
T	F	T	T	T　t	F	t　f　f　T	
T	F	T	F	T　t	F	t　f　f　T	
T	F	F	T	F　f	F	t　t　t　T	
T	F	F	F	F　f	F	t　t　t　F	
F	T	T	T	T　t	T	f　f　f　T	
F	T	T	F	T　t	T	f　f　f　T	bad line
F	T	F	T	T　t	T	f　f　t　T	
F	T	F	F	T　t	T	f　f　t　T	bad line
F	F	T	T	T　t	T	t　f　f　T	
F	F	T	F	T　t	T	t　f　f　T	bad line
F	F	F	T	T　f	T	t　t　t　T	
F	F	F	F	T　f	T	t　t　t　F	

Using the first bad line in the above table as a clue, I want to tell a story which meets the three tests for a good counterexample to the original argument:

(1)　In the bad line, 'A' was false. To apply this to our argument, we will substitute "The corn crop this year is the largest ever harvested" for 'A'. Since we are interested in the world in which 'A' is false, that means in this particular case that we are interested in the world in which it is *not* the largest corn crop ever harvested. This will be a key element in our counterexample.

(2)　In the same bad line, 'B' was assigned a "T". We see that 'B' is "The price per bushel of corn will hold steady". Accordingly, in our counterexample we will want this sentence to be true.

(3)　In the bad line 'C' was also true. Looking back we find that in our particular argument, the truth of 'C' is the truth of "The price per bushel of corn will drop". So if we are to continue using this bad line as a clue to the construction of counterexamples to the argument, we will want to incorporate this truth into our example as well.

But wait just a moment. We by now are planning on incorporating three elements into our counterexample: (1) This year's corn crop is not the largest ever harvested. (2) The price per bushel of corn will hold steady. (3) The price per bushel of corn will drop. We got these three elements from interpreting the bad line in the truth table. But these three elements are not logically compatible. The price for corn cannot possibly hold steady and drop, at the same time. (Now it is true that nothing has been said to indicate exactly what period of time we are talking about, but in fairness to the original argument, it seems that we are talking about the same time for prices to hold steady and to drop.) Thus, this particular bad line turns out to be a flop. The particular combination of truth values occurring in this bad line does not represent a genuine logical possibility. If all the bad lines work out like this one, we are in trouble with our counterexample. This illustrates the advantage of finding *all* the bad lines when you are constructing counterexamples.

Looking now at the second bad line as a clue to the construction of a counterexample, I see that 'A' is still false. That means that I want a story in which it is false that this is the largest corn crop ever. I see that in this bad line 'B' is true, which means that in my story I want it to be true that the price per bushel of corn will hold steady. Next, I see that 'C' in this line is false, which means that I want it to be false that the price per bushel of corn will drop. So far so good. This line is working out better, since the first three elements of the story, based on the first three letters in the table, are logically compatible. Finally, we look at 'D' in the table, and we find that in this bad line 'D' is false, which means that it is false in my story that the price for beef will go up. This completes the information I can get from my bad line. If I tell a story with the above elements incorporated into one coherent whole, and which makes the original premisses all true while making the original conclusion false, I am successful.

Here is a successful attempt to build a story around the above bad line:

Suppose that the corn crop this year is not the largest ever harvested, but it is nevertheless fairly large. If it were the largest, then the price for corn would hold steady or drop, because of the large supply. However, since the crop is not the largest, there is not quite as large a supply as would be needed from this crop to hold the price down. Suppose also that the number of beef cattle is dropping slightly, so that not as much corn is needed to feed them. This results in a dropping off in the demand for corn, with the result that a somewhat smaller supply of corn will meet the demand. The net result is that the price for corn holds steady, not because of the large supply of corn but because of the dropping demand for it. Now it is true that if the price for feeding corn to cattle goes up, so does the price of beef, as the argument says, but we are supposing that the price for corn does not go up, but holds steady. As a result, we may suppose, the price of beef may remain steady also.

You should now take time to review the above counterexample, making sure that it meets all three tests for a good counterexample. Get in the habit of checking your counterexamples in that way.

I could have used the last bad line in the table instead of the second one in order to get the basic elements of a counterexample. If you look back at that bad line, you will see that it is similar to the bad line I did use above. The only difference is that in the last bad line we are considering the case (or world) in which the corn price drops rather than holding steady.

The technique described in this Section does not always work well. You have to use some imagination in filling out the details of your counterexample, and you have to be sure that your story logically hangs together. There is no automatic checking procedure for insuring that these requirements are met properly. Even if you were to study logic to the most advanced levels now known, there still would not be any automatic procedure available for checking to see if your story contained any hidden logical impossibilities. Moreover, some arguments do not contain many connectives. Such arguments very often do not have useful truth tables, and the techniques we have been discussing will not help much in coming up with a counterexample.

I will close this Section by illustrating this last sort of argument:

All mammals are warm-blooded. This animal is warm-blooded. Hence, it is a mammal.

I suppose an argument diagram for this argument would look like this:

A
B
—
C

(Even if you are clever and find a way to rewrite the argument so as to get more structure into it, you can still appreciate the point.) Using the technique given in this Section for coming up with counterexamples, we know only that our counterexample should be a story in which 'A' and 'B' are true and in which 'C' is false. Big surprise! What have we learned from the diagram and the truth table that we didn't already know about what our counterexample should look like? Nothing. In this case the diagram is useless because the original argument does not have the kind of structure that will show up in one of our diagrams.

Counterexamples, again

40.1 Use the techniques described in this Section to help come up with counterexamples for the following arguments.

a. If it is true that thirty out of every fifty college coeds have sexual intercourse outside of marriage, then it is very important to have birth control information available from the Student Health Service. In fact, it is very important to have this information available from the Student Health Service. Thus, we know thirty out of every fifty college coeds have sexual intercourse outside of marriage.

b. (Background information: Suppose you are on the committee which has to plan the release of the tickets for a big rock concert, and you are faced with a problem. The problem is that some of your helpers will be out of town next week, and the concert is coming up soon, so that if you will not give out the tickets until later, there won't be enough lead time before the concert. You are then presented with the following invalid argument by one of your helpers:)

If we have the ticket release next week, we'll be shorthanded for the release. But if we wait and have the release later, people won't be happy with us. In fact, people will be happy with us only if we are not shorthanded for the release, because they don't want to have to wait too long in line. So, people will not be happy with us.

(You can ignore the statement that people don't want to have to wait too long in line when you do this truth table.)

c. (This argument derives from the passage by George Gilder occurring in Section 3. The key idea in the argument is that women have a biological need to bear children in order to feel satisfied.)

If we don't discriminate against women in the job market, they will be encouraged to seek fulfillment in careers. But if women are thusly encouraged, large numbers will be unsatisfied. If large numbers of women are going around unsatisfied, that will make for a more unstable society in which all sorts of changes may easily take place. Hence we ought to discriminate against women who are looking for a job.

(Because the argument concludes what we *ought* to do, it is important to distinguish the conclusion from the statement that says we *do in fact* discriminate. Use one letter for the "ought" statement and another for the "do" statement.)

d. If we give equal pay for equal work to both men and women, then more women will be encouraged to join the work force, and unemployment among men will rise. However, if the male unemployment rate rises, the labor unions in this country will be generally unhappy. Thus, if we don't give equal pay for equal work, that will make the labor unions happy.

40.2 What will happen if you try to construct a counterexample using the techniques of this Section for an argument that has a diagram which has a bad line when the argument itself is valid?

40.3 Use the shortcut truth tables to provide clues for the construction of counterexamples for the following arguments. Construct the counterexamples.[1]

a. The state will increase its financial support for our university if and only if the priorities within the thinking of the legislature are shifted in favor of higher education. But if such a shift were to occur, the people who benefit from other state projects would complain bitterly. If the state does not increase its financial support for the university, tuition will have to be raised. So, tuition will have to be raised.

b. If a man is to have a role to play in society, that role must be determined by nature or by society. However, if his role is determined by nature, that role will be the role of the selfish hunter on the make. Hence, either society determines a role for a man or men will play the role of selfish hunter always on the make.

41

When are symbolic approximations good enough?

When I defined the horseshoe, I claimed that it was an approximation for the connective, "IF. . .THEN", even though the truth table typical of "IF. . .THEN" sentences contains question marks and the truth table for the horseshoe contains no question marks. You might be willing to accept the idea that it would

[1]See Exercise 39.2.

be nice to get rid of the question marks, but you have been given no explanation for why the horseshoe is defined in such a way that all the question marks from the "IF. . .THEN" table get changed into "T"'s in the definition for the horseshoe. I have merely claimed that it works better that way, given the goals we have in mind. The purpose of this Section is to explain why it works better to use certain approximations rather than others, and to help you see how to tell when an approximation is good enough to be useful.

There is one very important relationship between an argument and its symbolic diagram which is important to try to preserve when we do approximations: when an argument diagram is valid, *all* the arguments which fit the diagram *precisely* are valid. We rely on this relationship in order to prove arguments valid. But what happens when the diagram only *approximates* the structure of the actual English argument? Then the above important relationship might not hold, and we would be without a way to prove the English argument valid. It would be highly desirable if our approximate diagrams would be good enough approximations to allow us to infer the validity of English arguments which *almost* fit them. You may have noticed that in previous Sections I have asked you in the exercises to prove certain English arguments valid, on the *assumption* that their symbolic diagrams were good enough approximations to allow the inference from the validity of the diagram to the validity of the argument. That assumption was required because at that point you had no way of telling for sure when an approximation was good enough to warrant such an inference. Now let's see when the approximation is good enough to warrant that inference.

Suppose you have an argument written in ordinary English, which I shall call Argument 1, and another argument, which I shall call Argument 2, that is just like Argument 1 except that all the connectors in Argument 1 are replaced by our special symbols, and the appropriate rewriting has been done to get all the sentences in order. Here is an example:

ARGUMENT 1	ARGUMENT 2
It will rain only if the front passes through.	(It will rain) $\supset$ (the front will pass through).
The front will not pass through, because the high pressure system will not move.	$\sim$(the front will pass through) & $\sim$(the high pressure system will move).
It will not rain.	$\sim$(it will rain).

Now, consider what would happen if both the following conditions were met:

(1) Each of the premisses in argument 2 is true in at least all the lines of the truth table in which the corresponding premiss in argument 1 is true or questionable.

(2) The conclusion of argument 1 is true in at least all the lines of the truth table in which the conclusion of argument 2 is true.

I claim that the result would be as follows: if there are no bad lines in the truth table for argument 2, there will be no bad lines in the truth table for argument 1. In other words, if the diagram of argument 2 is valid, argument 1 will be valid! So if the claim I just made is correct, the satisfaction of conditions (1) and (2) above will guarantee that the approximation contained in argument 2 is good enough to prove the validity of the original argument.

Before I try to prove my claim, let's take a look at the example arguments given above, to see if conditions (1) and (2) are met by these arguments. Here are the relevant truth tables:

For argument 1:

It will rain.	The front will pass through	The high pressure system will move.	(prem) It will rain ONLY IF the front will pass through.	(prem) NOT (the front will pass through) BECAUSE (NOT (the high pressure system will move)).	(concl) NOT (it will rain).
T	T	T	?	F	F
T	T	F	?	F	F
T	F	T	F	F	F
T	F	F	F	?	F
F	T	T	?	F	T
F	T	F	?	F	T
F	F	T	?	F	T
F	F	F	?	?	T

For argument 2:

It will rain.	The front will pass through.	The high pressure system will move.	(prem) (It will rain)⊂(the front will pass through).	(prem) ~(the front will pass through) & ~(the high pressure system will move).	(concl) ~(it will rain).
T	T	T	T	F	
T	T	F	T	F	F
T	F	T	F	F	F
T	F	F	F	T	F
F	T	T	T	F	T
F	T	F	T	F	T
F	F	T	T	F	T
F	F	F	T	T	T

Verify that condition (1) is satisfied by these two tables. This requires checking every line of the tables in which a "T" or a "?" appears under a premiss in argument 1. Make sure that the corresponding slot in the table for argument 2 contains a "T". If you do that, you will see that this condition is satisfied. Condition 2 is also satisfied, since the tables for the conclusions of the two arguments are identical. Thus, if my claim is correct, that means if argument 2's diagram is valid argument 1 is valid. (In other words, if my claim is correct, the symbolization of argument 1 indicated by argument 2 is "good enough".) So, we turn now to a proof that my claim is indeed correct.

What I have to show is this: When conditions (1) and (2) are satisfied, and argument 2 (the symbolic argument) is valid, then argument 1 (the nonsymbolic one) is guaranteed to be valid. To show this, I'm going to assume for the sake of argument that conditions (1) and (2) are satisfied, and that argument 2 is valid, as was the case in our example above. On the basis of this assumption, I will then show that argument 1 must be valid as well. If these assumptions lead to the result that argument 1 is valid, that proves what I want to prove.

Assuming, then, that conditions (1) and (2) are satisfied, and that argument 2 is valid, I can first prove the following:

(a)   One can validly infer each premiss of argument 2 from the corresponding premiss of argument 1.

(b)   From the premisses of argument 2 one can validly infer the conclusion of argument 2.

(c)   From the conclusion of argument 2 one can validly infer the conclusion of argument 1.

(a) is true because condition (1) says that each premiss of argument 2 is true when the corresponding premiss of argument 1 is true. (b) is true because we are assuming argument 2 is valid. (c) is true because condition (2) says that the conclusion of argument 1 is true when the conclusion of argument 2 is.

Therefore, on the basis of my assumptions, I know (a), (b), and (c) are all true. But (a), (b), and (c) form a chain. Starting with the premisses of argument 1, and using (a), I can validly infer the premisses of argument 2. Then, using (b), I can validly infer the conclusion of argument 2. And finally, using (c), I can infer the conclusion of argument 1. Putting this all together, I can start with the premisses of argument 1 and end up with the conclusion of argument 1 by a series of valid steps of inference. Thus, argument 1 must be *valid* under the conditions assumed above. But this is exactly what I had to prove, using my original assumptions. Thus, my claim has been proved.

To summarize, we have just seen that an English argument which is symbolized in such a way that conditions (1) and (2) are satisfied can be proved valid by means of a validity proof of its symbolic approximation. (Argument 1 is the original English argument. Argument 2 is the symbolic version.) Since the truth table test for argument 2 amounts to a test of the *diagram* of argument 2, we can say that if the diagram of argument 2 is valid, argument 1 is valid. Conditions (1) and (2) jointly give us a test of when a symbolization is "good enough" to be useful in proving the validity of an English argument.[1]

This result can be used to justify the definition of the horseshoe. Let argument 1 be the argument given above in our example. Notice that it contains the connector, "ONLY IF". We would like to have a symbol which will be a satisfactory approximation for "ONLY IF". Let it be the horseshoe, but let's assume just for the moment that we are unsure how to define the horseshoe in order to make it the most useful approximation for "ONLY IF". We know that we have to get rid of the question marks in the table which goes with "ONLY IF", but we haven't

---

[1]Conditions (1) and (2) are not actually necessary to having a "good enough" approximation. A weaker, but more complex set of conditions could be substituted for (1) and (2); however, (1) and (2) provide a good working test of the reliability of symbolic approximations. An approximation diagram which does not satisfy (1) and (2) is unlikely to be good enough, although it is theoretically possible for it to be good enough.

decided on whether to convert them to "T" 's or to "F" 's, or to some mixture of "T" 's and "F" 's. Now suppose we decide to define the horseshoe as follows:

A	B	A ⊃ B
T	T	T
T	F	F
F	T	T
F	F	F

Look what that does to the relationship between the table for "It will rain ONLY IF the front passes through" and the table for "It will rain ⊃ the front will pass through":

It will rain.	The front will pass through.	It will rain ONLY IF the front passes through.	It will rain ⊃ the front will pass through.
T	T	?	T
T	F	F	F
F	T	?	T
F	F	?	F

Notice especially the last line here. If we use this definition for the horseshoe, and apply it to argument 2, condition (1) will no longer be satisfied, because of the last line above. In other words, if we used the new definition for the horseshoe rather than the old definition, we would not be able to use the proof of validity for argument 1 which I worked so hard to give earlier in this Section.

The same thing is true of *any* other definition of the horseshoe which might possibly be thought to make the horseshoe an approximation for "ONLY IF". Any such definition will have to have an "F" in the second line, as usual. But if you try to put anything but a "T" in any of the other slots under the horseshoe, you will ruin the possibility of satisfying condition (1) in an argument like argument 1 above. That would be a shame, because that would make it very unlikely that using a symbolic form could prove the English argument valid. A new definition of the horseshoe would not make the horseshoe a good enough approximation for "ONLY IF", as it occurs in the above argument diagram.

Similar remarks can be made about "ONLY IF" as it occurs in a great many other common argument diagrams, and also about "IF. . .THEN" as it occurs in a great many common argument diagrams. That's why I claimed that it works out better to give the horseshoe the definition I gave it.

But it would be wrong to leave you with the impression that the standard definition of the horseshoe never causes problems. There are examples of arguments where the horseshoe as we have been using it does not work very well as an approximation, because conditions (1) and (2) are not both satisfied in these cases. Here is such an example:

It will not rain tomorrow.
_____
It will rain tomorrow ONLY IF the sun shines brightly all day tomorrow.

I would think that normally a person would say this argument is invalid. However, when we symbolize it, using the horseshoe as an approximation, it comes out having a *valid* approximation diagram:

A	B	(prem) ~A	(concl) A ⊃ B
T	T	F	T
T	F	F	F
F	T	T	T
F	F	T	T

(Let 'A' stand for "It will rain tomorrow".)

There are *no* bad lines in this table!

Our English argument is *not* valid even though this diagram is valid because the English argument does not fit the diagram exactly—the conclusion of the English argument is an "ONLY IF" statement which is only approximated by the horseshoe statement. In fact, condition (2) is not met by this diagram. To see that condition (2) is not met, examine the relation between the conclusions of the symbolic argument and the English argument in a truth table:

It will rain tomorrow.	The sun will shine brightly all day tomorrow.	It will rain tomorrow ONLY IF the sun shines brightly all day tomorrow.	It will rain tomorrow ⊃ the sun shines brightly all day tomorrow.
T	T	?	T
T	F	F	F
F	T	?	T
F	F	?	T

There are three lines of this table which show why condition (2) will not be satisfied, because the three lines in which the horseshoe statement is true do *not* correspond to lines in which the "ONLY IF" statement is definitely true. Since condition (2) fails, we have no grounds for thinking the symbolization of the argument is good enough to yield accurate results.

The approximations which I have suggested in earlier Sections of this book are designed to yield results that meet conditions (1) and (2) most of the time. That is ultimately the best justification for using the suggested approximations rather than other approximations which might be developed. However, as we have just seen, these approximations do not work all the time. If, in dealing with a specific argument, you are worried that your approximations might not be good enough, you can use conditions (1) and (2) as tests. If these conditions are met, you are safe in inferring the validity of the English argument from the validity of its symbolic approximation. If the conditions are not both met, you are no doubt in hot water, and it would be good to try to find a better approximation.

When are symbolic approximations good enough?

41.1 Determine for each of the following arguments whether the use of standard approximation techniques yields a symbolic argument diagram that meets conditions (1) and (2).

a. It is cold outside.
   If it's cold outside, the children should stay in.
   ───
   The children should stay in.

b. Percy won because he was better known.
   Percy was better known because he was the incumbent.
   ───
   Percy won and he was the incumbent.

c. Percy won because he was better known.
   Percy was better known because he was the incumbent.
   ───
   Percy won because he was the incumbent.

d. The country is safe only if Jones won.
   Jones won because she is better known.
   ───
   The country is safe.

e. We will hire a new manager, because we want greater efficiency.
   ───
   We will lose money unless we hire a new manager.

f. We will lose money unless we hire a new manager.
   We will lose money.
   ───
   We will not hire a new manager.

41.2 Which of the symbolic argument diagrams for the above arguments are valid? (Show your calculations.)

41.3 Which of the above arguments can be proved valid by appeal to the symbolic argument diagrams which one obtains when one uses the standard approximation techniques? Explain.

41.4 Which of the symbolic argument diagrams for these arguments contained bad lines? If conditions (1) and (2) are met, and the symbolic diagram contains bad lines, does that mean the original English argument is invalid?

7

Symbolic Propositional Logic—Natural Deduction

42

Introduction to natural deduction: horseshoe elimination

In this chapter we will take a look at one system of rules for constructing valid arguments in a step-by-step way, valid arguments which move from premises to intermediate conclusions to ultimate conclusions slowly and carefully. There are many such sets of rules known today, and as we go along you may well think of some alternative rules which could be added to make our system work more easily for particular cases. However, I ask that you learn to operate exactly within just the rules laid down here in order to avoid confusion and the possibility of error.

Our system will contain ten basic rules and a few extra, supplementary rules added here and there for convenience. These ten basic rules will allow you to construct, step-by-step, an argument from any symbolic premisses to any symbolic conclusion which validly follows from those premisses according to a truth table. Such arguments will look like this:

```
1  |
2  |   premisses
3  |  _____
   |
.  |   .
.  |   .
.  |   .
   |
   |   intermediate conclusions
   |   .
   |
   |   .
   |
   |   .
   |   final conclusion
```

Here the idea is to write the premisses above the horizontal line, numbering each one, and to arrive at the ultimate conclusion on the bottom. Intermediate conclusions will be written below the horizontal line, in the order in which we come up with them, and these conclusions may be used to help us to arrive at later intermediate conclusions, and the ultimate conclusion. However, no later conclusion may be used to help in obtaining an earlier conclusion—the argument proceeds logically from the top down and never from the bottom up.

The rules of our system will specify exactly what range of conclusions may be written at any stage of the development of such an argument. I will refer to an argument written in the above format, following the rules, as a *derivation* of the conclusion from the premisses. All derivations will be *valid* because the rules will be designed to guarantee that each step of the argument is itself a valid argument. Thus, a derivation will constitute a proof of validity for the symbolic diagram whose conclusion is derived from its premisses. (However, I will not *prove* that the system of rules in fact produces only valid diagrams, although each rule will be justified informally at the time of its introduction.)

Each of our rules will be stated in terms of a simple valid pattern of reasoning whose validity depends solely on structure revealed by our special connective symbols. In this Section we consider just two such valid patterns, both of which deal with the horseshoe. The first pattern is illustrated by the following two obviously valid diagrams:

$$X \supset Y \qquad\qquad X$$
$$\underline{X} \qquad\qquad \underline{X \supset Y}$$
$$Y \qquad\qquad\quad Y$$

Some books refer to this pattern of inference by its traditional name, *modus ponens,* but I shall call it by a more recent name, *horseshoe elimination.* Perhaps you can see why this pattern of inference has the name it has, since the conclusion is obtained by, so to speak, "eliminating" the horseshoe from one of the premisses. Of course, for the elimination of the horseshoe from one premiss to be valid, one needs the other indicated premiss as well.

Do not confuse the above valid pattern with the following *invalid* patterns:

X ⊃ Y   Y
Y   X ⊃ Y
————   ————
X   (Invalid)   X   (Invalid)

Of course, the valid horseshoe elimination pattern may occur even when the propositions represented by 'X' and 'Y' above are complex. In such a case, it may be possible to symbolize these propositions more fully. For example, the following diagram might more fully represent a particular argument fitting the above valid diagram:

(C & ~D) ⊃ (G ∨ H)
C & ~D
————————————————
G ∨ H

Let me illustrate the use of the rule of inference in a more complex case. Suppose you have the following four premisses to work with:

(1) A ⊃ B
(2) A
(3) B ⊃ C
(4) C ⊃ D

From these premisses, by repeated use of the pattern described above we can derive a conclusion in a series of steps, each of which is valid:

1. | A ⊃ B
2. | A
3. | B ⊃ C
4. | C ⊃ D
——————————————
5. | B   1, 2
6. | C   5, 3
7. | D   6, 4

Here I have written all the premisses above the horizontal line. Then, below the horizontal line, I have written a series of conclusions which can be derived from these premisses by repeated use of the pattern. Thus, in line 5 I wrote 'B' to indicate that the conclusion 'B' can be obtained from lines 1 and 2 by applying the pattern. Then, once having indicated that 'B' follows from the first two premisses, I can use 'B' to get 'C' by applying the rule once again. Line 6 comes from lines 5 and 3. That is, the argument which leads to line 6 looks like this:

3.   B ⊃ C
5.   B
————————————
6.   C

It is not hard to see this fits the same pattern. Line 7 is obtained from lines 4 and 6 by following the pattern again. Thus, the array of lines 1 through 7 indicates in a compact way a *chain* of arguments which lead first to 'B', then to 'C', and finally to 'D', as conclusions.

I will draw a *vertical* line to form a framework behind which the premisses and the series of conclusions can fit, as I did above. This will become important later. I will also write the line numbers which are used to get each conclusion to the right of the conclusion, as I did above. Finally, I will write the name of the rule I used to get each conclusion to the right of the line, since after a bit we will have several rules to work with. I will abbreviate "horseshoe elimination" by "⊃ E". Thus, the complete derivation, showing how to get 'D' from the premisses 1 through 4, should look like this:

1.	A ⊃ B	
2.	A	
3.	B ⊃ C	
4.	C ⊃ D	
5.	B	1, 2, ⊃ E
6.	C	5, 3, ⊃ E
7.	D	6, 4, ⊃ E

I write the official version of the ⊃ E rule as follows:

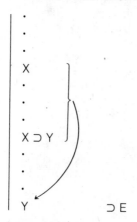

⊃ E

In words, this rule simply means: whenever you have any formula 'X' in a derivation (either as a premiss or as an intermediate conclusion) and you also have a formula 'X ⊃ Y' in that derivation behind the same vertical line, you are allowed to conclude the formula 'Y' at any time later on behind the same vertical line. The patterns of three dots indicate merely that there can be other formulas also occurring in the derivation besides the ones in which we are interested. These other formulas would be irrelevant to what we are doing.

Whenever I write a rule in its official form, as I did with ⊃ E, and the rule requires that two or more formulas already have been established before one can use the rule to get a conclusion, as where both 'X' and 'X ⊃ Y' are required before one can draw the conclusion 'Y' using ⊃ E, *it is to be understood that the order in which the already established formulas appear is irrelevant.*

Thus, I will not consider it necessary to write an alternate version of ⊃ E to cover this situation:

1. | X ⊃ Y
2. | X
   |———————
3. | Y   1, 2, ⊃ E

Here's another example of the employment of the horseshoe elimination rule:

1. | (A & B) ⊃ C
2. | A & B
   |———————
3. | C   1, 2, ⊃ E

Note that in this example, the formula which plays the role of 'X' is complex; that formula is 'A & B'. It doesn't matter how complex the formulas are, so long as they fit the pattern laid out in the rule.

The rule of ⊃ E is one of the ten basic rules in the derivation system which will be developed in this chapter. Although those basic rules by themselves are sufficient for obtaining all the conclusions a truth table can show we ought to be able to derive, it is sometimes frustratingly awkward to obtain some of those conclusions using just the ten rules. We can make our derivations shorter and more natural by adding more rules to the system, but each such addition makes the total rule set harder to remember. I will follow a policy of moderation, then, adding only a few extra, nonbasic or "derived" rules. These will be mentioned as we go along, when related to the basic rule under consideration. For most of the derived rules, I will eventually produce proofs showing that the conclusions obtained in one step by use of a derived rule could have been obtained in a more roundabout way using only the ten basic rules.

Our first derived rule will be known as "negative horseshoe elimination", abbreviated "N ⊃E". Its pattern is given below, and its validity is fairly obvious. (If you doubt the validity of the pattern, check it with a truth table.)

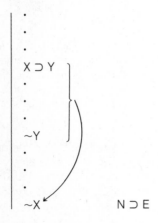

X ⊃ Y

~Y

~X   N ⊃ E

(The traditional name for this rule is "modus tollendo tollens".) Its operation is illustrated in the following derivation.

1.	A	
2.	A ⊃ ~B	
3.	(D ∨ E) ⊃ B	
4.	~B	1, 2, ⊃ E
5.	~(D ∨ E)	3, 4, N ⊃ E

Note how the negation sign in line 5 above applies to the whole formula, 'D ∨ E', since it is that whole formula which plays the role of 'X' in the rule.

Since we as yet have only one of our basic rules, I cannot *prove* that N ⊃E is theoretically superfluous; however, after we have the basic rules for handling negations you will be able to prove for yourself that we don't really need to have the rule N ⊃E in order to get all the conclusions we ought to be able to get. Still, N ⊃E is a natural pattern of inference which we use in everyday life and which is also quite useful in our system.

Introduction to natural deduction: horseshoe elimination

42.1 Derive the conclusion 'B' from the premisses 'D', 'A', and 'D ⊃ B'.

42.2 Derive the conclusion 'C' from the premisses 'A ∨ B' and '(A ∨ B) ⊃ C'.

42.3 Derive the conclusion 'D & E' from the premisses '(A ∨ B) ⊃ C', 'A ∨ B', and 'C ⊃ (D & E)'.

42.4 Derive the conclusion 'C' from the premisses 'A ⊃ (B ⊃ C)', 'B' and 'A'.

42.5 Derive the conclusion 'P & ~R' from the premisses 'C ∨ Q', 'A ⊃ (P & ~R)', and '(C ∨ Q) ⊃ A'.

42.6 Symbolize the following arguments and derive their symbolic conclusions from their premisses. (Hint: In problem (c) use the horseshoe in symbolizing the "unless" premiss—if necessary, refer back to the discussion of how to symbolize "unless".)

  a. If I stop dating Jerry, he'll soon forget about me. If he soon forgets about me, he won't be sad long. If he won't be sad long, he has nothing to complain about. I'm going to stop dating Jerry. So, he has nothing to complain about.

  b. Your contract is legally enforceable only if it's in written form and signed by both parties. But if you got a court order enforcing your contract, it must be legally enforceable. You did get the court order. Hence, your contract is in written form and is signed by both parties.

  c. There will be a revolution unless something is done to improve living conditions among the poorer classes. But nothing will be done to improve those conditions. Hence, there will be a revolution.

42.7 What conclusions can be derived (approximately) from the following sets of premisses?

    a. Premisses: If I go to work for you I'll stay longer than a year only if I get good raises. I'll go to work for you.

    b. Premisses: If we increase our net income by raising prices, our sales volume will eventually drop and the number of distributors willing to carry our product will decline. If our sales volume drops off eventually, and the number of our distributors declines, our net income will drop later on. We are going to increase our net income for now by raising our prices.

42.8 Derive '~C' from 'C ⊃ (D & A)', 'E', and 'E ⊃ ~(D & A)'.

42.9 Derive 'A' from '~D ⊃ A', 'D ⊃ (B ∨ C)' and '~(B ∨ C)'.

43

Ampersand and triple bar
introduction and elimination

Not many valid argument diagrams can be constructed using merely ⊃ E and N ⊃ E. In this Section I will introduce two additional rules both pertaining to the ampersand and two rules pertaining to "≡". There will be rules for introducing additional ampersands or triple bars into a derivation, called "ampersand introduction" and "triple bar introduction": "& I" and "≡ I" for short; and rules for getting rid of unwanted ampersands or triple bars, called "ampersand elimination" and "triple bar elimination": "& E" and "≡ E" for short.

    In order to understand and remember the ≡ E and ≡ I rules, it is helpful to consider the fact that

      'X ≡ Y' has the same truth values as '(X ⊃ Y) & (Y ⊃ X)'

Thus, the "≡" is closely related to the "⊃". Accordingly, the following argument diagram will be valid

A ≡ B
A
―――
B

just as if the "≡" were a "⊃". But, *unlike* the situation with "⊃", the following diagram is also valid:

A ≡ B
B
―――
A

(This fact makes sense if one remembers that "≡" amounts to *two* horseshoes rather than one.)

Another way to look at this matter is to remember that the formula 'X ≡ Y' is true when and only when 'X' and 'Y' are equal in truth value. Thus, when 'A ≡ B' is a premiss along with 'B', one may conclude that 'A' is also true.

These observations lead to establishing the following rule for ≡ E. This rule, in its two versions, is a basic rule of our system.

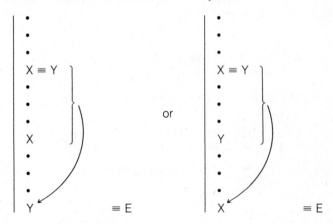

The following argument diagram illustrates the use of the second version of ≡ E:

1. | A ≡ ~(B & C)
2. | (B & C) ⊃ D
3. | ~D
   ―――――――――――――――――
4. | ~(B & C)   2, 3, N ⊃ E
5. | A   1, 4, ≡ E

Note that line 4 is needed before one can use ≡ E on line 1. One may *not* conclude 'A' is true from line 1 alone, since 1 merely says 'A' has the same truth value as '~(B & C)' has.

So much for ≡ E. The introduction rule for "≡" is closely related. Recall that "≡" amounts to two horseshoes as explained above. Our ≡ I rule will capitalize on that by allowing you to conclude 'A ≡ B' once you have proved or been told that 'A ⊃ B' *and* that 'B ⊃ A'. Formally, the ≡ I rule looks like this:

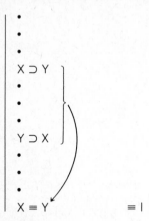

X ⊃ Y

Y ⊃ X

X ≡ Y   ≡ I

This rule is the third of our basic set of ten. Note that the conclusion authorized by this rule can only be drawn after there are *two* premisses or intermediate conclusions available above; that is, one cannot legitimately infer 'A ≡ B' from 'A ⊃ B' alone.

The use of ≡ I may be illustrated by working from the following set of premisses:

P1.   The U.S. can increase defense spending only if spending for other programs is cut or there is a tax increase.

P2.   Moreover, if we do cut spending for other programs, or increase taxes, we can indeed increase defense spending.

Symbolizing these premisses as 1 and 2 below, and using ≡ I, we get the conclusion 'I ≡ (C ∨ T)':

1. | I ⊃ (C ∨ T)
2. | (C ∨ T) ⊃ I
3. | I ≡ (C ∨ T)   1, 2, ≡ I

At present, we can't do much with ≡ I, but as more rules are added to the system ≡ I will become more useful.

Having added ≡ E and ≡ I to ⊃ E and N ⊃ E we turn now to the rules & I and & E.

To get the idea behind & I, consider the pattern present in the English argument:

God loves everyone, no matter how young or old.
God cares for everyone, no matter how young or old.

God loves and cares for everyone, no matter how young or old.

Obviously, the pattern here looks like this:

A
B
―――
A & B

Equally obvious is the validity of such a pattern, as can be verified by a simple truth table. The same pattern will be valid, no matter how complex the propositions to be substituted for 'A' and 'B'. Hence, the following rule will be a good one:

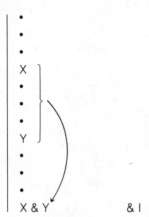

X & Y   & I

Here is an example of & I at work:

1.	A ⊃ B	
2.	A ⊃ C	
3.	A	
4.	B	1, 3, ⊃ E
5.	C	2, 3, ⊃ E
6.	C & B	4, 5, & I

An English argument which is well approximated by the above derivation could look like this:

> If the driver was drunk, he should be arrested. Also, if he was drunk, the accident was unnecessary. He was, in fact, drunk. So, it follows that he should be arrested. And it follows that the accident was unnecessary. Thus, the accident was unnecessary and he should be arrested.

You should be able to find that each line of the derivation fits something said in this English argument chain.

In a way the & I rule is trivial and obvious. When we actually give arguments in English we often take bigger steps than the ones taken above in our example. But the rule is nevertheless valid, and useful.

The ampersand elimination rule is equally trivial and obvious. The idea it represents is very simple: once you know that 'A & B' is true, you can say that 'A' is true; you could also say that 'B' is true in this circumstance. This leads immediately to another basic rule, the & E (ampersand elimination) rule:

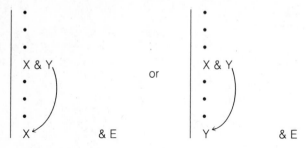

Here is an example of this rule in action:

1.	$(B \lor C) \supset (A \supset D)$	
2.	$A \& (B \lor C)$	
3.	$B \lor C$	2, & E
4.	$A \supset D$	1, 3, $\supset$ E
5.	A	2, & E
6.	D	4, 5, $\supset$ E

Here's an English argument which is approximated by the steps taken in this last example:

> If Jones or Smith works on the plumbing, the job will be finished on time only if we use Frank as foreman. But the job must be finished on time and we will use Jones or Smith on the plumbing. That means Jones or Smith will work on the plumbing. But that in turn means that the job will be finished on time only if Frank is the foreman. However, the job must be finished on time, as we said earlier. Thus Frank will have to be the foreman.

Make sure you see how each of these sentences corresponds to one line of the derivation in the example.

Of course you can use any one of these rules as many times as you want in a derivation, and you can use the rules in any order you please. Just be sure each and every line of the derivation below the horizontal line is justified by using one of the rules in application to earlier lines of the derivation.

Ampersand and triple bar
introduction and elimination

43.1 Copy the following derivations. For each line below the premises indicate how that line can be derived from earlier lines by listing to the right the earlier line number(s) and the rule that can be used to get that line.

a. 1. | A & B
   2. | B ⊃ (C & D)
   3. | C ⊃ E
   ---
   4. | B
   5. | C & D
   6. | C
   7. | E

b. 1. | A ⊃ (D ⊃ R)
   2. | A & D
   3. | S
   ---
   4. | A
   5. | D ⊃ R
   6. | D
   7. | R
   8. | R & S

43.2 Derive 'E' from the premisses '(A & C) ⊃ (B & E)', 'A', and 'C'.

43.3 Derive 'E' from the premisses '(R & S) ⊃ E', 'R', 'A ≡ S', and 'A'.

43.4 Derive 'G' from the premisses 'E & A', 'A ⊃ (C & D)', and 'D ≡ G'.

43.5 Derive '(B & C) ≡ A' from the premisses '(B & C) ⊃ A', 'C & D', and 'A ⊃ (B & C)'.

43.6 Derive '~C ≡ D' from the premisses '(A ⊃ B) & (~C ⊃ D)', 'A', and 'B ≡ (D ⊃ ~C)'.

43.7 Derive 'C' from the premisses '~D', 'B ⊃ A', '~E ⊃ (A ⊃ B)', 'E ⊃ D', and '(A ≡ B) ⊃ C'.

43.8 For each of the following arguments, do the following: symbolize the argument; derive the conclusion of the argument from its premisses, as symbolized; read the derivation to yourself, step by step, substituting the English expressions for the appropriate capital letters in the derivation; as you read, try to see how the derivation makes sense, in English.

a. Although it is technologically possible to entirely eliminate air pollution from factories, the expense required for doing so is great enough to make the cost of running the factories prohibitive. But if this cost is prohibitive, manufacturing companies and their customers, the public, will resist efforts to force such stringent requirements on the business community. Thus, even though the entire elimination of air pollution from factories is technologically possible, the public will resist efforts to force such strict requirements on business.

b. The proper reaction occurred if and only if the chemicals were measured very carefully before mixing and the room temperature was just right. However, if the room temperature was right for the reaction to take place, it was difficult to work with the chemicals because they are unstable at the temperature. If they were unstable at that temperature, they gave off toxic fumes. The proper reaction did occur. Hence, the chemicals gave off toxic fumes.

44

Some rules about the wedge

The first rule presented in this Section may appear odd to you, but it is necessary for our system of basic rules to be complete enough to construct all the valid argument diagrams which can be built around our five special symbols. The new rule is the basic wedge introduction rule, $\vee$ I. Keep in mind during the discussion that the wedge symbol does not correspond exactly in meaning to all occurrences of the English word "or". Its truth table defines the wedge. Here, then, is the rule:

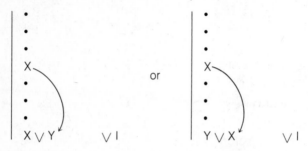

The rule $\vee$ I allows the following kinds of moves to be made:

1. | A
2. | A $\vee$ B   1, $\vee$ I

1. | A $\vee$ ~C
2. | (Q ≡ R) $\vee$ (A $\vee$ ~C)   1, $\vee$ I

1. | B
2. | ((Q & R) ≡ S) $\vee$ B   1, $\vee$ I

In other words, once you have any formula, 'X', you can then conclude 'X $\vee$ Y', where 'Y' is *any* formula whatsoever. That's right—*any* formula whatsoever! (You can also conclude that 'Y $\vee$ X', where 'Y' once again is any formula whatsoever.) In particular, the formula 'X' need not be related in any special way to the formula 'Y' for the rule to work.

To many people, this rule seems foolish and invalid at first sight. How can it be valid to argue from 'A' to 'A $\vee$ B', when 'B' has nothing to do with 'A'? In order to answer this question it is important to remember that 'A $\vee$ B' is automatically true whenever 'A' is true. No matter what 'B' is. Remember also the definition of a valid argument diagram—one with no bad lines in its table. Check out the preceding diagrams with tables. The tables have no bad lines.

It is very misleading in an ordinary conversation to say something of the form 'A $\lor$ B' when all the time you know that 'A' is true. If you know that it's Monday, but you go around saying "Either it's Monday or it's raining", you're misleading people about how much you know, and you are talking as if there were some special relation between the rain and the day of the week. But we are not concerned with the practicality of saying something nor are we concerned with all English "or"'s. Instead, we are concerned with what counts as evidence for the truth of something made with a "$\lor$". And the truth of 'A' does *guarantee* the truth of 'A $\lor$ B'. Thus the argument

$$\frac{A}{A \lor B}$$

is valid, even though odd. (The truth of 'A' would not necessarily guarantee the truth of 'A OR B', since "OR" need not mean "$\lor$" in English.)

You'll have to take my word for it that this rule, the sixth of our ten basic rules, is useful in constructing certain sorts of arguments. Perhaps you can see in general why it's nice to have a rule for introducing new occurrences of "$\lor$" into an argument chain.

Next, we will deal with wedge elimination, which is a bit more complex. This is a rule frequently used in mathematics and in common life to construct arguments from premisses containing "or". Here is an example of the main idea of this rule in an English argument:

> If I stop studying now and go to bed, I will do badly on the test tomorrow. But if I stay up all night and study, I will be very tired tomorrow. So, no matter what I do, I will do badly on the test tomorrow or I will be very tired.

In order to deal with this argument in symbols so as to show that it is valid, we need to make explicit the unstated premiss

> I will either stop studying now and go to bed or I will stay up all night and study (but not both).

(Without some such added premiss the argument won't be valid.) With the addition of this premiss, the argument can be approximately symbolized as follows:

$$\frac{[(A \,\&\, B) \lor (E \,\&\, F)] \,\&\, \sim[(A \,\&\, B) \,\&\, (E \,\&\, F)]}{(A \,\&\, B) \supset D} \\ (E \,\&\, F) \supset G$$
$$\overline{D \lor G}$$

A: I will stop studying now.

B: I will go to bed.

D: I will do badly on the test.

E: I will stay up all night.

F: I will study all night.

G: I will be very tired tomorrow.

You might say that 'E' amounts to '$\sim$B', but that is not strictly correct, since not going to bed is not the same as staying up all night. (One could manage to not go to bed by managing to die, but dying seems to me not to be the same thing as staying up.) So, caution is required.

The next step is to show how we will be able to derive the desired conclusion from the given premises, using a sensible rule of wedge elimination, $\lor$E.

1. | [(A & B) $\lor$ (E & F)] & ~[(A & B) & (E & F)]
2. | (A & B) $\supset$ D
3. | (E & F) $\supset$ G
---
4. | (A & B) $\lor$ (E & F)   1, & E
5. | D $\lor$ G   2, 3, 4, $\lor$ E

What is going on here? Line 4 tells us that I will stop studying and go to bed, or I will stay up all night and study. This line basically states the alternatives. Line 2, however, tells us that 'D' will happen if I stop studying and go to bed, while line 3 tells us that 'G' will happen if I stay up and study. Because line 4 tells us what the alternatives are, and lines 2 and 3 tell us what will happen under each alternative, we can conclude that either 'D' or 'G' is true, as shown in line 5.

Note that we must *not* eliminate the wedge in line 4 by concluding 'A & B' at line 5. Nor may we conclude 'E & F'. (I cannot tell from line 4 which alternative is correct; so I may not arbitrarily conclude at line 5 that one specific alternative will be true.) Since line 4 lists two alternatives, and there is no way to tell which of them is true, we must explore *both*. The first leads to 'D'. The second to 'G'. Hence 'D $\lor$ G'.

This correctly suggests the following basic rule of wedge elimination:

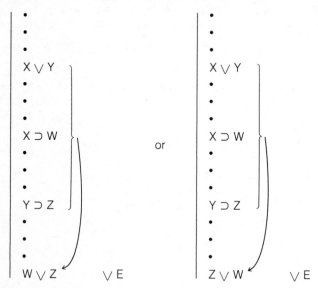

Here is a typical *misuse* of $\lor$ E:

1. | A $\supset$ C
2. | B $\supset$ D
---
3. | C $\lor$ D   1, 2, $\lor$ E ?

The difficulty with the preceding derivation is that one cannot conclude 'C $\vee$ D' from the given premisses unless one knows 'A $\vee$ B' is true, which one does *not* know in the above case. Remember that $\vee$ E requires *three* known lines before it allows you to draw any conclusion.

Sometimes, when there are two known alternatives, X and Y, each can be shown to lead to the *same* outcome, as with the following set of premisses:

P1: If I stay up and study all night, I'll do badly on the test tomorrow.
P2: On the other hand, if I go to bed now, I'll also do badly on the test.
P3: Those are the only alternatives.

Symbolizing and using our rule $\vee$ E we obtain:

1. (A & B) ⊃ C
2. D ⊃ C
3. (A & B) $\vee$ D
4. C $\vee$ C   3, 1, 2, $\vee$ E

So far, so good, but obviously 'C $\vee$ C' is equivalent to just plain 'C', and we should be able to find a way to conclude 'C' rather than having to stop with 'C $\vee$ C'. For this, we need a simple new basic rule called redundant wedge elimination, R $\vee$ E.

- •
- •
- •
- X $\vee$ X
- •
- •
- •
- X   R $\vee$ E

$\vee$ E and R $\vee$ E will serve as our basic rules for eliminating wedges. (It is possible to have only one slightly different basic rule for wedge elimination, but I find it more convenient in practice to have two such rules, since $\vee$ E is especially useful in its present form.)

Here is an example illustrating the proper use of several of our rules:

1. (A & B) ≡ (C $\vee$ D)
2. A
3. C ⊃ ~E
4. D ⊃ ~E
5. A ⊃ B
6. B   2, 5, ⊃ E
7. A & B   2, 6, & I
8. C $\vee$ D   1, 7, ≡ E
9. ~E $\vee$ ~E   8, 3, 4, $\vee$ E
10. ~E   9, R $\vee$ E

It is necessary to include line 9 in the above derivation if the rules are to be followed precisely: one cannot go directly from 8 to 10 by $\lor$ E or R $\lor$ E.

The procedure of $\lor$ E followed by R $\lor$ E is known in mathematics as "proof by cases". When in a mathematics problem one can show that there are only a fixed number of alternatives ("cases") and then can show that each alternative leads to the same result, one can conclude that result is thereby proved correct.

Our rules so far are set up to handle only two alternatives at a time, but once we have more of our basic rules, problems involving more alternatives can be solved using the same basic strategy. Of course we could add more wedge elimination rules to the system to take care of the case of three alternatives, and the case of four alternatives, and so on, but I will refrain from doing that, to keep the rule set more manageable.

There is, however, one additional supplementary rule regarding the wedge, not to be included in our basic set of ten, but especially handy and natural, and ultimately derivable from our basic set. I call this the rule of "elimination of alternatives", or "EA" for short.[1] The fundamental idea behind this supplementary rule is illustrated in the following simple valid diagram:

$$A \lor B$$
$$\underline{\sim B}$$
$$A$$

(If you doubt the validity of such a diagram, check it with a truth table.)

Generalizing this pattern into a rule, we obtain the following:

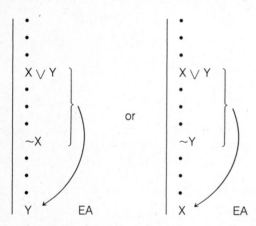

This rule can be viewed as an alternative wedge elimination type of rule, usable whenever the necessary premises are available. The following correct derivation serves as a further illustration:

---

[1]This rule has sometimes been known by other names, such as "disjunctive syllogism".

```
1.  │ D ⊃ (A & B)
2.  │ D ∨ ~C
3.  │ ~(A & B)
    └──────────
4.  │ ~D                1, 3, N ⊃ E
5.  │ ~C                2, 4, EA
```

Here is a somewhat more complex example:

```
1.   │ A ∨ (B ∨ C)
2.   │ A ⊃ D
3.   │ (G & H) ≡ ~(B ∨ C)
4.   │ G
5.   │ G ⊃ H
     └──────────
6.   │ H                4, 5, ⊃ E
7.   │ G & H            4, 6, & I
8.   │ ~(B ∨ C)         3, 7, ≡ E
9.   │ A                1, 8, EA
10.  │ D                2, 9, ⊃ E
```

Now that we have added one basic rule of ∨ I and two for ∨ E, plus EA, a great many more interesting derivations can be constructed.

Some rules about the wedge

44.1 Derive 'C & D' from the premisses 'A ∨ B', 'A ⊃ (C & D)', and 'B ⊃ (C & D)'.

44.2 Derive 'C' from the premisses 'E ⊃ (A ∨ B)', 'A ⊃ (C & D)', 'B ⊃ (C & D)', and 'E'.

44.3 Derive 'C ∨ D' from the premisses 'A ∨ B', 'A ⊃ C', and 'B ⊃ C'.

44.4 Derive 'C' from the premisses '(~A ∨ B) ≡ C', 'A ⊃ E', and '~E'.

44.5 Derive 'D & A' from the premisses 'B ⊃ D', '((A & ~B) ∨ C) ⊃ D', 'A', and 'A ⊃ ~D'.

44.6 Derive 'G ≡ H' from the premisses 'B ∨ E', 'A ≡ ~C', 'E ⊃ A', '(D ∨ ~C) ⊃ (G ⊃ H)', 'B ⊃ A', and 'H ⊃ G'.

44.7 Derive 'D' from the premisses '(B & E) ∨ A', 'A ⊃ C', 'B ≡ (C ∨ D)', and '~C'.

44.8 Why is it not a good idea to add to our system a rule which would allow the derivation of '~Y' from the premisses 'X ⊃ Y' and 'X'?

45

Horseshoe introduction— conditional proof

You may have noticed that we have elimination and introduction rules at this point for "&", "$\vee$", and "$\equiv$", but we only have elimination rules for "$\supset$". I have waited until now to discuss the horseshoe introduction rule, $\supset$ I, because this rule employs a powerful new device which initially can be hard to accept. However, we can no longer put off the $\supset$ I rule, sometimes called the rule of conditional proof.

Let me illustrate the use of this rule by first presenting an example of its approximate use in an English argument. Suppose Herman and Helga have been discussing next year's prospects for their favorite baseball team, the West Gulch Vultures, who have suffered in the past from a weak pitching staff, and Herman puts forward the following argument:

> If the Vultures can sign Tom Slinger for next season, the team will improve its pitching staff tremendously. If the pitching staff improves significantly, the team will do much better next season. So, if Tom signs with the Vultures for next season, the team will do better.

This appears to be a valid argument which can be approximated by the following diagram:

$$S \supset P$$
$$\underline{P \supset B}$$
$$S \supset B$$

You can check out this diagram with a truth table if you wish; the table contains no bad lines. However, we as yet have no way to derive this conclusion from its premises. (We need the $\supset$ I rule to do that job.)

One might try to explain the validity of the Vulture argument by making the following analysis:

> Suppose, just for the sake of argument, that the Vultures do sign Slinger for next season. Then, according to the first claim in the original argument, the team will improve in its pitching staff tremendously. But, then we get the result that the Vultures will do much better next season. This all shows that if Slinger does sign, the Vultures will do much better next season.

Note how the above analysis works. It attempts to show that the premises of the original argument imply the whole statement that

(1)　IF Tom Slinger signs with the Vultures for next season, THEN the team will do better.

In order to show this, the analysis *assumes,* temporarily, that

(2)　Tom Slinger does sign with the Vultures for next season.

Then, *on the basis of* this temporary assumption, the analysis argues that

(3)　The Vultures will do better next season.

That is, in order to show the original argument is valid, the analysis adds a temporary additional premiss, (2), to the argument, and then, using this additional premiss in conjunction with the original premisses, a new temporary conclusion (3) is obtained. After doing this, the analysis in effect says that since you can get (3) from assuming (2) together with the original premisses, you can get the whole statement: "IF (2) THEN (3)" from the original premisses alone.

We need a system for keeping all this straight when writing out a derivation. Fortunately, there is a fairly easy way to do it. I will illustrate how it can be done in the case of the Vulture argument. First, we set up the premisses as usual. Our goal will be to obtain the conclusion, 'S ⊃ B', as the bottom line in the derivation. That is, we want something like this:

```
1. |  S ⊃ P
2. |  P ⊃ B
   |_____
   |
   |
   |
   |
   |  S ⊃ B
```

The question is, How do we fill in the missing steps? We know from our analysis in English that the missing part of the derivation must somehow report the fact that we can derive 'B' from the original premisses combined with the assumption 'S', for this fact is our justification for saying 'S ⊃ B' follows from just the original premisses alone. We handle this very easily, by filling in the blank area of the above derivation like this:

```
1. |  S ⊃ P
2. |  P ⊃ B
3. |  | S
   |  |____
4. |  | P          3, 1, ⊃ E
5. |  | B          2, 4, ⊃ E
6. |  S ⊃ B        3-5, ⊃ I
```

In this complete derivation, we start out with the original premisses, listed above as lines 1 and 2. We then construct a new subderivation which has the same premisses as the original derivation, plus one more premiss, 'S'. Then we derive a conclusion, 'B' inside the subderivation. Once we have this conclusion, we eliminate our temporary assumption, 'S', by building it into the conclusion in line 6 in front of the horseshoe. In this way we show that 'S ⊃ B' follows from the original premisses.

The justification for writing line 6 is the whole set of lines 3 through 5. This whole set of lines is a subderivation inside the main derivation. I indicate that lines 3 through 5 constitute a derivation by writing them behind their own vertical line. Inside this subderivation I follow all the usual rules. Once the subderivation is finished at line 5, I end the vertical line which marks off the subderivation. I then write line 6 behind only one vertical line, to indicate that it is obtained from just the original premisses listed as lines 1 and 2.

When we end the subderivation headed by 'S' and build 'S' into the next line, 'S ⊃ B', that is referred to as *discharging* the premiss 'S'. 'S' was just a temporary assumption anyway, added to the original premisses for the purpose of making it easier to see what would follow if the original premises were true. When line 6 is written, there is no more need for 'S', and so it is discharged.

Here is another example of this type of derivation:

1. | A ⊃ B
2. | B ⊃ (C ∨ D)
3. | | A
4. | | B   1, 3, ⊃ E
5. | | C ∨ D   2, 4, ⊃ E
6. | A ⊃ (C ∨ D)   3-5, ⊃ I

In this derivation we see that 'A', together with lines 1 and 2, allows the derivation of 'C ∨ D'. We then say this shows that lines 1 and 2 by themselves imply 'A ⊃ (C ∨ D)'.

In general, the rule for horseshoe introduction can be stated as follows:

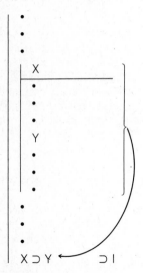

Note that you *always* build the discharged temporary assumption, 'X', into the conclusion 'X ⊃ Y' when using this rule.

There is no limit to the number of subderivations which may occur in one larger derivation. Consider the following proof:

1. | (A ∨ B) ⊃ C
2. | D ≡ C
3. | | A
4. | | A ∨ B   3, ∨ I
5. | | C   1, 4, ⊃ E
6. | A ⊃ C   3-5, ⊃ I
7. | | D
8. | | C   2, 7, ≡ E
9. | D ⊃ C   7-8, ⊃ I
10. | (A ⊃ C) & (D ⊃ C)   6, 9, & I

Each subderivation is marked off by its own vertical line. When that line ends, the subderivation ends. Above, 3 was assumed to help get 6, while 7 helps get 9.

My students often ask me what *justifies* making a particular temporary assumption. The answer is that *nothing* justifies it. Temporary assumptions are exactly what they are called: assumptions. They very well may not be true, and when you write one you are not claiming that it follows logically from the original premisses. You may assume anything at all, simple or complex, silly or wise. However, there *is* a catch. Once you assume something, you must start a new framework, marking the assumption as an added premiss by a horizontal line, and indicating by means of a new vertical line that a new subderivation has begun. *The conclusions you obtain by using your temporary assumption will all appear behind the vertical line associated with that assumption, as a warning that the assumption is still being made, except when you use a specific rule such as ⊃ I which allows you to take something obtained by the help of a temporary assumption and move it out from behind that vertical line.* (So far, ⊃ I is the *only* rule which allows you to remove anything from behind a vertical line. And ⊃ I allows such removal only under specific, legitimate conditions.) So you can assume anything you want, but you will not find just any arbitrary assumption to help much.

Thus, if asked to derive 'B' from 'A ⊃ B', 'C ≡ A', and 'C', it is not *wrong,* but it is *foolish* to start out like this:

1.	A ⊃ B	
2.	C ≡ A	
3.	C	
4.	A	
5.	B	1, 4, ⊃ E

This derivation has not yet succeeded in showing 'B' follows *from the original premisses,* since 'B' in line 5 is behind *two* vertical lines, indicating that the temporary assumption 'A' has not yet been discharged. This derivation *can* be completed, as follows:

1.	A ⊃ B		
2.	C ≡ A		
3.	C		
4.	A		
5.	B	1, 4, ⊃ E	
6.	A ⊃ B	4-5, ⊃ I	(Compare this to line 1.)
7.	A	2, 3, ≡ E	
8.	B	6, 7, ⊃ E	

However, you should be able to see that lines 4, 5, and 6 are unnecessary. The above derivation is "correct" but very inefficient. *As a general rule of strategy, you should not assume anything without knowing in advance what rule you will use to discharge the assumption in order to get something useful from it.* (So far, the only time you would ever sensibly assume anything is when you need to derive a horseshoe statement.)

There's another troublesome feature of subderivations. People are tempted to cheat when using them, like this:

1. | A ⊃ B
2. | C ≡ A
3. | | C
4. | | A   2, 3, ≡ E
5. | B   1, 4, ⊃ E

This derivation is incorrect, because 'B' was obtained by the help of line 4, and thus 'B' should be behind the same vertical lines 'A' is behind. The reason is simple: We used 'C' to prove 'A' above. So 'C' is indirectly being used to prove 'B' as well. The vertical line that goes with 'C' must be continued, to indicate 'C' is still being assumed true. The above derivation may be corrected to read as follows:

1. | A ⊃ B
2. | C ≡ A
3. | | C
4. | | A   2, 3, ≡ E
5. | | B   1, 4, ⊃ E
6. | C ⊃ B   3-5, ⊃ I

Here we legitimately take 'B' out from behind the vertical line by using ⊃ I, which requires us to build our temporary assumption into the antecedent of our conclusion. Of course it would be illogical to change line 6 above to 'B'. It *must* read 'C ⊃ B'.

The preceding discussion illustrates a general point about subderivations which I will discuss by using the following complex derivation outline:

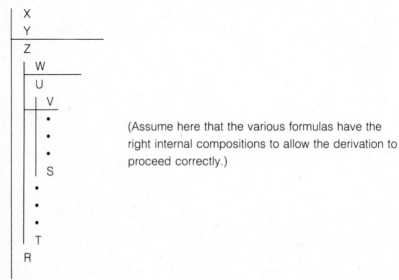

(Assume here that the various formulas have the right internal compositions to allow the derivation to proceed correctly.)

Above, when we are working on proving 'Z', 'X' and 'Y' alone are available. On the other hand, when we are working on proving 'U', we may treat our situation as though 'X', 'Y', 'Z', and 'W' are all directly above 'U'; that means we can use our rules to get 'U' as though we had the following arrangement above 'U'.

X
Y
Z
W

When we are working on the slot marked by 'S', we have, in effect, the following formulas above 'S': 'X', 'Y', 'Z', 'W', 'U', and 'V'. But when we are working on the 'T' slot, we do *not* have 'X', 'Y', 'Z', 'W', 'U', 'V', . . . , and 'S' with which to work. Rather, we *do* have, in effect

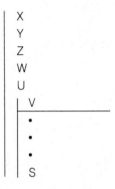

That is, 'V' and all the formulas under it through 'S', are *not* treated as behind the same vertical line as 'T'. Similarly, when working on 'R', only 'X', 'Y', and 'Z' are behind its vertical line.

The vertical lines show graphically which premisses are still being assumed at any point in the derivation. So long as a vertical line continues, the premisses appearing at its top are still being assumed true. It would not be reasonable to use 'S' to get 'T' as though 'S' and 'T' were behind the 'W' vertical line, for the appearance of 'S' behind the 'V' vertical line indicates that 'V' was still being assumed true when 'S' was obtained. If you want to use 'S' to get 'T' you must either continue the 'V' vertical line, or use a special rule like ⊃ I which specifically allows you to bring out something from under assumption 'V'.

To sum up, when you need to justify some particular conclusion in a derivation, you must consider which earlier formulas in the derivation count as though they were behind the vertical line closest to the left of your conclusion. The preceding discussion shows that *those formulas which count are those which appear immediately behind any vertical line which passes by anywhere to the left of your conclusion.* This principle is important because it tells you which formulas may be cited in justifications for any given conclusion. ⊃ I is the only rule discussed so far which allows you to cite formulas which do not count as being immediately behind the same vertical line which the conclusion is immediately behind.

Applying this principle to the above derivation outline, you can see that 'V' and 'S' do *not* count as behind the vertical line left of 'T', because they do not appear *immediately* behind any vertical line which passes by 'T' on the left. 'X', 'Y', and 'Z', on the other hand, *do* count as being behind 'T''s vertical line. Accordingly, 'X', 'Y', and 'Z' may be cited in justifying 'T', even though they are not actually behind the line immediately to the left of 'T', for they count as though they were behind that line. But 'V' and 'S' may not be cited, except when using a rule like ⊃I which contains reference to a subderivation. (Thus far, ⊃ I is the only such rule.)

Turning to another topic, if you are a bit clever, you may notice that the ⊃ I rule allows us to discharge *all* the premises in an argument. For example, the following will be a legitimate derivation.

1.	B & C	
2.	B	1, & E
3.	(B & C) ⊃ B	1-2, ⊃ I

In this case, lines 1 and 2 constitute a subderivation occurring under no premisses at all. The vertical line to the far left has no premisses at the top. If you like, you could write this derivation like this:

1.		
2.	B & C	
3.	B	2, & E
4.	(B & C) ⊃ B	2-3, ⊃ I

I leave that up to you.

There is nothing wrong with the above derivation, even though it has no premisses. Its conclusion, '(B & C) ⊃ B', is surely true, no matter what, as can be seen by doing a truth table, which will show that '(B & C) ⊃ B' is true in *every* line of the table—'(B & C) ⊃ B' is said to be a *tautology*. One needs no premisses to base the truth of '(B & C) ⊃ B' on. Any conclusion which can be reached from no premisses at all in our system of rules will be a tautology, true in every logically possible world.

Here is a final example of a derivation from no premisses at all:

1.		
2.	C	
3.	C ∨ D	2, ∨ I
4.	C ⊃ (C ∨ D)	2-3, ⊃ I

Again, the conclusion obviously has to be true, no matter what.

As indicated earlier, a subderivation can contain another subderivation inside it. The vertical lines always indicate what's going on by marking off one subderivation from another, as in this example:

```
1.  │ (A ⊃ B) ⊃ (C & D)
2.  │ D ⊃ B
3.  │ ┌─ D
4.  │ │ ┌─ A
5.  │ │ │ B              2, 3, ⊃ E
6.  │ │ A ⊃ B            4-5, ⊃ I
7.  │ │ C & D            1, 6, ⊃ E
8.  │ │ C                7, & E
9.  │ D ⊃ C              3-8, ⊃ I
```

We see that lines 4 and 5 constitute one subderivation, occurring inside the subderivation 3 through 8. Since line 4 occurs *inside* the derivation headed by line 3, and line 3 occurs *inside* the main derivation, the two original premisses in lines 1 and 2 are also premisses for the first subderivation, which in turn means that lines 1, 2, and 3 represent premisses for the second subderivation. This justifies the use of lines 2 and 3 to get line 5. It may interest you to notice that the assumption made at line 4 is never used inside the subderivation headed by line 4. Line 4 serves only as a dummy assumption which plays no actual role in obtaining line 5. Line 4 comes into play only when it is discharged at line 6.

There is no substitute for practice in mastering the ⊃ I rule—or any of the other rules for that matter. So here are the exercises.

Horseshoe introduction—conditional proof

45.1 Fill in the justifications that go to the right of each intermediate and final conclusion in the following derivations. (These derivations are all correct; so your job is just to figure out where each line comes from. It would also be a good idea to pay attention to the strategies employed.)

a.
```
1.  │ A ⊃ B
2.  │ B ⊃ (C & D)
3.  │ ┌─ A
4.  │ │ B
5.  │ │ C & D
6.  │ │ D
7.  │ A ⊃ D
```

b.
```
1.  │ (A ∨ C) & D
2.  │ (D ⊃ B) ⊃ E
3.  │ ┌─ D ⊃ B
4.  │ │ E
5.  │ │ D
6.  │ │ B
7.  │ │ B & E
8.  │ (D ⊃ B) ⊃ (B & E)
```

c.
```
1.  │ (A ∨ B) ⊃ C
2.  │ ┌─ A
3.  │ │ A ∨ B
4.  │ │ C
5.  │ A ⊃ C
```

d.
```
1.  │ C & D
2.  │ ┌─ A
3.  │ │ C
4.  │ A ⊃ C
```

(Derivation (d) illustrates a general point: once you know that something is true, like 'C', you thereby can prove that anything else, like 'A', can be placed in front of 'C' with a horseshoe.)

e. 1.
   2. | A ⊃ B
   3. | | A
   4. | | B
   5. | A ⊃ B
   6.   (A ⊃ B) ⊃ (A ⊃ B)

45.2 Each of the following "derivations" contains at least one serious blunder. Find the blunders.

a. 1. | A ⊃ (B & C)
   2. | B ⊃ D
   3. | | A
   4. | | B & C   1, 3, ⊃ E
   5. | | B   4, & E
   6. | A ⊃ B   3-5, ⊃ I
   7. | D   2, 5, ⊃ E

b. 1. | A ⊃ (B & C)
   2. | B ⊃ D
   3. | | A
   4. | | B & C   1, 3, ⊃ E
   5. | | C   4, & E
   6. | C   3-5, ⊃ I

c. 1. | A ⊃ (B & C)
   2. | B ⊃ D
   3. | | B & C
   4. | | B   3, & E
   5. | | D   2, 4, ⊃ E
   6. | (B & C) ⊃ D   3-5, ⊃ I
   7. | B   3, & E
   8. | D   2, 7, ⊃ E

d. 1. | A ⊃ (B & C)
   2. | B ⊃ D
   3. | | A
   4. | | B
   5. | | D   2, 4, ⊃ E
   6. | | B & C   1, 3, ⊃ E
   7. | | C   6, & E
   8. | A ⊃ C   3-7, ⊃ I

45.3 Derive '(A & B) ⊃ C' from '(A & B) ⊃ E' and 'E ⊃ C'.

45.4 Derive 'P ⊃ Q' from '(P ∨ R) ⊃ Q'.

45.5 Derive 'P ⊃ (Q ∨ R)' from 'P ⊃ Q'.

45.6 Derive 'P ⊃ (Q ⊃ R)' from 'P ⊃ R'.

45.7 Symbolize each of the following arguments as accurately as possible and then derive the symbolic version of the conclusion from the symbolic premisses, thus verifying the validity of the arguments and showing how to get the conclusion in a step-by-step way from the premisses.

    a. If the amendment is adopted, a great many laws will have to be changed. If a great many laws will have to be changed, the legislators will work overtime. The legislators will work overtime only if they are paid extra. If the legislators are paid extra the voters will be angry. Thus, if the amendment is adopted, the voters will be angry.

    b. If we improve the quality of our product or if we lower its price, we will gain a competitive advantage in the marketplace. The board of directors has decided to go for quality improvement. If the board decides in favor of quality improvement, we will in fact improve the quality of our product. So we will gain a competitive advantage in the marketplace.

    c. If the workers go on strike next week, we will have to shut down production of the new model, and we will be unable to fill the back orders for the old model. If we shut down production of the new model, we will lose about one million dollars in gross sales, unless we can begin production again within one month. (Use "∨" for "unless".) If we are unable to fill the back orders for the old model, our distributors will drop our product. So, if the workers go out on strike next week, our distributors will drop our product, and we will lose one million dollars in sales unless we will get back into production within one month.

46

Repetition and the rules for negations

Now that subderivations have made their appearance, our system of rules has much greater power. However in order to make subderivations do all the work for us which they should, we can introduce a rule which allows us to get a formula from one place in a derivation to appear later on in the derivation. The rule is very simple, and is obviously valid:

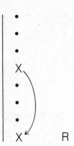

This rule is called the rule of repetition.

A trivial example of the rule R in action is the following:

1. | A ∨ B
   |‾‾‾‾‾‾‾‾‾‾‾‾
2. | A ∨ B   1, R

We will not normally use the rule so trivially. Nevertheless, you can see that the argument diagram which results from the use of the rule R is valid. After all, if the premiss, 'A ∨ B' is true, then the conclusion, 'A ∨ B' must also be true!

The above trivial example of the use of R can, however, be turned quickly into a somewhat less trivial example:

1. |
   |‾‾‾‾‾‾‾‾‾‾‾‾‾‾‾‾‾‾‾‾
2. | | A ∨ B
   | |‾‾‾‾‾‾‾‾‾‾‾
3. | | A ∨ B   2, R
4. | (A ∨ B) ⊃ (A ∨ B)   2-3, ⊃ I

This derivation shows that the formula '(A ∨ B) ⊃ (A ∨ B)' is derivable from no premisses at all and is thus a tautology. An English sentence which has approximately this form is the sentence "If it will rain or snow, then it will rain or snow". Not very interesting, perhaps, but true enough, on a truth-functional interpretation.

There are some more common uses of R involving subderivations. Here is one such example:

1. | (D & B) ⊃ R
2. | B
3. | | D
   | |‾‾‾‾‾‾‾‾‾‾‾
4. | | B   2, R
5. | | D & B   3, 4, & I
6. | | (D & B) ⊃ R   1, R
7. | | R   5, 6, ⊃ E
8. | D ⊃ R   3-7, ⊃ I

In this example the reiterations are not really necessary; one could have done the job without them, but some people prefer to see the things they are working with close together. Since lines 1 and 2 above can be treated as being behind the same vertical line as 'D' is, it is perfectly legitimate to copy them down inside the subderivation. I prefer not to use R except when it actually saves steps in getting a desired conclusion, as it sometimes does. For example, we can use R to get 'P ⊃ Q' from the premiss 'Q':

```
1.  │ Q
2.  │ │ P
3.  │ │ Q        1, R
4.  │ P ⊃ Q      2-3, ⊃ I
```

The preceding derivation becomes very difficult without R:

```
1.  │ Q
2.  │ │ P
3.  │ │ │ Q
4.  │ │ │ Q & Q    1, 3, & I
5.  │ │ │ Q        4, & E
6.  │ │ Q ⊃ Q      3-5, ⊃ I
7.  │ │ Q          1, 6, ⊃ E
8.  │ P ⊃ Q        2-7, ⊃ I
```

The existence of this strategy does illustrate, however, that R need not be a basic rule. We can always do something like the above to repeat a formula. I'm sure, though, that you'll prefer to use R, which becomes our third derived rule.

There is one principal mistake associated with the rule R, illustrated below:

```
1.  │ (A ⊃ B) ⊃ C
2.  │ B
3.  │ │ A
4.  │ │ B        2, R (This time used properly.)
5.  │ A ⊃ B      3-4, ⊃ I
6.  │ A          3, R (This time used improperly.)
7.  │ B          5, 6, ⊃ E
```

The problem with the second use of R above arises because R allows you to copy a formula when and only when that formula is copied behind the *same* vertical line. Our understanding about what counts as being behind the same vertical line, explained in the previous Section, allows you to count 'B' in line 2 as being behind the same vertical lines as 'A' in line 3. But our understanding does not allow line 3 and line 5 to count as being behind the same vertical lines. If the above "derivation" were all right, that would mean that line 6 *validly* follows from lines 1 and 2. But a moment's thought should show that it does not, for there is no way one could know on the basis of lines 1 and 2 alone that 'A' is true. This observation shows why line 6 must not be allowed.

So much for R. Enjoy it, but with care.

We are now ready for the last of the ten basic rules, reductio ad absurdum (meaning "reduction to absurdity"), known as "RA" for short. The addition of this extremely powerful rule makes our system sufficiently complete that for any argument diagram which truth tables can prove valid there exists a derivation starting with that diagram's premisses and ending with the conclusion.

RA relies on the notion of a *contradiction*. For our purposes, a contradiction is defined as any formula which is false in *every* line of its truth table. Thus, 'A & ~A' is a contradiction, for it remains false no matter whether 'A' itself is true or false. There are infinitely many different contradictory formulas, such as '~(A ∨ ~A)', '~(P ⊃ P)', and so on, but RA is concerned only with the contradictory pattern 'A & ~A' since this is the easiest type of contradiction to recognize.

RA essentially says that any *valid* argument which ends up with this contradiction as conclusion *must* have at least one false premiss. That at least one of the premisses of such an argument must be false follows from the definition of "valid argument" plus the fact that the contradiction cannot be true. After all, if all the premisses of a valid argument are true, then the argument is sound, and its conclusion will automatically be true as well. But if the conclusion of the argument is a contradiction, the conclusion cannot be true. Hence, not all the premisses are true in a valid argument which has a contradiction as its conclusion.

Here is an example of a set of premisses which leads by way of a valid argument to a contradiction:

If I go to the store, I'll spend too much money. But I'm a compulsive shopper and I'll go to the store. I won't spend too much money, because I can't afford to.

In symbols, these premisses can be approximated as follows:

A ⊃ B   C & A   ~B & ~D

You shouldn't have too much trouble seeing how to get the conclusion 'B & ~B' from those premisses by means of a valid argument. The same line of argument will also be valid in English and will lead to the conclusion that I'll spend too much money and I won't spend too much money. Once this contradictory conclusion is reached, it is obvious that at least one of the premisses is false. Crudely speaking, the premisses fight each other. They don't fit together. But there is no way of telling from the argument *which* of the premisses is false. We only know they can't *all* be true. Maybe more than one of them is false; maybe not.

Even though the argument which leads to the contradiction above does not show us which premiss is false, we can say at least this much: there are three premisses in all; if we continue to *assume* the first *two* are true, we can then say, *on the basis of that assumption,* (and only on that basis), that the *third* premiss is false. Similarly, if we were to assume instead that the last two premisses are true, we could conclude on the basis of that assumption that the first premiss is false. Of course, our assumption may be incorrect. Perhaps one of the premisses we assume to be true really isn't true. Nevertheless, the logical point remains: so long as we continue to assume that a particular set of premisses is true, and the addition of an extra premiss then leads to contradiction, we can validly conclude that the extra premiss is false.

The previous discussion was intended to help you see the validity of the rule of RA which we can now set out in its most general schematic form:

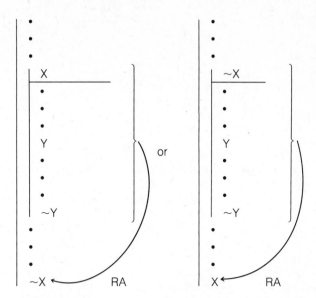

Of course, the *order* in which 'Y' and '~Y' appear in the above rules is irrelevant. The idea in both versions of the rule is very closely related to the discussion of contradictions given earlier.

As usual, the best way to get a firm grasp of this rule is to look at examples of its use, and then use it yourself. Here are some examples:

1.	A ⊃ B	(These are the premisses for the argument
2.	C & A	about shopping used above.)
3.	~B & ~D	
4.	A	2, & E
5.	B	1, 4, ⊃ E
6.	~B	3, & E
7.	~(~B & ~D)	3-5 and 6, RA

Notice that the way the vertical lines are arranged here tells an important story. The final conclusion of the argument is obtained on the assumption that the first two premisses are true. That is, line 3 is proved false only on the assumption that lines 1 and 2 are true. That is why line 7 is behind the vertical line that frames lines 1 and 2. In short, when you use the rule of RA, you put the premiss you want to *disprove* at the top of a subderivation. You then get the pieces of the contradiction to appear under this premiss.

Let me illustrate this point again, this time by using the same three basic premisses as before, but putting the last two premisses at the top of the whole derivation, which will mean that these will now be the preferred premisses which are not questioned. On the basis of these premisses, we will be able to derive the conclusion that the other one of the three premisses is false by using our rule.

```
1.  |  C & A
2.  |  ~B & ~D
3.  |  |  A ⊃ B
4.  |  |  A              1, & E
5.  |  |  B              3, 4, ⊃ E
6.  |  |  ~B             2, & E
7.  |  ~(A ⊃ B)          3-5 and 6, RA
```

Finally, we could choose to put the one remaining pair of our three premisses at the top in order to derive the negation of the other one of the three. (I leave it to you to verify this.) Thus, given any two of our three shopping premisses, we can easily show that the remaining premiss is false, since the addition of any one of the three to the other two leads to a contradiction.

The rule RA is used extensively in mathematical proofs, as well as in ordinary life. In mathematics, the rule is usually known by another name—"indirect proof". If you want to prove that an even number greater than two is never prime, you assume the opposite and then derive a contradiction. The final result is that you get the conclusion you want. The trick is to assume the opposite of what you really want to prove, and then to derive a contradiction from that assumption.

This strategy makes many proofs easy which would otherwise be difficult or impossible. The great benefit derived from this strategy for proving is that the added temporary assumption gives you one extra premiss to work with. You can add *any* assumption that you want, at any time that you want, in order to derive the contradiction. So, by choosing your assumption *carefully,* you can often get desired conclusions quickly.

There are several very useful derived rules regarding "~". The first of these, "double negation elimination", works like this:

```
1.  |  ~~A
2.  |  A              1, ~~E
```

The validity of such an argument is clear, of course, but it is somewhat amusing to see how this same conclusion would have to be obtained without the use of a derived rule:

```
1.  |  ~~A
2.  |  |  ~A
3.  |  |  ~~A          1, R
4.  |  A               2 and 3, RA
```

The above derivation shows why ~~E is theoretically superfluous, since RA will do the desired job in several steps. To save those steps, though, ~~E is handy.

Actually, I will officially specify a more general version of ~~E than you might have guessed from the above:

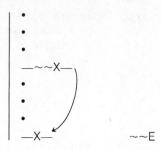

$\sim\sim$E

This rule specification looks different from any rule description we've had before, in virtue of the long dashes in it.

These long dashes signify that the formula '$\sim\sim$X' might appear *inside* a longer formula, such as 'A ⊃ $\sim\sim$B', but the rule of $\sim\sim$E may still be used to eliminate the double negation signs:

1. | A ⊃ $\sim\sim$B
2. | A ⊃ B   1, $\sim\sim$E

This works only because '$\sim\sim$X' is always equivalent to 'X', and this is justified by our rules only because of the long dashes occurring within the rule specification. All our other rules apply to *whole lines only,* and none of them may be applied to pieces of lines as $\sim\sim$E may be.

Even this last feature of $\sim\sim$E is not strictly needed within our system. We could have obtained the conclusion 'A ⊃ B' above without using $\sim\sim$E at all:

1. | A ⊃ $\sim\sim$B
2. | A
3. | $\sim$B
4. | $\sim\sim$B   1, 2, ⊃ E
5. | B   3 and 4, RA
6. | A ⊃ B   2-5, ⊃ I

Obviously, though, $\sim\sim$E makes things more natural and direct.

Moreover, since '$\sim\sim$X' is fully equivalent to 'X', we can reverse the process to get a new derived rule, $\sim\sim$I, which you will rarely use. I include it mainly to illustrate the point that *some* rules are actually reversible:

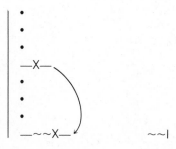

$\sim\sim$I

What justifies this rule? The truth table equivalence of '~~X' and 'X' guarantees that the substitution of '~~X' for 'X' will not change the value of any formula containing 'X'. A little imagination can make ~~I useful once in a while:

1. | ~A ∨ B
2. | A
   |————————————
3. | ~~A   2, ~~I
4. | B   1, 3, EA

(Try doing that one without ~~I. Then try it without ~~I or EA.)

Students commonly misuse ~~E in cases where there are two negation signs in a row applying to *different* formulas:

1. | A ≡ ~(~B & C)
   |————————————
2. | A ≡ (B & C)   1, ~~E misused

Our ~~E rule does *not* allow this move, since it requires *both* negation signs to apply to just *one* formula, 'X'. In fact, the above argument diagram is *invalid,* and we would not want ~~E to generate it.

The final derived rule for this Section, the rule of contradiction, comes quite directly from RA, although it looks insane:

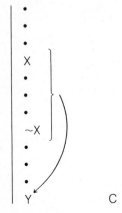

In words, what this rule says is simple: from a contradiction, *anything* validly follows. The validity of such arguments has been discussed previously. (See Section 33.) Moreover, we can use RA to derive the relevant conclusion in such cases. For example, if we assume 'A & B' and '~(A & B)', we can derive the arbitrarily chosen formula 'R ⊃ (S ≡ T)', as follows, without using the derived rule C:

1. | A & B
2. | ~(A & B)
   |————————————
3. | | ~(R ⊃ (S ≡ T))
4. | | A & B   1, R
5. | | ~(A & B)   2, R
6. | R ⊃ (S ≡ T)   3-4 and 5, RA

Obviously, in order to get any conclusion one might want to appear in line 6 above, all one need do is assume its opposite in line 3. Rule C merely shortens the process, and allows you to proceed in one step, which is often very handy:

1.	A & B	
2.	~(A & B)	
3.	R ⊃ (S ≡ T)	1, 2, C

This completes the set of rules focussed specifically on the negation sign.

Repetition and the rules for negations

46.1   Pick out each error in the following "derivations".

a.
1.	A ≡ ~(~B & C)	
2.	Q ⊃ A	
3.	A	2, ⊃ E
4.	~(~B & C)	1, 3, ≡ E
5.	B & C	4, ~~ E
6.	C	5, & E

b.
1.	A ⊃ (B & ~C)	
2.	A ∨ D	
3.	~B ⊃ D	
4.	B & C	
5.	(B & ~C) ∨ ~B	2, 1, 3, ∨ E
6.	B & ~C	5, ∨ E
7.	~B	5, ∨ E
8.	~C	6, & E
9.	~B & ~C	7, 8, & I
10.	~B & ~C	4 and 9, RA

46.2   Write the justification for each step in the following correct derivations.

a.
1.	~(P & A)
2.	Q ≡ (R ⊃ P)
3.	R
4.	Q & A
5.	Q
6.	R ⊃ P
7.	P
8.	A
9.	P & A
10.	~(P & A)
11.	~(Q & A)

b.
1.	A ∨ ~B
2.	A ≡ (C & D)
3.	~D ⊃ B
4.	A
5.	C & D
6.	D
7.	A ⊃ D
8.	~B
9.	~~D
10.	D
11.	~B ⊃ D
12.	D ∨ D
13.	D

46.3   Derive '~A' from the premisses 'A ⊃ B' and '~B', using RA rather than N ⊃ E.

46.4   Derive '~(A ∨ B)' from the premisses 'A ⊃ C', 'B ⊃ C', and '~C'.

46.5   Derive '(A ∨ B) ≡ ~C' from the premisses '(A ∨ B) ⊃ ~C', '~C ⊃ (D & E)', and 'D ⊃ A'.

46.6   Derive '~A & ~B' from the premisses 'C' and '(A ∨ B) ⊃ ~C'.

46.7   Derive '~B ⊃ ~A' from the premiss 'A ⊃ B'.

46.8   Derive 'A ⊃ B' from the premiss '~B ⊃ ~A'.

46.9   Derive '(A ⊃ B) ≡ (~B ⊃ ~A)' from no premisses at all.

46.10 Derive '~(A & B)' from the premiss '~A & ~B'.

46.11 Derive '~(A ∨ B)' from the premiss '~A & ~B'.

46.12 Derive '~(A ∨ B) ≡ (~A & ~B)' from no premisses at all.

46.13 Derive the following well-known tautologies from no premisses:
   a. ~(A & ~A)
   b. A ∨ ~A
   c. (B ⊃ ~B) ⊃ ~B
   d. (~C ∨ D) ⊃ (C ⊃ D)

47

More about derived rules of inference

Now that our basic rule set is complete, we can use it to justify the derived rules we've already seen, as well as to generate new derived rules. I'll begin by justifying N ⊃ E.

Examine the following correct derivation, which employs only basic rules.

1.	A ⊃ B	
2.	~B	
3.	A	
4.	B	1, 3, ⊃ E
5.	~B	2, R
6.	~A	3-4 and 5, RA

It is not hard to see this derivation accomplishes in six lines exactly what can be accomplished by N ⊃ E in three lines. Obviously, this same derivation method would work, no matter how complex the formulas 'A' and 'B' might become. Thus, we have a justification showing that N ⊃ E is just as valid as our basic rules are. N ⊃ E does not allow the derivation of anything not already derivable using only the ten basic rules.

The rule EA can be similarly justified:

1.	$A \lor B$	
2.	$\sim A$	
3.	A	
4.	$\sim B$	
5.	A	3, R
6.	$\sim A$	2, R
7.	B	4-5 and 6, RA
8.	$A \supset B$	3-7, $\supset$ I
9.	B	
10.	B	9, R
11.	$B \supset B$	9-10, $\supset$ I
12.	$B \lor B$	1, 8, 11, $\lor$ E
13.	B	12, R $\lor$ E

The complexity of the above derivation amply illustrates why the derived rule EA is very useful. (EA came in two versions. The other version is justifiable in much the same way.)

Next, I'd like to add four more derived rules, which will complete the system of rules to be discussed in the present chapter. All four of these rules have the same basic purpose, namely, to convert a formula containing one connective into a formula containing a different connective. In that regard, these rules differ from the introduction and elimination rules we have been studying so far.

The first pair of derived rules allows one to convert certain formulas containing a horseshoe to certain formulas containing a wedge, or vice versa. I call the first one "horseshoe to wedge conversion", or "$\supset$ T $\lor$" for short; the second one is the reverse conversion, $\lor$ T $\supset$:

These may be checked for validity with truth tables. (You should stop and do that now to convince yourself.) I will show below how ⊃ T ∨ can be derived from our other rules, but I will leave the analogous task regarding ∨ T ⊃ as an exercise.

1.	X ⊃ Y	
2.	~(~X ∨ Y)	
3.	X	
4.	Y	1, 3, ⊃ E
5.	~X ∨ Y	4, ∨ I
6.	~(~X ∨ Y)	2, R
7.	~X	3-5 and 6, RA
8.	~X ∨ Y	7, ∨ I
9.	~(~X ∨ Y)	2, R
10.	~X ∨ Y	2-8 and 9, RA

The second pair of derived rules, traditionally known as DeMorgan's laws, deal with conversions between "&" and "∨". If you give some thought to what these rules say, they ought to seem obviously valid to you, but of course they may be checked with truth tables if you are suspicious. I will refer to these rules by the more descriptive names, "ampersand to wedge" and "wedge to ampersand", or "& T ∨" and "∨ T &", respectively. Here they are:

The following derivation shows &T∨ is theoretically superfluous (but handy). One can get '~A ∨ ~B' from '~(A & B)' without using & T ∨:

1.	~(A & B)	
2.	A	
3.	B	
4.	A & B	2, 3, & I
5.	~(A & B)	1, R
6.	~B	3-4 and 5, RA
7.	A ⊃ ~B	2-6, ⊃ I
8.	~A ∨ ~B	7, ⊃ T ∨

Although I used the derived rule ⊃T∨, the above derivation nevertheless shows that &T∨ can be replaced by only basic rules, since ⊃T∨ can itself be replaced by only basic rules. I leave it to you to show how ∨T& can be derived from our other rules.

This completes the system of rules of inference for this chapter. It is important that you realize this system is not the only possible one governing formulas built around the connectives. As you might have guessed, there is an infinite variety of valid alternative rules one might adopt. However, much of ordinary reasoning whose structure is adequately represented by our argument diagrams can be duplicated fairly naturally by following the particular set of rules presented here. This particular rule set is my own modification of a set devised by Gerhard Gentzen, while the format of the derivations, utilizing the vertical and horizontal lines, is due to Frederick B. Fitch. Various logicians have employed each of the rules I use, in varying combinations with other rules.

There is an important limitation to our system of rules: it can only handle argument diagrams which are built around our preferred connectors. You should recall that some valid English arguments cannot be adequately symbolized since their validity depends on more detailed structure which is lost in our symbolic notation. Our rules of inference will fail to prove these arguments valid. Our rules can prove valid only those arguments with symbolic diagrams which are either exact or are close enough approximations (in the sense of Section 41) and which can themselves be proved valid by means of a truth table. (In the next chapter, we will work on symbolization and derivation techniques to handle a much wider range of valid arguments.) Since truth tables reveal logical relations between whole propositions, and the smallest units in our symbolic argument diagrams are signs for whole propositions, the logic we are constructing in this chapter is known as "propositional" (or sometimes, "sentential") logic.

Nevertheless, by learning to work intelligently with the system of rules found in the present chapter, you should acquire a sense of how one might work through a set of premises and figure out, step-by-step, what conclusions validly follow from those premises.

More about derived rules of inference

47.1 How can you tell that there is not supposed to be any way of deriving 'A' from '~(A & B)' in our system? (If you could find a way of correctly making this derivation, using our rules, what would that show about our system of rules?)

47.2 Would the following rule be a good derived rule? Explain.

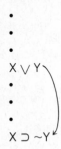

47.3 Derive the conclusions of the following argument diagrams from their premisses, using appropriate derived rules discussed in this Section, plus other rules from earlier Sections as needed.

a. $\underline{(A \And B) \supset C}$
   $C \vee \sim(A \And B)$

b. $\underline{\sim A \vee B}$
   $\sim(A \And \sim B)$

c. A
   $\underline{A \supset \sim(C \And \sim D)}$
   $\sim C \vee D$

d. $\underline{A \vee B}$
   $B \vee A$

e. (No premisses)
   $A \vee \sim A$

f. $C \supset D$
   $A \equiv (\sim C \vee D)$
   $\underline{A \supset B}$
   B

g. $\underline{A \supset \sim B}$
   $\sim(A \And B)$

h. $\sim(A \And B)$
   $\underline{\sim(\sim B \vee R)}$
   $\sim A$

i. $\underline{B}$
   $A \supset B$

j. $A \vee (B \And \sim C)$
   $\underline{\sim B}$
   $C \supset A$

k. $(C \vee B) \And \sim(E \supset G)$
   $\underline{\sim G \supset \sim B}$
   C

l. $\sim C \And D$
   $\underline{(B \equiv D) \supset \sim(C \supset E)}$
   $\sim B$

m. (No premisses)
   $\sim(A \supset B) \supset A$

n. (No premisses)
   $\sim(A \supset B) \supset \sim B$

47.4 Justify the claims that $\vee T\supset$ and $\vee T\And$ are derived rules which do not add any new conclusions to those that already can be obtained in the system without them.

47.5 For each new rule introduced in this Section, state in ordinary English a one-step argument which would fit the pattern of argumentation represented by that rule.

48

Review of propositional logic— strategy and applications

Successful derivations generally don't just happen; they are planned. In fact, the basic planning strategy for derivation construction parallels the strategy for most human problem-solving situations: one must keep one eye on the starting points—the original premisses and other formulas available to serve as prem- isses—and the other eye on the goal—the conclusion. In keeping with that gen- eral advice I often work derivations backwards, at least partially, writing the de- sired conclusion at the bottom, and leaving a blank space in which to write the intermediate steps. Then I write just above the conclusion the formula or outline of a series of formulas which would give me the conclusion I want if I could manage to derive that formula or set up that series correctly.

Accordingly, if I am trying to derive '(A & B) ⊃ C' from 'A ⊃ (D ∨ E)', 'D ⊃ ~B', and '~C ⊃ ~E', after I write the premisses, I scan them to look for any obvious moves that go directly to the conclusion, and finding none, I im- mediately write the last line of the unfinished derivation, and fill in a framework for doing a horseshoe introduction:

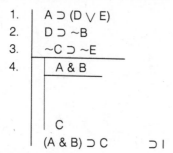

This gives me a general strategy to follow. (It may not work, but at least I have a plan to try.)

My next step will be to shift attention completely away from the bottom line, and direct all my attention to asking how to derive 'C' from lines 1 through 4. I might notice, for example, that line 3 tells me that I could get 'C' (basically by N ⊃ E) if I could get 'E'. Suppose I decide to adopt that strategy. I would then fill in the relevant parts of the derivation as follows:

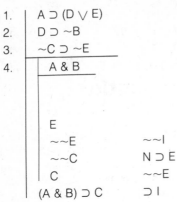

1. | A ⊃ (D ∨ E)
2. | D ⊃ ~B
3. | ~C ⊃ ~E
    _____
4. | | A & B
    | |
    | |
    | | E
    | | ~~E   ~~I
    | | ~~C   N ⊃ E
    | | C   ~~E
    | (A & B) ⊃ C   ⊃ I

Now my task has once again taken new shape. I shift my attention toward deriving 'E' from lines 1 through 4, forgetting about what will happen once 'E' is obtained. Perhaps I will now be able to see how to get 'E'. Probably 3 will be useless, since 3 tells me a condition under which 'E' is false, not under which 'E' is true. But what about 1? That looks more promising. If I had 'A' (which I can get easily from 4), I can get 'D ∨ E'. If 'D' is then ruled out, I will be left with 'E'. Can I somehow rule out 'D'? Yes, using line 2 with 4. So I'm finished! All that remains is to write it up in orderly fashion:

1. | A ⊃ (D ∨ E)
2. | D ⊃ ~B
3. | ~C ⊃ ~E
    _____
4. | | A & B
5. | | A   4, & E
6. | | D ∨ E   1, 5, ⊃ E
7. | | B   4, & E
8. | | ~~B   7, ~~I
9. | | ~D   2, 8, N ⊃ E
10. | | E   6, 9, EA
11. | | ~~E   10, ~~I
12. | | ~~C   3, 11, N ⊃ E
13. | | C   12, ~~E
14. | (A & B) ⊃ C   4-13, ⊃ I

Of course this derivation is not the only possible correct one. You may be able to find another. But the general strategy of working backwards at first before working forwards, as illustrated here, often works.

In addition to the general problem-solving strategy of working backwards, there are several other rules of strategy which often work, or which can at least tell you where to start looking.

(1) When trying to derive something of the form 'X ⊃ Y', try ⊃ I. Thus, start out by assuming 'X', whatever 'X' may happen to be. Alternatively, try ∨ T ⊃, which requires no assuming.

(2) There are only two rules which allow you to get any results by making temporary assumptions: RA and ⊃ I. *Never* make a temporary assumption unless you plan to use RA or ⊃ I to end the subderivation created by the assumption.

(3) If all else fails, and you can't find any way to derive the conclusion you're after, try an indirect proof—try assuming the opposite of the conclusion you want, and then use RA.

(4) Most of the rules do not allow the introduction of any capital letters into a derivation which were not already present in the premisses. Thus, if your desired conclusion contains a new capital letter, you'll have to use ∨ I, RA, C, or ⊃ I somewhere in your derivation, since these are the only rules which can possibly introduce a new letter.

(5) If you are trying to derive something of the form 'X ∨ Y', it is not likely ∨ I will work, although it's worth a quick try. In most interesting arguments with conclusions of this form, the last step will be an application of ∨E, & T∨, ⊃T∨, or RA, rather than ∨I. The reason for this is simple: the only time you can prove 'X ∨ Y' by first proving just plain 'X' is when 'X' really does follow logically from the original premisses. If the original premisses do not imply 'X', you're wasting your time trying to derive 'X' from them.

(6) Whenever your premisses contain something of the form 'X ∨ Y', consider whether ∨ E and R ∨ E may be helpful.

(7) If you can figure out how to get a contradiction of the form 'X & ~X' from your premisses, you can prove *anything,* easily, using C or RA. Contradictory premisses lead validly to any conclusion whatsoever.

Now that you have a complete system of rules for deduction relating to the five special connectives, you have a second method for proving symbolic argument diagrams valid. (The first method employs truth tables.) You know that a correct derivation of a particular conclusion from a given set of premisses proves the corresponding argument diagram valid. This method has at least two great advantages over the truth table method: (1) The truth table method becomes very cumbersome when applied to arguments involving more than four different capital letters. (2) The method of derivation is psychologically more satisfying and natural feeling, for it reveals *how* the conclusion follows, and not merely *that* the conclusion follows.

However, if you lose sight of what the symbols mean and why the rules work, then you will lose the second benefit described above. (You will also lose much of your ability to do derivations correctly since mistakes are much more difficult to prevent when you become a mere symbol-pusher.) For instance, my students sometimes try maneuvers like this within a derivation:

$$
\begin{array}{ll}
\vdots & \\
A \lor B & \\
A & \lor E
\end{array}
$$

That this amounts to a silly mistake can be readily seen once one considers that "∨" and "&" mean radically different things, with the result that "& E", but not

"$\lor$ E", works like the above diagram. Merely remembering that "$\lor$" is a rough equivalent of "OR" ought to prevent such mistakes.

One final reminder: failure to derive a desired conclusion from given premisses does *not* prove the relevant argument diagram invalid. Perhaps a derivation does exist, but you just have not found it. This statement remains true even if you are able to derive the *opposite* of the desired conclusion from the premisses, for it may be possible to derive both the desired conclusion and its opposite. (This would happen if the premisses were contradictory.) Before a failure to derive the desired conclusion could count as an invalidity proof, you would have to know somehow that every possible derivation strategy had been tried. (There are ways of knowing this, but I will not discuss them here.) Moreover, even if you once did know that a particular diagram was invalid, that would not by itself prove invalid any particular English argument fitting the diagram. (Recall that invalid diagrams often fit valid arguments.) To prove a particular English argument invalid, use a counterexample or an absurd logically analogous argument. (See Sections 25 and 26.)

Review of propositional logic— strategy and applications

48.1 Derive the conclusions of the following argument diagrams from their premisses.

a. $\dfrac{(A \& B) \lor (A \& C)}{A \& (B \lor C)}$

b. $\dfrac{(A \& B) \supset C}{A \supset (B \supset C)}$

c. $\dfrac{A \equiv B}{(A \supset B) \& (B \supset A)}$

d. $\dfrac{(A \lor B) \supset C}{\begin{array}{l} \sim B \supset D \\ \sim D \end{array}}$ ... C

e. $\dfrac{\begin{array}{l} A \supset (B \& C) \\ B \supset D \\ \sim D \end{array}}{\sim A}$

f. $\dfrac{\begin{array}{l}(A \& B) \supset C \\ D \supset A\end{array}}{(D \& B) \supset C}$

g. $\dfrac{\begin{array}{l} A \supset (B \supset C) \\ A \supset (D \supset E) \\ A \& (B \lor D) \\ \sim C \end{array}}{E}$

h. $\dfrac{\sim A}{A \supset B}$

i. $A \supset (B \lor C)$
$\dfrac{\sim B}{A \supset C}$

j. $\sim A \lor B$
$A \supset B$

k. $\sim A \lor B$
$(A \& B) \supset C$
$\dfrac{A}{C}$

l. $A \supset \sim (B \supset C)$
$A \supset B$

m. $(A \supset B) \& (\sim C \supset \sim D)$
$\dfrac{(\sim B \lor D) \& \sim (C \& \sim A)}{A \equiv C}$

48.2 Symbolize each of the following arguments as accurately as possible and derive the conclusion of each from its premises. Note that you are proving the argument diagrams valid by doing this. On the assumption that your diagrams are accurate enough, this means you are proving these arguments valid. You may then wonder whether the premises of these arguments are true, since if they are, the controversial conclusions of these arguments are true also, at least to the extent that they are adequately symbolized.

a. If God is perfectly good, there is evil in the world only if God cannot prevent it or God doesn't know about it. But God cannot prevent evil only if God is not completely powerful, and God doesn't know about it only if God is not completely knowledgeable. There is evil in the world. So, if God is completely powerful and completely knowledgeable, God is not perfectly good.

b. If there is to be any ultimate explanation of why the physical universe as a whole exists, that explanation must be given by reference to some being which exists independently of the physical universe. However, if an explanation of the latter sort is given, the being to which it refers cannot be part of the physical universe, nor can that being itself stand in need of an explanation outside itself for why it exists as it does. If that being causes events to occur in the physical universe in the way that one thing inside the universe causes other things inside the universe, then that being is part of the physical universe. So either there is no ultimate explanation of why the physical universe as a whole exists, or that explanation is given by reference to a being that does not itself stand in need of an explanation outside itself and that does not cause events to occur in the physical universe in the same way one thing inside the universe causes other things inside the universe.

48.3 For each argument below, produce a derivation to prove its symbolic diagram valid, or produce a proof of invalidity using appropriate techniques.

a. The universe exhibits a very complex pattern of organization, which could only have come about by design. Hence, there must be a God.

b. On the average, blacks score lower than whites on the standard written intelligence tests. Thus, it is easy to see that blacks are on the average less intelligent than whites.

c. If we are experiencing high unemployment along with inflation, our situation is different from any previously encountered. A new economic analysis is called for, if our situation is different from any previously encountered. In fact we are experiencing high unemployment; we are also experiencing inflation. Thus, a new economic analysis is called for.

d. Every girl who goes to college wants to get married. Some people who want to get married are desperate for a mate. Thus, every girl who goes to college is desperate for a mate.

e. In the past increased unemployment has generally preceded a decline in prices. So, we may conclude that increased unemployment tends to cause price declines.

f. An all-out nuclear war would wipe out most of the world's population, and make the world unfit for human habitation in many places. The only way to avoid nuclear war in the long run is to allow the Russians to have their own way in world politics. Hence, that is what we should do.

g. If the university requires each student to take courses in a wide variety of areas, many students will not be adequately prepared to work in a specialized job when they graduate. But if the university does not make such a requirement of the students, many will graduate without a college-level understanding of the basic environment in which they live. So either many graduating students will not be adequately prepared for specialized jobs, or they will lack a college-level understanding of their environments.

h. For Thanksgiving, I'll either go home or go visit my friends in Chicago. If I go home, I'll fight with my father, but I'll enjoy being with my mother. I'm not going to fight with my father any more, because I can't stand it. So, I'll go visit my friends in Chicago for Thanksgiving.

48.4 Derive each of the following tautologies from no premises.
  a. $R \supset (R \supset R)$ b. $(A \supset B) \equiv (\sim A \vee B)$
  c. $(A \supset B) \equiv \sim(A \,\&\, \sim B)$ d. $(A \,\&\, B) \supset (A \equiv B)$
  e. $[A \vee (B \,\&\, C)] \equiv [(A \vee B) \,\&\, (A \vee C)]$

48.5 Find interesting conclusions, validly derived, for each of the following sets of premises using all the given premises in each case. (Don't restrict yourself merely to using & I to conjoin the premises.)

  a. $A \vee (B \,\&\, C)$ b. $A \supset (B \supset C)$
      $\dfrac{\sim B}{?}$ $\dfrac{B \,\&\, \sim C}{?}$

c. ~(A & B)   d. ~A ∨ B
  ~A ≡ C   (B ⊃ D) & A
  <u>~C ⊃ B</u>   <u>(D ⊃ R) & P</u>
                          ?

e. A ⊃ ~B   f. ~(A & B) ⊃ (C ∨ D)
  C ⊃ B   E ⊃ (~A ∨ ~B)
  <u>~C ⊃ D</u>   <u>~E ⊃ ~(A & B)</u>
  ?   ?

48.6 Write three new derived rules which could be added to our system without making it *too strong*. The system will be too strong if it allows the creation of invalid derivations.

8

Predicate Logic

49

Propositional logic, syllogisms, and first-order predicate logic

Symbolic propositional logic has two limitations, already noted: (1) Because they are nontruth-functional some English connectors can only be approximated by our five symbolic connectives. (2) Some English arguments have more logical structure than can be revealed by the symbolic diagrams available in propositional logic. We will now take steps to partially overcome the second (and only the second) of these limitations by greatly increasing the amount of structural detail an argument diagram can represent. Of course this will mean we need some new symbols and some new ways to use them. However, we will also retain the old symbols in approximately their old meanings. The symbolic techniques to be presented in this chapter may be viewed as an expansion of propositional logic rather than a replacement for it.

A great many valid English arguments which cannot be diagramed in sufficient detail to be proved valid within propositional logic use in a crucial way words like "all", "every", "some", and "there are". Such words are not connectors, and are thus not representable within the diagrams of propositional logic. Nevertheless, they are like connectors in so far as they importantly affect the logical relationships between propositions.

The argument

All apples are fruits.
Some of the things in my refrigerator are apples.
Some things in my refrigerator are fruits.

is such an argument. It seems to contain no connectors, but yet it seems obviously valid. If we symbolize it, we get the uninformative

A
B
C

This difficulty can be solved in part by developing symbolizing techniques which *break up* the atomic sentences in an argument, using special symbols for "all" and "some", known as *quantifiers*. I will use "∀" for "all" and "∃" for "some". (The backwards "E" is used for "some" because "some" in the relevant sense works like "there exists", and the "∃" suggests existence. For example we could have said "There exist things in my refrigerator which are apples" in place of the premises above.)

However, use of "∀" and "∃" must be coupled with further symbolic devices before our sample argument above can be handled adequately. For trying something like

> ∀A
> ∃B
> ___
> ∃C

obviously doesn't help much to clarify that argument.

What more is needed? We need to connect, in symbols, the idea that "apples" appears in both premisses, "things in my refrigerator" appears in one premiss and the conclusion, and so on. Now, "apple" and "things in my refrigerator" are not whole clauses and do not express whole propositions. So we are going into the *internal* structure of the clauses here, into the relations between the main component phrases within the clauses, to put the matter grammatically. We must introduce symbolic recognition of these phrases in addition to representation for "all" and "some" in our argument diagram. But of course we do not want to define a particular new logical symbol to always mean "apple", for we already have such a symbol—the word "apple". What we want instead is a symbolic device to merely mark the places where "apple" occurs in the argument. (More precisely, to mark where the concept of an apple occurs in the argument, since the same concept may be expressed by differing words.)

We are getting close to what we will be using if we construct diagrams like this one for the sample argument:

> ∀A's are B's.   (A's are apples, B's are fruits, C's are things in my
> ∃C's are A's.   refrigerator, in this diagram.)
> _____
> ∃C's are B's.

Notice how the double occurrences of 'A', 'B', and 'C' reveal logical connections within the argument. (Of course, in this "diagram", the capital letters no longer stand for propositions. They stand for concepts instead.)

At least loosely speaking, the main phrases which are symbolized above are called "subjects" and "predicates" in grammar. We might think of 'A' as the subject term and 'B' as the predicate term in the first formula above; then, 'C' is the subject term and 'A' is the predicate of the second premiss. (Strictly, this is not quite grammatically correct, since the quantifiers have to be taken into account in a completely accurate story about subject terms; but it's close enough to the truth for our purposes.) Because the symbolism we are developing here recognizes predicate/subject differences it is called *predicate* logic.

The ancient Greek philosopher, Aristotle, was the first to deal with such argument diagrams, in his theory of categorical syllogisms.[1] This theory con-

_____
[1]Aristotle did not use quite the same symbolic techniques as these, but his techniques were analogous to these. See his *Prior Analytics*.

cerns the various predicate logic argument diagrams which may be constructed by using formulas of the following four general forms as premisses and conclusions:

∀X's are Y's.

∃X's are Y's.

~(∃X's are Y's).

∃X's are not Y's.

Propositions expressed in one of the above ways are called "categorical", and a categorical syllogism is an argument whose premisses and conclusion are categorical propositions.

I mention syllogisms because syllogistic logic has had a long and influential career. Many thousands of logic students, perhaps millions, have studied it in some detail. But I will not present syllogistic logic in this volume, for it has unnecessary limitations which have been overcome in the last one hundred years by the development of a more complete system of predicate logic, which we will study instead. Anything which syllogistic logic elucidates can also be brought out by predicate logic.

It is not difficult to see the limitations of categorical syllogistic logic, since it deals with only the four types of proposition listed above (including, however, any sentence which can somehow be rephrased so as to express a proposition belonging to one of these four types).

We have been seeing how some interesting logical structure might be developed by working with "∀", "∃", and subject-predicate structure. Such structure will be the subject matter for the rest of this chapter.

Propositional logic, syllogisms, and first-order predicate logic

49.1 Write a new argument which is clearly valid, but which cannot be proved valid using symbolic propositional logic. Why can your argument not be proved valid with propositional logic?

49.2 Using notational devices introduced in this Section, diagram the following sentences:
   a. Every college student wants to get rich.
   b. No college student wants to get rich.
   c. There are college students who want to get rich.
   d. Some college students do not want to get rich.
   e. All male college students want to get rich.
   f. All male college students are male.

49.3 What is the main difference between predicate logic and propositional logic?

50

Single quantifier symbolization

In pursuing the development of predicate logic, we will follow the same strategy as that followed in the development of propositional logic. First, we will look at individual propositions expressed in English and talk about how to diagram them. This task will be aided and legitimized by reference to Venn diagrams and small test universes in the following Sections, playing something of the same explanatory role as that of truth tables in the discussion of symbolization within propositional logic. Of course, *argument* diagrams will consist of strings of diagrams of individual propositions, as before. Once you can construct argument diagrams in predicate logic, you will be given some new rules of inference to use in connection with these diagrams. Interestingly, it takes only four additional basic rules of inference to complete the system.

You were already informally introduced in the previous Section to the predicate logic symbolization of individual propositions. Our first job is to continue that introduction more formally.

Let's start with the simplest cases. We know we want to be able to diagram subject-predicate structure; so we will need some symbols for subjects and others for predicates. I'll use single small letters of the alphabet for subjects and single capital letters for predicates. Thus, if "f" stands for the subject named by "Fred", and "P" stands for the predicate concept expressed by "is a pharmacist", we can diagram the proposition

Fred is a pharmacist

by writing

Pf

Note that the order in which the subject and predicate expressions appear in the diagram reverses that in which they appear in English. The idea is simple: think of the predicate as *applying to* the subject. Thus, the predicate is given first; then the thing to which it applies.[1]

We will continue to use single capital letters to stand for whole propositions, as well as for predicate concepts. You will be able to tell the difference between these uses in two ways: (1) If a capital letter is followed by a small letter, the capital letter stands for a predicate.   (2) If a capital letter has parentheses after it—like this: 'P( )'—to indicate a slot where a small letter can appear, then again the capital letter stands for a predicate. Otherwise, a single capital letter stands for a whole proposition. Single capital letters standing for predicates will be called *predicate letters*.

We can still use the connectives from propositional logic. Thus, if we want to symbolize the thought that Fred is *not* a pharmacist, we can use the negation sign:

~Pf

---

[1]Historically, this way of writing the diagram came about because symbolic logic copied the techniques of writing used in mathematics, where one finds notation like 'Fx', referring to some function of the variable 'x'.

The negation sign here applies to 'Pf', not merely to 'P', since it is 'Pf' which expresses a whole proposition in this case, and "~" so far applies only to expressions for whole propositions.

Similarly, if we want to say that Fred is a pharmacist and also a joker, we could write something like this:

Pf & Jf

Again, "&" is used to connect expressions for two whole propositions, as before.

Note the repeated use of 'f' above to signify Fred. This is not strictly necessary. We could write

Pf & Jg

letting both 'f' and 'g' stand for Fred. But just as in propositional logic, it is generally more revealing to use repeated letters where possible, to display logical connections. Whenever the same letter, whether capital or small, is used within one argument or sentence diagram, it is understood to refer to the same thing throughout. Thus,

Pf & Pf

does not properly symbolize "Fred is a pharmacist and a joker", since one may not use "P( )" for the concept of being a joker and also for the concept of being a pharmacist in one diagram.

Small letters will be called *names,* unless they come from the very end of the alphabet: small 'x', 'y', and 'z' will be called *variables* instead, and will have a use somewhat distinct from that assigned to names. The difference in function between a name and a variable roughly approximates the difference between a proper name, like "John Johnson", and a pronoun, like "he", in English—the proper name always stands for the same thing (at least in a given discussion), while the reference of the pronoun can shift from one thing to another. More exactly, a name refers to just one thing in a given context, like the expression "2" does in mathematics, while a variable ranges over a whole collection of things, like 'x' does in algebra, where it ranges over a whole collection of numbers, and can take on any one of them as its value.

When do we use variables instead of names? In predicate logic, if we want to write a diagram for a proposition, we use variables in place of names *only* when we are also using "∀" or "∃", the quantifiers. Suppose, for example, that we want to say that *every* pharmacist is a joker. Then it will not do to write

∀ (Pf & Jf)

since this would say something like all Fred pharmacists are Fred jokers, or maybe all Fred is a joker pharmacist. We need to eliminate Fred from the symbolization, and replace his name with a more general symbol which can range over all the pharmacists rather than referring to just one of them. That will be the function of a *variable.* Suppose we use 'x' for the purpose. We want to say that all x's that are pharmacists are also jokers. That seems just like saying that all x's are such that if an x is a pharmacist *then* that x is a joker. This yields the following diagram:

∀x(Px ⊃ Jx)

The above diagram is correct. Note the three occurrences of 'x' in it. The parentheses indicate that the quantifier applies to the whole formula 'Px ⊃ Jx'.

The "∀" is known as the *universal quantifier*.

Similarly, if we wanted to diagram the proposition that every joker is a pharmacist, we could write

∀x(Jx ⊃ Px)

(Compare this diagram to the previous one to be sure you see the difference.) Or, if we want to diagram the proposition that there is a joker who is also a pharmacist, we could write

∃x(Jx & Px)

Read this formula as follows:

Some x is such that it is a joker and a pharmacist.

It would have been wrong to write

∃x(Jx ⊃ Px)

above, since *this* formula says roughly some x is such that *if* it is a joker, *then* it is a pharmacist. In other words, this latter diagram does not capture what we want because it does not say there actually is a joker or that there actually is a pharmacist.

Let's consider a bit more just what it means to say, "Some x is such that it is a joker and a pharmacist", which was the interpretation of the formula, '∃x(Jx & Px)'. First, note that in order for some x to be both a joker and a pharmacist, it is necessary for that x to exist and for that existing x to be both a joker and a pharmacist. So when we use "∃" as a quantifier, we are committing ourselves to the existence of something. That's why we can use the quantifier to say *there is* a joker who is also a pharmacist. This quantifier, accordingly, is known as the *existential quantifier*.

Secondly, when we say that some x is such that it is both a joker and a pharmacist, we are not saying that it is the *only* such x. There may be others. Therefore, when we write '∃x(Jx & Px)', we are saying that there is *at least one* x such that it is a joker pharmacist. (We are not, however, saying that there definitely is more than one.)

If I say in ordinary English, "Some jokers are pharmacists", I clearly imply there is *more than one*. This sentence is therefore different from the English sentence, "Some joker is a pharmacist", which the above formula fits well. Since our formula does not say there is more than one joker pharmacist, it is not quite an accurate diagram for the sentence which implies there is more than one. Thus, if there being more than one has importance for the argument in which the sentence appears, it will be necessary to modify the formula given above in some way so as to make it clear that there is more than one. (See Section 56.) We generally won't bother to worry about that, since it usually doesn't make any difference for the validity of the arguments we will deal with. Nevertheless, there are ways to ensure that the formula does actually say that there is more than one such x, if we wish it to.

Also, if I say in ordinary English, "Some jokers are pharmacists", I probably mean to imply that *not all* jokers are pharmacists. Again, we find this sentence differing in meaning from the sentence, "Some joker is a pharmacist" which our formula accurately diagrams, for this latter sentence does not give a clue about whether there are jokers who are not pharmacists. Here again, our formula corresponds to the proposition expressed by the latter sentence rather than the former one, since our formula does not imply that there are jokers who are not pharmacists. In other words, our formula would be true so long as at least one joker is a pharmacist, even if all jokers are pharmacists. In this regard, then, our formula does not correspond well to the usual meaning of the sentence "Some jokers are pharmacists". I will not normally worry about this, and will thus use the formula to symbolize this English sentence, as well as the sentence which the formula more accurately symbolizes. However, if it is important in a given argument, you can always add the thought that not all jokers are pharmacists to '∃x(Jx & Px)' as follows:

∃x(Jx & Px) & ~∀x(Jx ⊃ Px)

To help keep all this straight, it is useful to think of the existential quantifier as saying something equivalent to "some", but a bit more elaborately spelled out, namely, to think of it as saying "there exists at least one". On this reading of the quantifier, '∃x(Jx & Px)' clearly says that there exists at least one x such that it is both a joker and a pharmacist, as the preceding paragraphs have pointed out. If it then becomes necessary to construct a formula which says more than this, there might be ways to add to this formula whatever else needs to be said.

Another technical point may cause some worry for some people. You may have noticed that I have suddenly started using our old connectives to connect expressions which do not strictly speaking express propositions—expressions containing variables, like 'Jx' and 'Px'. Although neither 'Px' nor 'Jx' by itself expresses a proposition, because 'x' is a variable rather than a name, both 'Px' and 'Jx' would become expressions for propositions if a name referring to something were substituted for the variable. That is, substitution of 'f' for 'x' above would yield expressions of propositions about Fred. This makes 'Px' and 'Jx' close enough to expressing propositions that it will be sensible to allow the propositional connectives, "&", "⊃", "~", etc., to apply to them. In general, I will allow the connectives between expressions containing variables provided that these expressions would turn into propositional expressions when the variables are replaced by names which refer to things. The regular truth table definitions of the connectives will apply once the variables have all been replaced by names in an expression like 'Jx & Px', or 'Px ⊃ Fx', or '~Px'.

An expression like 'Jx & Px', containing occurrences of variables which are *not* governed by any quantifiers, are not propositional diagrams. (Such an expression amounts to much the same thing as saying in English, "It's a joker and a pharmacist", when nothing in the context gives any clue about who or what is being talked about.) In order to convert such an expression into a diagram of a proposition, all the ungoverned occurrences of variables must be replaced by names or else must be made to be governed by a quantifier. Thus

∃xPx & Jx

does not diagram a proposition, since the third occurrence of 'x' in it is not governed by any quantifier. (The leading quantifier governs only up to "&".) This situation may be remedied in various ways. Here is a list of them, along with what each says:

∃x(Px & Jx).   There is at least one pharmacist joker.

∃xPx & Jf.   There is at least one pharmacist and Fred is a joker.

∃xPx & ∃xJx.   There is at least one pharmacist and there is at least
                       one joker (not necessarily the same people). The
                       repeated use of 'x' does not mean that the same thing
                       is being referred to each time. In this way 'x' and the
                       other variables differ from names.)

None of above is inherently more correct than any of the others. Choose the one which says the appropriate thing for the context in which you are working.

Let's look at some more examples. Examine each of the following carefully, paying special attention to parentheses:

There are no pharmacists.	∼∃xPx
There are no joker pharmacists.	∼∃x(Px & Jx) or ∀x(Px ⊃ ∼Jx)
No pharmacists are jokers.	∀x(Px ⊃ ∼Jx) or ∼∃x(Px & Jx)
If there are pharmacists, then jokers exist too.	∃xPx ⊃ ∃xJx
If Fred is a joker, then there are joking pharmacists.	Jf ⊃ ∃x(Jx & Px)
Fred is a joker and there are pharmacists who are jokers.	Jf & ∃x(Px & Jx)
If all pharmacists are jokers, then Fred is a joker.	∀x(Px ⊃ Jx) ⊃ Jf
Everything is a pharmacist!	∀xPx

Do *not* write diagrams like this one:

∃x∃x(Px ⊃ Jx)

The problem here lies in the fact that there is no way to tell which quantifier goes with which later occurrence of 'x'. (Or maybe the first quantifier is just superfluous.) Any occurrence of a quantifier will always be followed immediately by an occurrence of a variable, and that same variable must occur later on in the formula, governed by that quantifier. Thus,

∃xPy

makes no sense, and is "illegal". And if there are two quantifiers in one propositional diagram, they must either govern different variables, such as they do in

∃y∃x(Px & Jy)

or they must not overlap in the parts of the formula they govern, as is illustrated in the nonoverlapping formula

∃xPx ⊃ ∃xJx

We need to look a bit at diagrams beginning with more than one quantifier, such as

∃y∃x(Px & Jy)

In translation, this formula says there is at least one y and there is at least one x such that x is a pharmacist and y is a joker. Since 'x' and 'y' could refer to different individuals, this amounts to saying

∃xPx & ∃yJy   or   ∃xPx & ∃xJx

Similarly, '∀x∀y(Px & Jy)' says in translation, that everything is a pharmacist and everything is a joker. One could just as well have written '∀xPx & ∀yJy'.

But things are not always so simple. Some multiple quantifiers cannot be so neatly separated from one another. I leave the discussion of those cases to Section 56.

The part of a diagram governed by a quantifier is said to be in the *scope* of that quantifier. Thus, in the present Section we restrict our attention to propositions which can be diagramed without having two quantifiers overlap in scope.

As a summary, it may be helpful if I list some of the more common English expressions along with their usual diagrams. (But be careful, since the usual diagram may not always work. Be aware of the meaning you want to capture in your diagram.)

English format	Usual correct diagram	Common mistaken diagram
All A's are B's.	∀x(Ax ⊃ Bx)	∀x(Ax & Bx)
Some A's are B's.	∃x(Ax & Bx)	∃x(Ax ⊃ Bx)
No A's are B's.	~∃x(Ax & Bx) or ∀x(Ax ⊃ ~Bx)	∃x(Ax & ~Bx) ~∀x(Ax ⊃ Bx)
Some A's are not B's.	∃x(Ax & ~Bx)	~∃x(Ax & Bx)
All A's are B's and C's.	∀x(Ax ⊃ (Bx & Cx))	∀x(Ax & (Bx & Cx))
Some A's are B's and C's.	∃x(Ax & Bx & Cx)	∃x(Ax ⊃ (Bx & Cx))
If all A's are B's, then c is a B.	∀x(Ax ⊃ Bx) ⊃ Bc	∀x(Ax ⊃ (Bx & Bc))
Everything that is an A or B is a C.	∀x((Ax ∨ Bx) ⊃ Cx)	
All A's except for B's are C's.	∀x((Ax & ~Bx) ⊃ Cx)	
(Stronger:)	∀x(Ax ⊃ (~Bx ≡ Cx))	

If you ever need more names or variables than there are available letters of the alphabet, use numeral subscripts: 'x_1', 'x_2', 'x_3', etc., will all be variables, and 'f_1', 'f_2', 'f_3', etc., will all be names.

Single quantifier symbolization

50.1 Pick out from among the following strings of symbols just those which could symbolize a proposition correctly. Explain what's wrong with the others.

a. ∀xPx ⊃ Gx	g. Pf & Gy	m. ∃xPx & ∀xGy
b. ∀x(Px ⊃ Gx)	h. Pf & Gc	n. (Pf & Gy) ⊃ ∀xFx
c. ∀xPx ⊃ ∃xGy	i. Fx & Gc	o. ∀x(Px ⊃ (Gx ∨ Qx))
d. ∀x∀yPx	j. ∃x(Px & Gy)	p. K ⊃ ∀xPx
e. ∀x∀yPx & Gy	k. ∃x(Px & Gx)	q. ∀xK
f. ∀x∀y(Px & Gy)	l. ∃x(Px & ∀xGx)	r. ∀x(K ⊃ Px)

50.2 Symbolize the following propositions in as much detail as possible, using the following translation scheme:

j: Joe
k: Karen

S( ): ( ) is a supervisor
E( ): ( ) is an employee
H( ): ( ) is hard-working
L( ): ( ) is lazy

a. All supervisors are hard working.
b. All supervisors are hard working but Joe is lazy.
c. If Joe is a supervisor then he is an employee.
d. There are employees who are not supervisors.
e. Some supervisors are lazy, but Karen is hard working.
f. All lazy employees are supervisors.
g. No lazy employee is a supervisor.
h. If Karen is a supervisor, Joe is an employee who works hard.
i. Although Joe and Karen are supervisors, they are both lazy employees.
j. No lazy employees are hard-working supervisors.
k. If all supervisors are lazy, the Karen is not a supervisor.
l. Unless she's a lazy supervisor, Karen is hard working.
m. Only hard-working employees are supervisors.
n. All the employees except the supervisors are lazy.
o. The only employees who are lazy are the supervisors.
p. None of the lazy employees are supervisors.

50.3 Provide your own translation schemes and symbolize in as much detail as possible the structure of the following propositions:
a. Everyone who came to the party was a bore.
b. None of the large cats leads an active life for more than two hours a day.
c. If you're a lazy overpaid supervisor, you're vulnerable to being fired.
d. The only countries which give voting privileges to women are large.
e. All my books except the math texts were purchased at Zimbo's or at Klein's.
f. Someone ate supper here last night.
g. All the king's horses and all the king's men couldn't put Humpty together again.

     h. The United States contains fifty states.

     i. If Zimbo's is the largest store in town, everyone shops there sooner or later.

     j. Sometimes tornadoes touch down in town, but some of them only affect the countryside.

50.4 We now have symbols to represent certain quantifiers occuring in English, such as "all", "every", "some", and so on. However, certain other English quantifiers do not seem to have any adequate symbolic notation as yet within our system. For example, "many" is more than "some" but is not as many as "all", and has no adequate symbolic representation. List five other such quantifiers in English. (This begins to reveal some of the limitations of even our new symbolization techniques.)

51

Venn diagrams

Constructing a truth table for a diagram in propositional logic revealed in what sorts of worlds propositions fitting the diagram were true, and this knowledge provided a check of the correctness of a given diagraming attempt. We want to do something like truth tables now for predicate logic, since the symbolization process and the rules of inference become clearer when you know more about the conditions under which propositions fitting various diagrams are true.

Our technique for systematically laying out the truth conditions for predicate logic formulas is the invention of an English mathematician, John Venn (1834–1923). This technique works for many proposition diagrams which begin with one quantifier and contain no more than three predicate letters. With such severe limitations, it is obviously not applicable to the full range of predicate logic diagrams.

The key idea in Venn's technique is to use a circular disk to represent the group of all things to which a given predicate correctly applies—for example, the group or class of all things to which "bird" correctly applies is the group or class of birds. You are to use a different disk for each predicate appearing in a given diagram, and you will draw the disks overlapping one another so as to allow for the possibility that some things belong to more than one. Thus, to apply Venn's idea to '$\forall x(Px \supset Jx)$', we first need two overlapping disks, one to represent the class of things that are P's and the other the class of J's:

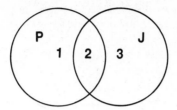

All things in region 1 of the P disk are things that are P's, but not J's. Things in region 2 are in both disks, and are thus P's and J's. Things in region 3 are J's but are not P's.

In order to make the above Venn diagram represent the truth conditions for '∀x(Px ⊃ Jx)' you will shade in a region to indicate that region is empty. Since '∀x(Px ⊃ Jx)' says that it is true for all things that they are J's if they are P's, region 1 must be empty, according to '∀x(Px ⊃ Jx)'. So,

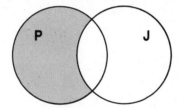

is a picture of what '∀x(Px ⊃ Jx)' says.

Aside from shading in regions which are empty, you will also need to place a small letter in regions which you know for sure are not empty. This does not affect the Venn diagram above, since '∀x(Px ⊃ Jx)' does *not* say there are some P's or that there are some J's. It only says that any P's which do happen to exist are also J's. Think of this English sentence, which might be symbolized by '∀x(Px ⊃ Jx)': "Everyone who studies will get at least a 'B' ". (Let 'P( )' stand for "is someone who studies", and let 'J( )' stand for "is someone who will get at least a 'B' ".) This sentence does not say there actually are studiers, or people who will get A's or B's. But other sentences do assert the existence of things, and for them you need to place the small letters on the appropriate disks.

An example of such a diagram is '∃x(Px & Jx)', which needs the following Venn diagram:

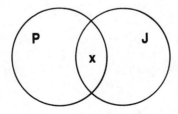

Since we don't know from '∃x(Px & Jx)' whether any regions are empty, no shading is required, and since we don't know the name of the thing in the middle region, we use an 'x'.

Some Venn diagrams are not very interesting. For example, the diagram for '∃xPx' looks like this:

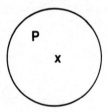

When you can tell from the proposition diagram that something exists, but you can't tell for sure which disk it should go on, put the same small letter in *each* of the possible regions where the existing thing might be, and draw a bar between each of these small letters to indicate that they go together as alternatives. Thus, '∃x(Px ∨ Rx)' gets the picture

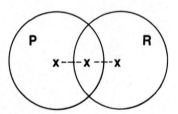

because the existing things could be in any of the three regions, and you know there is something in at least one of the regions.

When you know from the proposition's diagram that there is a particular named thing in one of the regions of a Venn diagram use the name instead of a variable. Thus 'Fa' gets the diagram

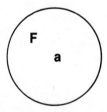

as does '∃xFx & Fa'. (In the case of the latter formula, you should not put both 'x' and 'a' in the 'F' disk, because you don't know from what the formula says whether 'a' names the *only* existing thing in the 'F' disk.) Similarly, 'Fa ∨ Ga' is diagramed by

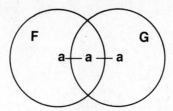

For completeness, it is sometimes necessary to enclose the disks in a box which represents the entire set of things under discussion:

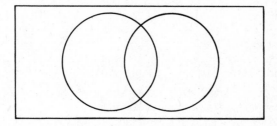

I usually leave out the box unless it is needed. Here's a case where it's needed. Diagram '∀xPx':

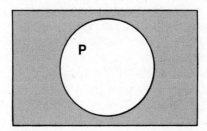

When there are three predicates involved, draw the three disks so as to allow for pictures of all possible combinations of empty and occupied regions:

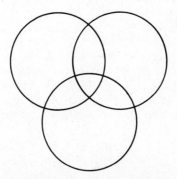

Accordingly, to picture '∀x(Px ⊃ (Rx & Qx))' write

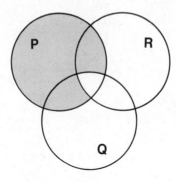

to show that every P thing must lie in the central region which is also part of the R and Q disks.

It is possible to construct Venn diagrams containing four disks, but these generally are too difficult to work with to be useful.

Our first main application of Venn diagrams will be for checking the symbolization of an English sentence for accuracy. The Venn diagram for the proposition expressed by a given English sentence and that for a proposed symbolization of that sentence must match if the proposed symbolization is to be considered correct. (This point parallels the claim made earlier regarding how the truth table for a proposition must match the table for its correct symbolization.)

To illustrate this test of correctness, try '∀x(Px ⊃ Rx)' as a symbolization of "A teacher is supposed to know his subject", letting 'P( )' stand for "is a teacher" and 'R( )' stand for "is one who is supposed to know his subject". The Venn diagram for the original English would look like diagram I below, while that for the formula would look like diagram II.

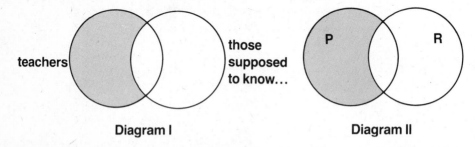

Diagram I   Diagram II

Since these diagrams match, the proposed symbolization is correct. On the other hand, if we use a similar symbolization for "A teacher is standing over there", things don't check out: (Let 'R( )' now represent "is standing over there".) The Venn diagram for '∀x(Px ⊃ Rx)' remains as above, Diagram II. But the Venn diagram for the English sentence looks like this:

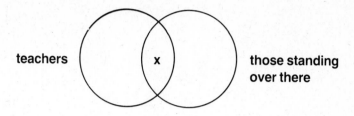

teachers   x   those standing
                         over there

Thus, the proposed symbolization in this latter case is wrong.

Unfortunately, some formulas and sentences may be too complex for you to diagram until you've had more experience, and some formulas such as '∃xPx ⊃ ∃xRx', containing two or three predicate letters, cannot be diagramed using just the Venn techniques so far discussed. Nevertheless, when you can do the diagram, it might be quite useful.

We have been looking at Venn diagrams for individual propositions. However, in much the same way that one may construct truth tables both for individual propositions and for whole arguments, Venn diagrams may sometimes be constructed for whole arguments as a test of validity.

When constructing one Venn diagram to handle a whole argument or argument diagram, begin by scanning the argument or argument diagram to see that in the entire argument or argument diagram there are no more than three different predicates or predicate letters. (If there are more than three, the Venn diagram cannot be completed.) Then draw the Venn diagram so that it contains a disk for each predicate (or predicate letter) which appears anywhere in the argument or argument diagram. Next, picture the first premiss in the arrangement of disks, using the usual techniques for dealing with individual sentences. Then, utilizing the same set of disks, try to picture the second premiss without erasing anything put into the picture from the first premiss. Continue in this same fashion for all remaining premisses. If you can picture all of them on one set of disks, the result will be a composite picture of what all the premisses, taken together, say. One then asks whether this picture guarantees the truth of the conclusion, on the same set of disks. If the conclusion has, so to speak, already been pictured once the composite picture of the premisses is completed, the argument or argument diagram is thereby proved valid. Otherwise, the Venn diagram leaves it open whether the argument is valid or not.

Let's look at an example in symbols. Let the argument diagram be

Pa
∀x(Px ⊃ Rx)
∀x(Rx ⊃ Gx)
Ga

Since there are three predicate letters in the argument diagram, we set up the Venn diagram with three disks and proceed to picture the first premiss:

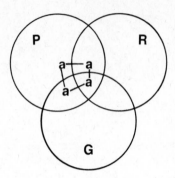

Adding the picture of the second premiss yields the following:

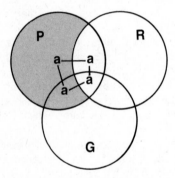

It is all right that two of our 'a' 's got eliminated, since they did not indicate definite inhabitants in those regions; and two of the 'a' 's tied to those eliminated still live on. In other words, we have not done anything to make the first premiss false. We have, however, narrowed down the places where the fellow named 'a' might be.

Now we put into the picture the third premiss, yielding this result:

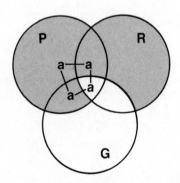

We lost another of our 'a' 's, but once again, that is no problem because there's still one left which was an alternate to those we've lost.

At this point, since all the premisses are pictured, we have a Venn diagram which tells us exactly what the sum of the premisses commits us to. Looking at the picture, we see that if the premisses are all true: (1) 'a' names an individual who is a P, a G, and an R; (2) there may or may not be things that are not P's but are both G's and R's; (3) there may or may not be things that are G's without being P's or R's.

Checking the composite diagram we see that it guarantees the truth of the conclusion. That is, if the composite Venn diagram of the premisses represents the real world, then we can tell from the diagram that the conclusion would be true, because in the composite diagram we see that 'a' must name something which is on the G disk. In other words, once we have a picture of what the world must be like to make the premisses true, we see that picture guarantees the conclusion is also true. Thus, we have proved the argument diagram valid. The same procedure could be used for picturing the argument expressed in English.

On the other hand, look at the composite Venn diagram for the following argument's premisses:

Only the brave deserve the fair.
Laura is brave.
_____
Laura deserves the fair.

Composite premiss diagram:

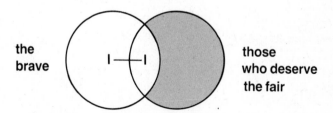

This diagram does *not* guarantee the truth of the conclusion, since Laura may be one of the brave who does not deserve the fair (represented by the leftmost 'j'). Thus, the diagram fails to prove the argument valid.

Caution: When a Venn diagram fails to prove an argument valid, do *not* conclude that the Venn diagram has proved the argument *invalid*. The situation here is exactly analogous to that involving truth tables. A failure to prove an argument valid does not thereby prove the argument invalid, for there might be redeeming features to the argument not captured by the diagram. However, recall that when a truth table for a symbolic argument *diagram* contained a bad line, that proved the *diagram* invalid (which meant that some but not all English arguments fitting the diagram are invalid). Similarly, a Venn diagram can prove a predicate logic argument *diagram* invalid, which means that some but not all English arguments fitting the diagram will be invalid.

One complication sometimes arises in connection with trying to produce a composite Venn diagram. Occasionally, the premisses contradict one another. In this case, the argument in question is automatically valid, but there will not be a composite Venn diagram. This happens in the following case:

∀x(Px ⊃ Rx)
Pa
~Ra
_____
∃xGx

The first two premisses may be pictured in the following way, but the third premiss does not fit with the picture.

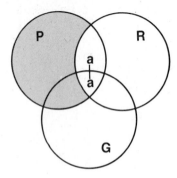

Thus, the argument diagram is valid no matter what the conclusion.

_____

Venn diagrams

51.1 Construct Venn diagrams to picture each of the following.
   a. ∃x(Px & Rx)   d. ∃x~Px   g. ∃x(Px & (Qx ∨ Rx)
   b. ∃x(Rx ∨ Qx ∨ Fx)   e. ∀x(Px & Rx) ⊃ Qx)   h. ∀x(Px ⊃ ~Rx)
   c. ~∃xPx   f. ∀x((Px ∨ Rx) ⊃ Qx)   i. ~∃x(Px & Rx)

51.2 Test each proposed symbolization below by comparing its Venn diagram with that of the English statement. State which symbolizations pass the test. When the diagrams for the English and the symbolization differ, display both. Use the following translation key:

T( ): is a teacher
R( ): is a responsible person
S( ): is a student

English statement	Proposed symbolization
a. No teacher is a responsible person.	$\forall x(Tx \supset \sim Rx)$
b. No teacher is a responsible person.	$\forall x(Rx \supset \sim Tx)$
c. No teacher is a responsible person.	$\sim \exists x(Rx \ \& \ Tx)$
d. Teachers and students are responsible people.	$\forall x((Tx \ \& \ Sx) \supset Rx)$
e. Teachers and students are responsible people.	$\forall x((Tx \lor Sx) \supset Rx)$
f. Some responsible people are both teachers and students.	$\exists x(Rx \ \& \ Tx \ \& \ Sx)$
g. Some responsible people are both teachers and students.	$\exists x(Rx \ \& \ (Tx \lor Sx))$
h. None but teachers are responsible persons.	$\forall x(Tx \supset Rx)$

51.3 Not all well-formed proposition diagrams beginning with one quantifier and containing three or fewer predicate letters have Venn diagrams. Construct two examples of such diagrams.

51.4 Test the following argument diagrams for validity, using the Venn technique.

a. $\exists x(Fx \ \& \ Gx)$
$\underline{\exists x(Gx \ \& \ Hx)}$
$\exists x(Fx \ \& \ Hx)$

b. $\forall x(Fx \supset Gx)$
$\underline{\exists xFx}$
$\exists xGx$

c. $\forall x((Fx \ \& \ Gx) \supset Hx)$
$\underline{\forall x(Fx \supset Gx)}$
$\forall x(Fx \supset Hx)$

d. $\underline{\exists xFx \ \& \ Fa}$
$\exists yFy$

e. $\forall x((Fx \lor Gx) \supset Hx)$
$\underline{Fa}$
Ha

f. $\forall x((Fx \lor Gx) \supset Hx)$
$\underline{\sim Fa \ \& \ \sim Ga}$
$\sim Ha$

g. $\underline{\exists xFx}$
Fa

h. $\exists xFx$
$\underline{\exists xGx}$
$\forall x(Fx \supset Gx)$

i. $\forall x(Fx \ \& \ Gx)$
$\overline{\sim \exists x(\sim Fx \lor \sim Gx)}$

j. $\forall x(Fx \ \& \ Gx)$
$\underline{\exists x \sim Fx}$
Ha

51.5 Choose one of the *invalid* diagrams from the preceding exercise and construct a *valid* argument in English which fits that diagram.

52

Existential import and domains of discourse

I have so far kept in the background two important matters affecting predicate logic symbolization. First, there is a problem that many, but by no means all, English sentences of the general type 'All A's are B's' and similar types such as 'All A's or B's are C's' have. They express propositions having what is known to logicians as "existential import".

A proposition of the type 'All A's are B's' has existential import when it implies that there actually do exist some A's. A proposition of the type 'All A's or B's are C's' has existential import if it implies there are things which are A's or B's. And so on, for similar types of propositions.

Most people seem to think the proposition normally expressed by "Everyone who came to the party had a good time" has existential import, for they would perhaps believe they had been lied to if they were later told no one at all showed up for the party. But "All trespassers will be shot" on the other hand seems to lack existential import, for it does not seem to imply there are or will be any trespassers.

This affects symbolization because '$\forall$x(Ax $\supset$ Bx)', for example, does *not ever* have existential import. Its Venn diagram *always* looks like this,

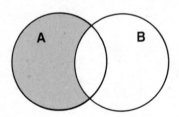

never like this:

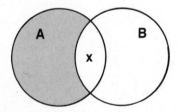

The Venn diagram for "Everyone who came had a good time", though, would probably look like this:

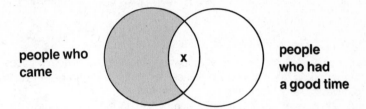

people who came   x   **people who had a good time**

If so, '∀x(Px ⊃ Gx)' cannot adequately symbolize the sentence.

There is an easy solution to this difficulty. When you want to include existential import in your symbolization of one of these sentences, do so by adding the existence claim explicitly. For our sentence about party-goers you could write

∀x(Px ⊃ Gx) & ∃xPx

to include the idea that there was at least one person at the party. This formula has the correct Venn diagram. (Check it out.) For our sentence about trespassers, however, you would presumably only write

∀x(Tx ⊃ Sx)

(In the above connection, predicate logic is superior to traditional categorical syllogistic logic, since the rules of traditional syllogistic logic assume existential import for *all* propositions of the form 'All A's are B's' and thus cannot deal adequately with the trespasser case.)

The second matter regarding symbolization which I wish to discuss in this Section has to do with *domains of discourse*. A domain of discourse is the range of objects, persons, or other entities which it is understood we are talking about in a particular section of discourse. For instance, when two friends are talking over the party they gave last weekend, they may well enter into discourse about the various guests. Within that discourse, one of them might truthfully say "Everyone had a good time". The domain of discourse for that discussion will naturally be the set of their party guests. The word, "everyone" is then understood to apply to that set only. If, however, we were strangely to insist on a larger domain of discourse, say, all people in the United States, then the word "everyone" would have to apply to this much larger set. Presumably, with the larger domain understood as the domain of discourse, the original statement becomes false, since it is false that everyone in the United States had a good time at the party.

Letting 'G( )' mean "had a good time (at the party)", we can symbolize "Everyone had a good time" as '∀xGx', *provided that* the understood domain of discourse is the set of party guests. We will be said in such a case to be "quantifying over" the set of party guests, since the universal quantifier, "∀", applies to all the party guests and to no one else.

However, if we change the understood domain of discourse, we will then be quantifying over a new set, and possibly '∀xGx' will no longer accurately represent the original thought. Letting the domain be all people in the world, '∀xGx' claims every one of those people had a good time at the party! In order to fix up the symbolization, if we want to insist on using this larger domain, we need to add a new predicate. Let 'P( )' mean "was a guest at the party". Then, we can write

$$\forall x(Px \supset Gx)$$

to accurately symbolize the original thought, using the larger domain consisting of all the people in the whole world. Once 'Px' is added it no longer hurts anything to have the larger domain.

Thus, whether or not a particular symbolization turns out to be correct will in many cases depend on the domain of discourse which is used in interpreting the symbolization. '∀xGx' was fine above so long as the domain was just the party guests, but when the domain was enlarged, '∀x(Px ⊃ Gx)' had to be used to get correct results.

How do you know what domain to use? This is largely a matter of practicality. If you are symbolizing propositions drawn from an extended section of discourse, choose a domain large enough to include all the things and people talked about in that section of discourse. This becomes crucial when you symbolize an entire argument, since the domain of discourse *must* be held constant for the entire argument. You may not switch domains between the various premisses or for the conclusion. First select a large enough domain; then do the symbolizing relative to that domain.

Existential import and domains of discourse

52.1 Every Venn diagram is always understood to have a box around it specifying the domain of discourse, even though we often do not actually draw the box. However, for this exercise, you are to always draw the box, and label it to indicate what domain of discourse is being used. The purpose of this exercise is to let you discover how changing the domain of discourse sometimes, but not always, changes the Venn diagram for a particular proposition. This should help you to see how domains of discourse function. Let 'P( )' and 'G( )' be as above.
   a. Draw a Venn diagram for "Everyone had a good time", assuming the domain consists of those who attended the party as guests. (Label the domain and the inner disk.)
   b. Draw the diagram again, for '∀xGx' (not for '∀x(Px ⊃ Gx)'), assuming the domain consists of all the people in the world.
   c. Draw it again, for '∀x(Px ⊃ Gx)', assuming the same domain.
   d. Draw it again, for '∀x(Px ⊃ Gx)', assuming the domain consists of all the people and all the animals in the world. Does it make any difference if we enlarge the domain to include animals?
52.2 What would happen if you choose as your domain all the males in the world, when you are symbolizing "Everyone had a good time", referring to the party goers? Is this a good choice of domain if you don't know of what

sex the party attenders were? What does this show about what might happen if you choose a domain that does not include all the relevant individuals?

52.3 For each statement below, two different domains of discourse are listed. For each, you are first to symbolize the statement in as much detail as possible, assuming the first domain of discourse; and then you are to symbolize it again, assuming the second domain. Provide translation keys for all your predicate letters.

   a. Context you are to assume: The speaker is referring to the men and women who work in a certain factory.

   Statement to be symbolized: The men are all suspicious of the women.

   First domain: All people throughout the world.

   Second domain: All the men and women who work in the factory in question.

   b. Context you are to assume: The speaker is referring to the people living in a certain community.

   Statement to be symbolized: There are some poor lawyers living around here.

   First domain: All lawyers.

   Second domain: All people living in the relevant community.

   c. Context you are to assume: The speaker is referring to the items in a package he just received in the mail.

   Statement to be symbolized: Everything got broken except for the handle.

   First domain: Items in the package.

   Second domain: All the objects in the whole world.

52.4a. If you were symbolizing an argument which had the two statements given below as premisses, what would be the smallest domain you could assume when symbolizing the first premiss?

   P1: Every student who came to the concert went home satisfied.
   P2: Some people who attended the concert were not satisfied.

   b. In what way would your answer above be affected if you knew the conclusion of the argument was "C: The concert was not a complete success"?

52.5 Consider the following argument:

   Everyone who played in the game was tall.
   Everyone in the world is tall.

   a. Is the above argument valid?
   b. Do you think the following argument diagram is valid? Explain.

   $\forall x T x$
   _____
   $\forall x T x$

     c. Let 'T( )' stand for "is tall". Assume the domain of discourse consists of all those who played in the game mentioned in the argument. Symbolize the premiss of the argument.

     d. Assume the domain consists of all the people in the world. Symbolize the conclusion of the above argument, using 'T( )' for "is tall".

     e. What do your answers to the above questions reveal about what can happen if you don't stick to using just one domain when symbolizing an argument?

52.6 Symbolize each of the following sentences twice, once with and once without assuming they have existential import. Give translation keys for all your predicate letters. State which sentences are likely to be uttered in contexts which would make it reasonable to suppose the propositions they express actually do have existential import.

     a. Everyone in this room should get out immediately.

     b. No one is allowed to smoke here.

     c. Anyone who falls behind in their payments will have to pay their entire debt immediately.

     d. The trees over there are dying.

     e. Anyone caught in the lava flow is dead.

     f. Everyone who studies hard in this course will get an "A".

53

∀ elimination and ∃ introduction

Although we certainly haven't exhausted the capability of predicate logic to incorporate detail into diagrams, you may be interested to see how one can use what has been presented so far to produce some valid argument diagrams. In this Section, I will present two new rules of inference which directly apply to our new quantifiers, and I will discuss how to modify our old rules to make them applicable to formulas which employ the symbolism of predicate logic.

First the modification of the old rules. When the rules of propositional logic were presented in Chapter 7, you were told that the 'X', 'Y', 'Z', etc., appearing in the rule descriptions could represent compound propositions. Accordingly, the rule & E, described as follows,

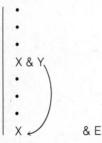

can be used when you wish to eliminate the "&" from even a compound formula such as 'P & (Q ⊃ R)'. In order to apply & E to our new predicate logic formulas, the same understanding still applies: 'X', 'Y', 'Z', etc., in the rule descriptions may stand for complex propositions. Thus, & E may now be used to eliminate "&" from the formula '∀xFx & ∃xGx'. The move might look like this:

```
        •
        •
        •
12. |   ∀xFx & ∃xGx
    |   •
    |   •
    |   •
17. |   ∃xGx            12, & E
```

Of course, the same sort of thing holds for *all* the rules of propositional logic. There is just one crucial thing to watch out for: the 'X', 'Y', and 'Z' in the rules of propositional logic always refer to *whole propositions,* and *never* to pieces of propositions. I emphasize this restricted application to whole propositions because it is easy to make the following kind of mistake:

```
        •
        •
        •
4.  |   ∀x(Gx & Fx)
    |   •
    |   •
    |   •
7.  |   ∀xGx            4, & E, mistakenly used
```

Even though the above step happens to be valid, it is *not* an appropriate application of & E, because 'Gx' and 'Fx' are not by themselves expressions for whole propositions, since they contain a variable without also containing a quantifier governing that variable.

You may wonder what harm can be done by applying the rules of propositional logic in the mistaken way illustrated above. The answer: if you use propositional logic rules on formulas which express mere pieces of propositions, you will *sometimes* produce invalid argument diagrams, even though in many cases the diagrams will turn out valid, as already shown. Here is an example where the diagram is invalid:

```
1. |   ∃xGx
2. |   ∀x(Gx ⊃ Fx)
3. |   ∀xFx            1,2, ⊃ E, mistakenly used
```

Given the importance of this restriction on the application of propositional logic rules in predicate logic, it might be worthwhile to briefly list some predicate logic expressions which do and do not express whole propositions:

Formulas which *do* express whole propositions
(when provided with an adequate translation)

Fa	∀x(Gx ⊃ Fx)	P ⊃ ∀xGx	∀xGx
Fa & Ga	∃x(Gx & Fx)	∀x(P ⊃ Gx)	∃x(Fx & P)
Fa ⊃ P	P & ∃x(Gx & Fx)	P & G	

Formulas which *do not* express whole propositions

Fx	Fx ⊃ P	Fx & P	∀x(Fy & Gx)
Fx & Gx	P ⊃ Gx	Fa & Gx	∃x(Fy & Gx)

Now that you know how to use rules from propositional logic within predicate logic, you can consider them part of predicate logic. You will then need only four more basic rules to complete the rule system for first-order predicate logic: ∀ introduction, ∀ elimination, ∃ introduction, and ∃ elimination. ("∀I", "∀E", "∃I", and "∃E" for short.) I will consider only ∀E and ∃I in this Section.

There are three important points we must agree to; otherwise the rules will not be valid. The first should come as no surprise: we must assume the same domain of discourse throughout any given derivation. The second point, however, may be a bit of a shocker: we must assume all the names which appear in our formulas refer to things within the relevant domain. That is, we cannot tolerate names which fail to refer to anything. Such names must never be used in symbolizing a proposition.[1] The third point, again somewhat troublesome, is that we must assume the domain contains at least one member. That is, the domain must not be completely empty.[2] (The domain may, however, consist entirely of fictional objects, such as elves or gremlins.)

Once the above restrictions are understood, the first new rule, ∀E, will probably seem very natural. Suppose you knew that '∀xFx' is true of a domain containing just four members, 'a', 'b,' 'c,' and 'd'. Clearly, you could validly infer 'Fa', 'Fb', 'Fc', and 'Fd', since '∀xFx' tells you that the concept signified by 'F( )' applies to every member of the domain. This is the basic idea behind ∀E.

Writing the above moves in our standard format yields

1.	∀xFx	
2.	Fa	1,∀E
3.	Fb	1,∀E
	etc.	

Of course, there will be more complex cases:

1.	∀x(Fx ⊃ Gx)	
2.	Fa ⊃ Ga	1,∀E

[1]It is possible to modify the rules in controversial ways to eliminate this restriction. The problem is complicated because there is not universal agreement about whether 'Fa' even has a truth value if 'a' does not refer to anything. Imagine someone who is an only child stating seriously and literally "My brother lives in Tennessee". Did he say something false?

[2]Presumably, it is possible to do away with this restriction, but it would complicate matters, and it isn't necessary. See N. Rescher, *Introduction to Logic* (New York: St. Martin's Press, 1964), Section 14.5 for an attempt. A logic which eliminates this restriction is called a *free logic*, because it is free from existence assumptions.

But all the cases are basically alike: you eliminate the leading quantifier, "∀", along with the variable immediately following it, and then substitute a name for that variable thoughout the remaining scope of the eliminated quantifier. (Recall that names are small letters, not including 'x', 'y', or 'z'.)

The rule will apply only to whole lines, just as almost all of our old rules did. Thus,

1. | P ⊃ ∀xFx
   |_____
2. | P ⊃ Fa   1,∀E misapplied

will not be legal.

Naturally, the rule ∀E doesn't apply to the existential quantifier, "∃". In general, it would not be valid to infer 'Fa' from '∃xFx', since "∃xFx' only tells you that at least one thing in the domain is an F, without telling you *which* thing is F. 'Fa' claims that it is the particular thing named by 'a' which is F. Thus, 'Fa' is too specific to be validly inferred from '∃xFx'. So, ∀E is only valid for "∀", not for "∃". The proper rule for existential quantifier elimination is more complex, and will be taken up later.

How do you know which name to use when applying ∀E? You are allowed to use any name you wish, so long as it names something in the domain. Generally, as a matter of practical strategy, you will choose a name which hooks up somehow with the rest of your argument to give you a desired conclusion. For example, to construct a derivation for the valid diagram of the famous valid argument

All humans are mortal.
Socrates is human.
_____
Socrates is mortal.

follow the following steps:

1. | ∀x(Hx ⊃ Mx)
2. | Hs
   |_____
3. | Hs ⊃ Ms   1,∀E
4. | Ms   2, 3, ⊃ E

Note that 's' was the name chosen in line 3, to make line 3 connect with line 2. If a different name had been used in line 3, it would not have been possible to get line 4 by ⊃ E. The rule of ∀E does not tell you which name to substitute for the variable, but it allows you to choose any name you wish (so long as that name refers to something in the domain).

Formally, the ∀E rule may be written as follows:

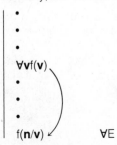

∀vf(v)

f(n/v)   ∀E

Here '**v**' stands for any variable, and 'f(**v**)' indicates a formula entirely within the scope of the universal quantifier, containing the variable, '**v**'; '**n**' stands for any name which refers to something in the domain, and 'f(**n/v**)' is the result of re-placing all the occurences of the variable '**v**' in formula 'f' with the name '**n**'— without altering formula 'f' in any other way. Thus, in applying this rule to '∀x(Fx ⊃ Gx)', '**v**' stands for 'x', 'f(**v**)' stands for 'Fx ⊃ Gx', and 'f(**n/v**)' would stand for 'Fa ⊃ Ga' if **n** stands for 'a', since you get 'Fa ⊃ Ga' from 'Fx ⊃ Gx' by replacing 'x' with 'a'. Read 'f(**n/v**)' as "The result of replacing '**v**' by '**n**' in 'f' ".

We can check this rule's validity by Venn diagraming. Treating 'f' as though it were a predicate letter, we get the following Venn diagram for the rule's prem-

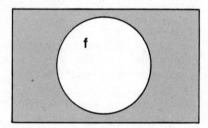

This diagram does not guarantee true the conclusion given by the rule *unless* we recall our background assumption that the name '**n**' stands for *something* in the domain.

With that background assumption built into the Venn diagram, we get the following picture of the rule's premiss plus background assumption;

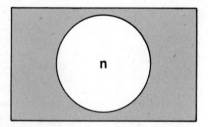

since there is no other place to put the thing which the name '**n**' refers to. Hence the conclusion of the rule is guaranteed true by the diagram after all, and the rule ∀E is justified.

The existential quantifier introduction rule, ∃I, is also fairly straightforward. One way to prove '∃xFx' is true to know first that 'Fa' is true. Since all names used in a derivation must refer to something in the domain, 'Fa' implies '∃xFx'. That will be our rule ∃I. Generalizing, we have

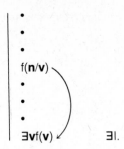

∃I.

It is not hard to see this process is exactly the opposite of ∀E, except that the bottom line begins with "∃" instead of "∀". In particular, 'f(**n/v**)' is related to 'f(**v**)' in exactly the same way as in the ∀E rule.

The validity of this rule may be established with a Venn diagram if we treat 'f' as a predicate letter. We then get the diagram

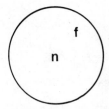

for the premiss in the rule, and this obviously establishes the truth of '∃vf(**v**)', since the truth of '∃vf(**v**)' is established once anything at all is in the 'f' disk.

No temporary assumptions are needed for either rule, but the rules ∀E and ∃I may be used within subderivations just as the rules of propositional logic could be used.

The following derivation illustrates both our new rules and also brings out how our understanding regarding the nonemptiness of the domain of discourse is built into the very fabric of our system.

1. | ∀xFx
2. | Fa   1,∀E
3. | ∃xFx   2,∃I

There is one twist to the working of ∀E and ∃I which you probably will not have noticed. When you use ∀E, *all* the occurrences of '**v**' disappear, being replaced by occurrences of '**n**' in the conclusion you obtain. But when you use ∃I, it is *not* necessary that all the occurrences of '**n**' disappear in the conclusion you obtain. Accordingly,

1. | Fa & Ga
2. | ∃x(Fx & Ga)   1,∃I

is a correct derivation. It *does fit* the format of the rule ∃I, since 1 here can be obtained by replacing all the occurrences of 'x' in 'Fx & Ga' by occurrences of 'a'. (Verify the validity of the above derivation for yourself with a Venn diagram.)

Examine carefully the following examples, looking for strategies you can use in constructing your own derivations.

1. | ∀xGx ⊃ Fa
2. | Hc ⊃ ∀xGx
3. | ∀x(Bx & Hx)
4. | Bc & Hc   3, ∀E (Why did I choose 'c'?)
5. | Hc   4, & E
6. | ∀xGx   2, 5, ⊃ E
7. | Fa   1, 6, ⊃ E
8. | ∃xFx   7, ∃I

1. | ∃x(Fx & Gb) ⊃ P
2. | Fa
3. | ∀yGy
4. | Gb   3, ∀E (Why choose 'b' here?)
5. | Fa & Gb   2, 4, & I (Getting ready for ∃I in line 6.)
6. | ∃x(Fx & Gb)   5, ∃I
7. | P   6, 1, ⊃ E

1. | (∃xFx ∨ ∃xGx) ≡ ∀x(Gx & Hx)
2. | Fa
3. | ∃xFx   2, ∃I (Why did I do this?)
4. | ∃xFx ∨ ∃xGx   3, ∨ I
5. | ∀x(Gx & Hx)   4, 1, ≡ E
6. | Gb & Hb   5, ∀E (Any other name could have been used just as well.)

1. | ~Ga   (The point of this one is to show how to get '~∀xGx' by RA.)
2. | | ∀xGx   (Getting set up to use RA.)
3. | | Ga   2, ∀E
4. | | ~Ga   1, R
5. | ~∀xGx   2-3 & 4, RA

Here are some *incorrect* moves utilizing the same premiss sets as above. Make sure you see why each is in fact a violation of the rules as formulated.

1. | ∀xGx ⊃ Fa
2. | Hc ⊃ ∀xGx
3. | ∀x(Bx & Hx)
4. | ∀xHx   3, & E, incorrectly applied
5. | Hc   4, ∀E

etc., as above in the first sample derivation.

```
1. │ ∃x(Fx & Gx) ⊃ P
2. │ Fa
3. │ ∀yGy
   ├─────────────────
4. │ Gb                   3, ∀E
5. │ Fa & Gb              2, 4, & I
6. │ ∃x(Fx & Gx)         5, ∃I, incorrectly applied
7. │ P                    1, 6, ⊃ E
```

```
1. │ (∃xFx ∨ ∃xGx) ≡ ∀x(Gx & Hx)
2. │ Fa
   ├─────────────────
3. │ Fa ∨ Ga             2, ∨ I
4. │ ∃xFx ∨ ∃xGx        3, ∃I, incorrectly applied
5. │ ∀x(Gx & Hx)         1, 4, ≡ E
6. │ Gb                   5, ∀E, incorrectly applied
```

It would also have been incorrect to write 'Gb & Hc' as line 6.

53.1 Write justifications for each nonpremiss line in the following derivations.

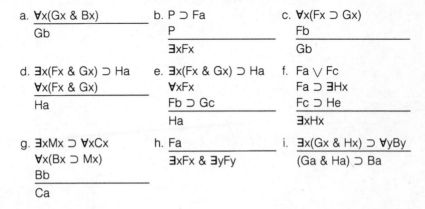

a. 1. │ (∃xFx ∨ ∀xGx) ⊃ Ha   b. 1. │ ∀x(Gx ⊃ Hx)
   2. │ ~∃xFx ⊃ Jb   2. │ ∃xHx ⊃ ∀xHx
   3. │ ~Jb   3. │ │ Ga
      ├──────────────   │ ├──────────
   4. │ │ ~∃xFx   4. │ │ Ga ⊃ Ha
   5. │ │ Jb   5. │ │ Ha
   6. │ │ ~Jb   6. │ │ ∃xHx
   7. │ ∃xFx   7. │ │ ∀xHx
   8. │ ∃xFx ∨ ∀xGy   8. │ Ga ⊃ ∀xHx
   9. │ Ha

53.2 Derive the given conclusions from the given premises.

a. ∀x(Gx & Bx)   b. P ⊃ Fa   c. ∀x(Fx ⊃ Gx)
   ─────────────   P   Fb
   Gb   ─────────   ───────
                         ∃xFx   Gb

d. ∃x(Fx & Gx) ⊃ Ha   e. ∃x(Fx & Gx) ⊃ Ha   f. Fa ∨ Fc
   ∀x(Fx & Gx)   ∀xFx   Fa ⊃ ∃Hx
   ───────────   Fb ⊃ Gc   Fc ⊃ He
   Ha   ─────────   ────────
                         Ha   ∃xHx

g. ∃xMx ⊃ ∀xCx   h. Fa   i. ∃x(Gx & Hx) ⊃ ∀yBy
   ∀x(Bx ⊃ Mx)   ──────────   ──────────────────────
   Bb   ∃xFx & ∃yFy   (Ga & Ha) ⊃ Ba
   ───────────
   Ca
```
```

j. ∃xDx ⊃ ∀x(Fx ⊃ Gx)   k. ∀x((Gx & Hx) ⊃ Cx)   l. ∀x(Hx ⊃ ~Gx)
    Fa   ∀x(Bx ⊃ (Gx & Hx))   Bc & Hc
    P ⊃ Dc   ~Ca   ∃xCx ⊃ Gc
    ――――――――   ――――――――   ――――――――
    P ⊃ Ga   ~Ba   ~∃xCx

53.3 Explain why the following could be a derived rule in our system.

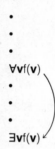

53.4 If we allowed names to fail to refer to anything, why would ∀E and ∃I fail
     to be reasonable rules of inference?

53.5 Suppose we were to allow empty domains. For instance, we might allow
     the domain to be the set of all human beings now alive who are over 200
     years old. Someone might then say, speaking of that domain, that everyone
     is blue-eyed. In symbols, using that domain, we get '∀xBx', for "Everyone
     is blue-eyed". Is this formula true? False? How do you tell? What does this
     show about the problems that would arise if we allowed empty domains?

53.6 Explain for each of the following formulas why it is not the sort that could
     symbolize a whole proposition.
     a. ∀x(Px ⊃ Gy)
     b. Bx & Fc

53.7 Verify the validity of 53.2a using a Venn diagram accompanied by an ex-
     planation. Which other argument diagrams in 53.2 may be proved valid by
     means of the Venn technique?

54

Universal quantifier introduction

To have a rule of universal quantifier introduction is to have a procedure for
validly deriving conclusions such as "All males are arrogant"—conclusions
which begin with a universal quantifier. What sort of rule is likely to do that job?
    To catch the scent of the rule we are going to use, begin by noticing that a

universally quantified formula diagrams propositions which are "about" the whole domain of discourse. "All males are arrogant", for example, might be symbolized by '∀x(Mx ⊃ Ax)' which says that all the things (or persons) in the domain are such that if they are male then they are arrogant. So, even if the domain contains females or tables, the proposition is still in some sense about all the entities in the domain, since it says of all these entities that if they are male then they are arrogant. This suggests that the way to derive something like '∀x(Mx ⊃ Ax)' would be to show first that 'Ma ⊃ Aa' is true, then 'Mb ⊃ Ab', then 'Mc ⊃ Ac', and so on, until *every* member of the domain had been considered and found to fit the pattern.

This procedure would be appropriate for small domains. However, some domains are infinitely large (such as the positive integers) and others are very large (such as the set of stars). We need a rule which will be practical even for large domains; so listing all the members of the domain one by one is out.

Instead of listing all the members of the domain and investigating each one separately, we will arbitrarily choose just one member and let it represent all the others. Thus, if we let our domain (for simplicity) be the set of males, and we wish to derive the conclusion that all men are arrogant, we will arbitrarily choose one of the males in the domain and check to see if it is true that he is arrogant. If it turns out that our arbitrarily chosen representative is arrogant, we will conclude that all men are arrogant.

You are probably thinking to yourself right now that the procedure just outlined sounds invalid. In fact it seems to be a paradigm example of hasty generalization. How can one validly conclude something about all men merely by examining one of them?

The rebuttal to the foregoing objection lies in specifying what will count as *arbitrarily* choosing one member of the domain for investigation. Suppose, for example, that you have the following (fictitious) information about men available for use as premises in your derivation:

P1. All men are born with a certain genetic structure called the Zerlof strand.   (∀xZx)
P2. Anyone with the Zerlof strand in their genetic structure is an arrogant person.   (∀x(Zx ⊃ Ax))

Now, in order to arrive at the conclusion that all men are arrogant (∀xAx), one might try to select Sam (named 's') for study, letting him represent the whole domain:

1.	∀xZx	
2.	∀x(Zx ⊃ Ax)	
3.	Zs	1, ∀E
4.	Zs ⊃ As	2, ∀E
5.	As	3, 4, ⊃ E

Note that from these same premises we could have derived 'Aj' (about Joe), or 'Aa', 'Ab', and so on, about *any* of the members of the domain, using *exactly* the same pattern of argument as that used above, except for using the appropriate name in place of 's'.

Under the conditions just outlined, Sam, Joe, or any other member of the domain could serve as an arbitrarily chosen member of the domain, representing all the other members. So the key to what counts as an arbitrary choice is this: *The choice is arbitrary only when the same general strategy of argument would have been possible using any of the other members of the domain.* It is therefore valid to complete the above derivation by adding line 6 at the bottom, which is the desired conclusion that all men are arrogant:

1.	∀xZx	
2.	∀x(Zx ⊃ Ax)	
3.	Zs	1, ∀E
4.	Zs ⊃ As	2, ∀E
5.	As	3, 4, ⊃ E
6.	∀xAx	5, ∀I (new rule)

The validity of the above diagram is verified by the following composite Venn diagram of the premisses:

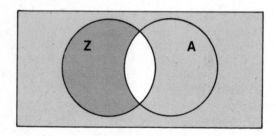

Clearly, this diagram guarantees the truth of '∀xAx', and thus the argument diagram with premisses 1 and 2 and conclusion 6 above is valid.

Our rule of ∀I thus turns out to be relatively simple:

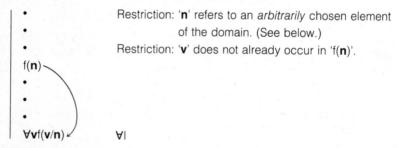

As in previous rules, 'f', 'n', and 'v' refer respectively to formulas, names, and variables. 'f(v/n)' stands for the formula you get by replacing all the occurrences of 'n' in 'f(n)' with occurrences of 'v'; the name 'n' thus disappears from the conclusion you obtain.

Exactly how can you tell when an element of the domain can be an arbitrarily chosen element? That, too, turns out to be simple: *At any given point in a derivation, a name 'n' refers to an arbitrarily chosen member of the domain* when and only when that name does not occur in any of the premisses which are still being assumed true at that point in the derivation. Recall that whenever a premiss (or a group of premisses) makes its first appearance in a derivation, a vertical line is begun immediately to its left. Until that vertical line ends, the premiss is still being assumed true.

Examples:

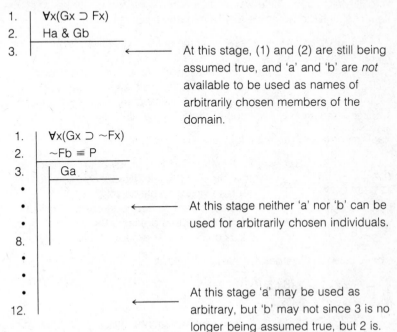

1. | ∀x(Gx ⊃ Fx)
2. | Ha & Gb
3. | ←——— At this stage, (1) and (2) are still being assumed true, and 'a' and 'b' are *not* available to be used as names of arbitrarily chosen members of the domain.

1. | ∀x(Gx ⊃ ~Fx)
2. | ~Fb ≡ P
3. | | Ga
   •
   •   ←——— At this stage neither 'a' nor 'b' can be used for arbitrarily chosen individuals.
   •
8. |
   •
   •
   •   ←——— At this stage 'a' may be used as
12. | arbitrary, but 'b' may not since 3 is no longer being assumed true, but 2 is.

The first restriction on ∀I prevents us from constructing invalid derivations like this one:

Invalid: 1. | Fa
          2. | ∀xFx   1,∀I without the first restriction

If the above were valid, it would be valid to argue that since I am poor, everyone is poor; or since Anne is rich, everyone is rich. The first restriction makes sure we don't use a name in ∀I which refers to some member of the domain about which we have specific information over and above the information we have about all the members of the domain.

The second restriction on ∀I ensures that the conclusion will be a well-formed expression for a proposition. Without the restriction, something like the following could happen:

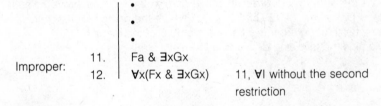

          •
          •
          •
          11. | Fa & ∃xGx
Improper:
          12. | ∀x(Fx & ∃xGx)   11, ∀I without the second restriction

You should recognize 12 as being improperly formed since it contains '∃x' within the scope of '∀x'. Formulas like '∀x(∃xGx & Fx)', '∃y(Fy & ∀yGy)', '∃y∃yFy', and '∀x∀x(Fx ⊃ Gx)' all have this same problem of overlapping scope, and none can be used to express a proposition.

So both restrictions on ∀I are quite important and must be carefully observed.

To illustrate in a simple way some of the power which our newly emerging system contains, consider the diagram for the following argument.

All the policies of the present government are directed toward the goal of preserving a stable society and a good business climate. But any policy which is directed toward preservation of stability in society automatically also tends to keep the less well-off groups in their disadvantaged position. Obviously, then, the present governmental policies tend to preserve the disadvantaged status of the poorer groups in our society.

A diagram for this argument might look like this:

∀x(Px ⊃ (Sx & Bx))	'P( )' stands for "is a policy of the present
∀x(Sx ⊃ Dx)	government"; 'S( )' for "is directed toward the goal
∀x(Px ⊃ Dx)	of preserving a stable society"; 'B( )' for "is
	directed toward . . . a good business climate";
	'D( )' for "tends to preserve the disadvantaged
	status of the poorer groups in our society".

The conclusion can be derived easily as follows:

1.	∀x(Px ⊃ (Sx & Bx))		
2.	∀x(Sx ⊃ Dx)		
3.	Pa	(Let 'a' signify a policy of the government.)	
4.	Pa ⊃ (Sa & Ba)	1, ∀E	(Apply 1 to that policy named 'a'.)
5.	Sa & Ba	3, 4, ⊃ E	
6.	Sa	5, & E	(That policy promotes stability.)
7.	Sa ⊃ Da	2, ∀E	(Apply 2 to that policy.)
8.	Da	6, 7, ⊃ E	(That policy tends to preserve the disadvantaged status.)
9.	Pa ⊃ Da	3-8, ⊃ I	
10.	∀x(Px ⊃ Dx)	9, ∀I	(Generalize to all policies of the government.)

Note the strategy. In order to prove '∀x(Px ⊃ Dx)', first prove something like 'Pa ⊃ Da' or 'Pb ⊃ Db', or 'Pc ⊃ Dc', where the name which occurs is one that can stand for an arbitrarily chosen member of the domain.

Examine the following derivations for useful strategies:

1.	$\forall x(Fx \supset Gx)$		
2.	$\forall xFx$		
3.	Fa	2, $\forall$E	('a' will name our arbitrarily chosen element.)
4.	Fa $\supset$ Ga	1, $\forall$E	(Apply line 1 to 'a'.)
5.	Ga	3, 4, $\supset$ E	
6.	$\forall xGx$	5, $\forall$I	(Since 'a' does not appear in 1 or 2 this is legitimate.)
7.	$\forall xFx \supset \forall xGx$	2-6, $\supset$ I	

1.	$\forall xGx \lor \forall xHx$	(This will be a $\lor$ E type of problem.)	
2.	$\forall xGx$	(Getting set up to prove '$\forall xGx \supset \forall x(Gx \lor Hx)$'.)	
3.	Ga	2, $\forall$E	('a' will name an arbitrarily chosen element.)
4.	Ga $\lor$ Ha	3, $\lor$ I	
5.	$\forall x(Gx \lor Hx)$	4, $\forall$I	(Note that 'a' disappears completely at this point.)
6.	$\forall xGx \supset \forall x(Gx \lor Hx)$	2-5, $\supset$ I	
7.	$\forall xHx$	(Getting set up to prove '$\forall xHx \supset \forall x(Gx \lor Hx)$'.)	
8.	Ha	7, $\forall$E	('a' will again name an arbitrarily chosen individual, because 1 and 7 do not contain 'a'.)
9.	Ga $\lor$ Ha	8, $\lor$ I	
10.	$\forall x(Gx \lor Hx)$	9, $\forall$I	
11.	$\forall xHx \supset \forall x(Gx \lor Hx)$	7-10, $\supset$ I	
12.	$\forall x(Gx \lor Hx) \lor \forall x(Gx \lor Hx)$	1, 6, 11, $\lor$ E	
13.	$\forall x(Gx \lor Hx)$	12, R $\lor$ E	

The following derivation illustrates a very important relationship between '∀' and '∃'.

```
1.  │  ~∃xFx
2.  │  │  Fa
3.  │  │  ∃xFx          2, ∃I
4.  │  │  ~∃xFx         1, R
5.  │  ~Fa             2-3 & 4, RA
6.  │  ∀x~Fx           5, ∀I
```

The justification for line 6 is important. 'a' counts as referring to an arbitrarily chosen individual, even though 'a' appears in 2, because 2 is no longer being assumed true in line 5. This technicality makes sense if you notice that we could just as well have used 'b' throughout the derivation in place of 'a', which is the rationale behind the idea of the arbitrarily chosen individual.

Given the above derivation, you will perhaps be able to see that the following *derived rule* is acceptable:

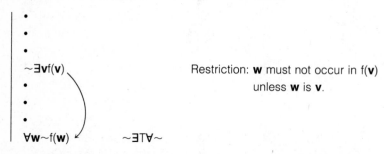

Restriction: **w** must not occur in f(**v**) unless **w** is **v**.

Analogously with ∨T& and similar rules from propositional logic, I name the above rule '~∃T∀~', since it allows you to convert formulas beginning with '~∃' into formulas beginning with '∀w~'. Of course, the rule applies no matter how complex 'f(**v**)' might be.

Conversion between '∀' and '∃' in the opposite direction can also be justified, as illustrated in the following derivation:

```
1.  │  ~∀xFx
2.  │  │  ~∃y~Fy           (Getting set up to derive '∃y~Fy' by RA.)
3.  │  │  │  ~Fa           (Getting set up to derive '∀xFx' by RA, for
                            above RA.)
4.  │  │  │  ∃y~Fy         3, ∃I
5.  │  │  │  ~∃y~Fy        2, R
6.  │  │  Fa              3-4 & 5, RA
7.  │  │  ∀xFx            6, ∀I (Legitimate because 3 has been
                            discharged.)
8.  │  │  ~∀xFx           1, R
9.  │  ∃y~Fy              2-7 & 8, RA
```

In other words, if you know not everything is fun, you thereby also know there is at least one thing which is not fun.

The derived rule which corresponds to the above maneuver, and which we can add to the system to avoid having to go through the above steps, looks like this:

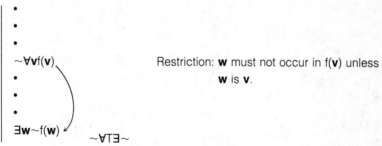

Restriction: **w** must not occur in f(**v**) unless **w** is **v**.

Note that both of these derived rules allow one to use a variable in the bottom line differing from that appearing behind the leading quantifier in the upper line, subject to the restriction mentioned. The reason for the restriction is to insure the bottom line is well formed. Without the restriction, the following could happen:

1. | ~∀x(Fx & ∃yGy)
2. | ∃y~(Fy & ∃yGy)   1, ~∀T∃~, without restriction

Line 2 above is not well formed since it contains overlapping scopes of two quantifiers each governing 'y'.

One final note about derived rules: so far, I have used the rules of propositional logic only in application to whole lines, just as in propositional logic. But what of ~~E and ~~I, which were set up to apply within lines too? They can continue to apply within lines. Thus, expressions such as '∀x(Fx ⊃ ~~Gx)' may be changed to '∀x(Fx ⊃ Gx)' in one step, by application of ~~E.

This has special importance now, since ~∀T∃~ generates '∃w~~f(w)' from '~∀v~f(v)', and ~∃T∀~ generates '∀w~~f(w)' from '~∃v~f(v)'. Application of ~~E to these results indicates that by ~∀T∃~ we get (in two steps)

'∃wf(w)' from '~∀v~f(v)'

if 'w' meets the restriction, and by ~∃T∀~ we get (in two steps)

'∀wf(w)' from '~∃v~f(v)'

through deletion of the double negations introduced by ~∀T∃~ and ~∃T∀~. The relationships between '∀' and '∃' will be explored further in the next Section.

Universal quantifier introduction

54.1 Pick out those of the following short derivations which are correctly done. Pick out the errors in those incorrectly done.

a. 1.	Fa		b. 1.	∀x(Fx & Gx)	
2.	∀xFx	1, ∀I	2.	Fa & Ga	1, ∀E
			3.	Fa	2, & E
			4.	∀xFx	3, ∀I

c. 1.	∀x(Fx & Ga)		d. 1.	Pb & ~∃xFx	
2.	Fa & Ga	1, ∀E	2.	~∃xFx	1, & E
3.	Fa	2, & E	3.	∀x~Fx	2, ~∃T∀~
4.	∀xFx	3, ∀I	4.	~Fb	3, ∀E

e. 1.	Pb & ~∀xFx		f. 1.	~∀xFx	
2.	~∀xFx	1, & E	2.	~∃xFx	1, ~∀T∃~
3.	~Fb	2, ∀E	3.	∀x~Fx	2, ~∃T∀~
4.	∀x~Fx	3, ∀I			

54.2 Derive each indicated conclusion from the given premisses.

a. ∀x(Fx ⊃ Gx)
   ∀x(Gx ⊃ ~Hx)
   —————————
   ∀x(Fx ⊃ ~Hx)

b. ∀x((Fx & Gx) ⊃ Hx)
   ∀x(Bx ⊃ (Fx & Gx))
   —————————
   ∀x(Bx ⊃ Hx)

c. Fa ⊃ ∀xGx
   ∀x(Gx ⊃ Hx)
   Fa
   —————————
   Hb

d. ∀x(Gx ⊃ ~Fx)
   ∀x(Hx ⊃ Gx)
   —————————
   ∀x(Fx ⊃ ~Hx)

54.3 Prove the following arguments have valid symbolic approximation diagrams. Assume the domain is the set of persons. Use the following translation key:

C( ): ( ) is a candidate
T( ): ( ) has promised to lower
       taxes
W( ): ( ) has promised not to
       curtail government
       services

F( ): ( ) is a fool
L( ): ( ) is a liar
D( ): ( ) deserves to be elected

a. All the candidates have promised to lower taxes without curtailing government services. Anyone who promises to lower taxes without curtailing government services is a fool or a liar. No one who is a fool or a liar deserves to be elected. So, no candidate deserves to be elected. (Hint: Use formulas beginning with '∀' for the last premiss and the conclusion.)

b. Only those candidates who promise to lower taxes deserve to be elected. But any candidate who promises to lower taxes is a liar. So, those candidates who deserve to be elected are liars.
(Hint: Use a formula beginning with '∀' for the first premiss.)

c. The only candidates who deserve to be elected are those who promise to lower taxes. However, no liar deserves to be elected. Unfortunately,

if a candidate promises to lower taxes, he's always a liar. Hence, none of the candidates deserves to be elected.
(Hint: All the formulas should begin with '∀'.)

d. Those candidates who promised to lower taxes are not liars, while those who promised not to curtail government services are. So, anyone who's not a liar either isn't a candidate, or hasn't promised not to cut back on government services.

e. Even though some of them are fools, a candidate deserves to be elected only if he is neither a liar nor a fool. George is a candidate who deserves to be elected. So, George is no fool.

54.4 Derive each indicated conclusion from the given premisses.

a. $\underline{\sim\forall x(Fx \;\&\; Gx)}$
 $\exists x\sim(Fx \;\&\; Gx)$

b. $\forall xFx \supset G$
 $\underline{\sim G}$
 $\exists x\sim Fx$

c. $\forall x(Fx \supset Gx) \supset \exists yBy$
 $\underline{\forall x\sim Bx}$
 $\exists x\sim(Fx \supset Gx)$

d. $\underline{\sim\exists x(Fx \;\&\; Gx)}$
 $\forall x\sim(Fx \;\&\; Gx)$

e. $\underline{\sim\exists x(Fx \;\&\; Gx)}$
 $\sim Fa \lor \sim Ga$

f. $\forall xHx \supset Fa$
 $\sim\exists x(Fx \;\&\; Gx)$
 $\underline{\sim Ga \supset \sim\forall xHx}$
 $\exists x\sim Hx$

54.5 Symbolize each of the following and derive the conclusions from the premiss sets. Specify your translation schemes and domains.

a. All students pay tuition, and all young people are relatively poor. So, if all those who pay tuition are young people, then all students are relatively poor.

b. Once some part of the code is broken, then all parts of the code are broken. The way in which proper names are handled is part of the code, and this part of the code has been broken. Another part of the code is the way numbers are handled. If that part of the code has been broken, the enemy knows how many troops we have available. Thus, the enemy knows how many of our troops are available.

c. It is not true that everything we do is determined by our genetic constitution. But if there are things we do which are not determined by our genetic constitution, then human reasoning has a role to play in choice-making. However, human reasoning has such a role only if choice-making is done deliberately at least sometimes. Therefore, human choice-making is at least sometimes done deliberately.

54.6 Derive the indicated conclusions from the given premiss sets. Construct an argument in English fitting each diagram.

a. ~∃x(Fx & Gx)
   _____
   ∀x((Fx & Gx) ⊃ Hx)

b. ~∃x~Fx
   _____
   ∀xFx

c. ∀xFx ∨ ∀xGx
   _____
   ∀x(Fx ∨ Gx)

d. ∀y(Py ⊃ (Qy & Sy))
   _____
   ∀y(Py ⊃ Qy)

e. ∀x(Mx ⊃ [∀y(Ny ⊃ Sy) ⊃ Ux])
   ∀x(Ux ⊃ [∀y(Ny ⊃ Fy) ⊃ Wx])
   _____
   ∀x(Nx ⊃ (Fx & Sx)) ⊃ ∀x(Mx ⊃ Wx)

55

Existential quantifier elimination

Our final basic rule, the existential quantifier elimination rule, or "∃E" for short, quite naturally applies to formulas which begin with "∃". However, it is not immediately obvious how such a rule should be formulated. If one knows, for instance, that there is a thief or two amongst the bank tellers, symbolized '∃xTx' (using the bank tellers as domain), one does not thereby also have grounds for concluding that any one particular teller is a thief, 'Ta' for instance. That is, the move

| ∃xTx
|‾‾‾‾‾
| Ta

is not valid, and ∃E cannot work like ∀E.

In order to give you something more to work with, let's also assume that all thieving bank tellers have personal debts to pay off, ∀x(Tx ⊃ Dx). Now you have grounds for validly concluding that there actually are bank tellers who have personal debts to pay off—that is, '∃xDx' validly follows from '∃xTx' and '∀x(Tx ⊃ Dx)', for even though you don't know who the thieves are, you do know that they have personal debts, whoever they are, and you do know there are some thieves. Our ∃E rule, in conjunction with the other rules we already have, should allow the derivation of '∃xDx' from the given premises. (Even though the desired conclusion begins with "∃", ∃E remains crucial because eliminating the

"∃" from '∃xTx' is crucial, since we need to get inside '∃xTx' to be able to hook up with '∀x(Tx ⊃ Dx)'. We will first eliminate "∃" from '∃xTx' and then later introduce "∃" into '∃xDx'.

Our purpose can be validly and sensibly accomplished by using the idea of an arbitrarily chosen individual again. Since '∃xTx' tells us some teller steals, but not which one, we can arbitrarily choose one to represent the thieves whoever they may be. Let's just arbitrarily suppose the teller named 'a' is a thief. Then, applying '∀x(Tx ⊃ Dx)' to 'a', we quickly find out 'Da'. But our conclusion here depends on our arbitrary assumption 'Ta'. So we could write the following derivation:

1. | ∃xTx
2. | ∀x(Tx ⊃ Dx)
3. | | Ta
4. | | Ta ⊃ Da   2, ∀E
5. | | Da   3, 4, ⊃ E
6. | | ∃xDx   5, ∃I

Up to this point, by assuming 'a' names a thief, we have concluded that 'a' also names someone with personal debts, and thus that there is a teller with personal debts. But a glance at the derivation, or a moment's thought about our logic so far, should reveal to you that we could have argued in exactly the same way if we had decided a different teller was a thief. Just substitute that teller's name for 'a' throughout, and you'll get it. Thus, 'a' names an arbitrarily chosen individual here in exactly the same sense in which individuals were arbitrarily chosen in the previous Section.

Note that the conclusion '∃xDx' is exactly what we wanted to derive. However, we've had to use an extra assumption to get it; so we're not yet finished. The extra assumption must be discharged, and that is where the arbitrariness of 'a' comes in, along with the first premiss.

The first premiss, line 1, tells us 'Ta ∨ Tb ∨ Tc ∨. . . .'. The subderivation beginning with line 3 tells us if 'Ta' is true then '∃xDx' is true. But since 'a' names an arbitrarily chosen individual, the subderivation beginning with line 3 can also reveal that if 'Tb' is true '∃xDx' is still true, and if 'Tc' is true '∃xDx' is still true, and so on for all formulas of the form 'T**n**'. Since we know from line 1 that 'Ta ∨ Tb ∨ Tc ∨ . . .' is true, and we can tell that *each* alternative 'Ta', 'Tb', 'Tc', etc., leads to the *same* result, '∃xDx', we know '∃xDx' is true no matter what name we use in 3. Hence 3 can be discharged, for line 1 by itself tells us that something of the form 'T**n**' is true.

The completed derivation looks like this:

1. | ∃xTx
2. | ∀x(Tx ⊃ Dx)
3. | | Ta
4. | | Ta ⊃ Da   2, ∀E
5. | | Da   3, 4, ⊃ E
6. | | ∃xDx   5, ∃I
7. | ∃xDx   1, 3-6, ∃E (new rule)

It is absolutely crucial that line 1 be available as a premiss for getting line 7. Without line 1 being assumed true, there would be no reason at all to discharge 3.

Thus, this "derivation" is incorrect:

1. | ∀x(Tx ⊃ Dx)
2. | | Ta
3. | | Ta ⊃ Da   1, ∀E
4. | | Da   2, 3, ⊃ E
5. | | ∃xDx   4, ∃I
6. | ∃xDx   2-5, ??   (No possible justification for discharging 2.)

The rule ∃E joins RA and ⊃ I in requiring that a subderivation be constructed for its operation. Here is the official version of ∃E:

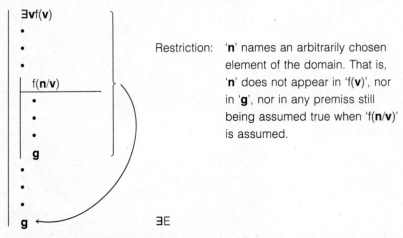

Restriction: 'n' names an arbitrarily chosen element of the domain. That is, 'n' does not appear in 'f(v)', nor in 'g', nor in any premiss still being assumed true when 'f(n/v)' is assumed.

In words, this rule says that the conclusion, indicated above by the 'g', can be brought out from under the temporary assumption, 'f(n/v)', provided that an earlier line of the derivation has established '∃vf(v)', and provided the restriction is met. 'g' is any well-formed formula whatsoever.

Why is the restriction so complicated? Let's see what disasters occur when any given piece of the restriction is dropped. First, suppose 'n' could appear in one of the still-active premisses. Then the following "derivation" would be legitimate:

1. | Fa
2. | ∃xGx
3. | | Ga
4. | | Fa & Ga   1, 3, & I
5. | | ∃x(Fx & Gx)   4, ∃I
6. | ∃x(Fx & Gx)   2, 3-5, ∃E (modified)

This "derivation" is undesirable because it is invalid! Knowing, for example, that a particular product, 'a', is cheap, symbolized by 'Fa', and that there exists a quality product, '∃xGx', does *not* validly yield the conclusion that there is a cheap quality product, '∃x(Fx & Gx)'. Yet the only thing which would block the above derivation from following our rule of ∃E is the restriction that '**n**' not appear in the premises active when 'Ga' appears. ('a' appears in line 1.)

Now let's drop instead the restriction that '**n**' not occur in '∃v**f**(**v**)'. Then the following illegitimate "derivation" would be allowed by our rule:

```
1. | ∃x(Fa & Gx)
2. |   | Fa & Ga
3. |   | ∃x(Fx & Gx)      2, ∃I
4. | ∃x(Fx & Gx)          1, 2-3, ∃E (modified)
```

This "derivation" is just as bad as the last one. In fact the same translation key can be used for this one as was used for the previous one, with the same results.

Finally, let's try just dropping the requirement that '**n**' not appear in '**g**'. Then the following so-called derivation would be sanctioned by our rule:

```
1. | ∃xGx
2. |   | Ga
3. |   | Ga          2, R
4. | Ga              1, 2-3, ∃E (modified)
```

Of course, this is ridiculous. I cannot legitimately conclude from the premiss that there are murderers ('∃xGx') that you ('a') are a murderer ('Ga').

All these examples illustrate what happens when the restrictions are violated. If you now go back and look at each example, asking yourself whether you could have obtained the same bottom line by using a name different from 'a' in place of 'x' in the subderivation, you will find that the answer is always "no". Hence in the above examples, 'a' is not arbitrary. That's why all those arguments are invalid. (Of course, one can sometimes violate the restrictions without producing an invalid argument diagram, but then that's just a matter of luck. We want rules which *never* produce an invalid diagram.)

The easiest way to make sure you've followed all the restrictions when using ∃E is to use in the temporary assumption a name which is new to the derivation, and then to keep that name out of the conclusion '**g**' by using ∃I to get '**g**', as I did in line 6 of the correct derivation about the thieves. This strategy may not always be appropriate, but it often works.

To illustrate the use of ∃E, and to obtain two new derived rules which will complete our entire rule set, I offer the two following derivations with commentary. To prepare for these, review ~∀T∃~ and ~∃T∀~. The derivations below provide justification for derived rules similar to those two derived rules already presented.

For the first derivation, assume there is something which is not an F. It should, and does, follow that not everything is an F:

1.	∃x~Fx		
2.	∀xFx		(Assume everything is an F to obtain a contradiction.)
3.	~Fa		(Let 'a' name the arbitrarily chosen thing that is ~F.)
4.	Fa	2, ∀E	(Line 2 immediately tells us that even 'a' names an F.)
5.	~∀xFx	3, 4, C	(I need a conclusion void of occurrences of 'a', as well as a conclusion which will yield a contradiction under line 2. So I choose the negation of 2.)
6.	~∀xFx	1, 3-5, ∃E	(Here's where we use ∃E. Line 5 was obtained from 3, but 3 contained reference to the arbitrarily chosen individual named by 'a'. Line 1 tells us that *something* is not an F, so we can discharge 3.)
7.	~∀xFx	2 & 6, RA	(We can't keep '∀xFx', since it leads to its own negation.)

Carefully examine the rationale for writing line 5.

The second sample derivation begins by assuming everything is non-F, and concludes, quite naturally, that there does not exist an F.

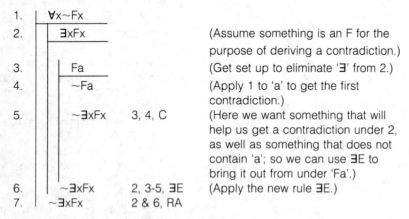

1.	∀x~Fx		
2.	∃xFx		(Assume something is an F for the purpose of deriving a contradiction.)
3.	Fa		(Get set up to eliminate '∃' from 2.)
4.	~Fa		(Apply 1 to 'a' to get the first contradiction.)
5.	~∃xFx	3, 4, C	(Here we want something that will help us get a contradiction under 2, as well as something that does not contain 'a'; so we can use ∃E to bring it out from under 'Fa'.)
6.	~∃xFx	2, 3-5, ∃E	(Apply the new rule ∃E.)
7.	~∃xFx	2 & 6, RA	

Here again the hardest line to come up with is probably line 5.

These two derivations provide some justification for the following two derived rules for converting "∃" to "∀" and "∀" to "∃", which I now state in official form and add to the system:

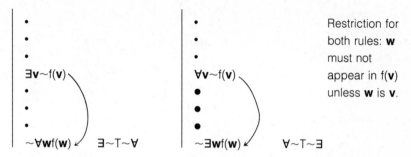

(Here, in the names of the rules, read the "T" as "to".)

Previously, after introducing ∼∀T∃∼ and ∼∃T∀∼, I remarked that we will interpret ∼∼E and ∼∼I to allow elimination and introduction of double negations inside predicate logic formulas. That allowed us in two steps to get

  '∃vf(**v**)' from '∼∀**v**∼f(**v**)'

using ∼∀T∃∼, and in two steps to get

  '∀vf(**v**)' from '∼∃**v**∼f(**v**)'

using ∼∃T∀∼. We can use the two new derived rules, ∃∼T∼∀, and ∀∼T∼∃ in much the same way to get in two steps

  '∼∀**v**∼f(**v**)' from '∃vf(**v**)'

and to get in two steps

  '∼∃**v**∼f(**v**)' from '∀vf(**v**)'.

In other words, we can reverse the order of derivation obtained earlier. The steps look like this:

1.	∃vf(**v**)			1.	∀vf(**v**)	
2.	∃**v**∼∼f(**v**)	1,∼∼I		2.	∀**v**∼∼f(**v**)	1, ∼∼I
3.	∼∀**v**∼f(**v**)	2, ∃∼T∼∀		3.	∼∃**v**∼f(**v**)	2, ∀∼T∼∃

Behold what this implies! Because we can move in either direction between

  ∃vf(**v**) and ∼∀**v**∼f(**v**)

and between

  ∀vf(**v**) and ∼∃**v**∼f(**v**),

we didn't really need to have *two* quantifiers after all. '∃**v**' could have been handled by '∼∀**v**∼' if we didn't want to use a special symbol "∃" for "some". Or, we could have not introduced "∀", using '∼∃**v**∼' instead.

Most people, however, find it much easier to symbolize ordinary English if they have both quantifiers available for use, and I will not take back either one of them, even though keeping both is redundant. (I did the same thing with the five propositional connectives, keeping all of them even though it is possible to prove some are redundant. One can in fact get by with just the "~" plus any *one* of the other five!)

We have now reached the point where the rest has to be up to you, in the exercises.

Existential quantifier elimination

55.1 Write the justifications which go to the right of the lines in each of the following correct derivations. (Study these for strategy.)

a.
1. | ∀xFx & ∃xGx
2. | ∃xGx
3. | | Ga
4. | | ∀xFx
5. | | Fa
6. | | Fa & Ga
7. | | ∃x(Fx & Gx)
8. | ∃x(Fx & Gx)

b.
1. |
2. | | ∃xFx
3. | | | Fa
4. | | | ∃yFy
5. | | ∃yFy
6. | ∃xFx ⊃ ∃yFy

c.
1. | ∀x~(Fx & Gx)
2. | ∃xFx
3. | | Fa
4. | | | Ga
5. | | | Fa & Ga
6. | | | ~(Fa & Ga)
7. | | ~Ga
8. | | ∃x~Gx
9. | ∃x~Gx

d.
1. | ∀x~(Fx & Gx)
2. | ∃xFx
3. | | Fa
4. | | | Ga
5. | | | Fa & Ga
6. | | | ~(Fa & Ga)
7. | | ~Ga
8. | | ∃x~Gx
9. | ∃x~Gx

55.2 The following incorrect derivations mimic 55.1c. and d. Pick out the illegal steps in each.

a.
1. | ∀x~(Fx & Gx)
2. | ∃xFx
3. | | Fa & Ga
4. | | ~(Fa & Ga)   1, ∀E?
5. | | ∃x~Gx   3, 4, C?
6. | ∃x~Gx   2, 3-5, ∃E?

b. 1. | ∀x~(Fx & Gx)
   2. | ∃xFx
   3. | | Fa
   4. | | | Ga
   5. | | | Fa & Ga   3, 4, & I?
   6. | | | ~(Fa & Ga)   1, ∀E?
   7. | | ~Ga   4-5 & 6, RA?
   8. | ~Ga   2, 3-7, ∃E?
   9. | ∃x~Gx   8, ∃I?

55.3 The derivation below is correct, but it might not appear to be so, since 'a' occurs in line 4 after occurring in line 3. Why does this occurrence in line 3 not ruin the possibility of treating 'a' as arbitrary for purposes of ∃E at line 7?

1. | ∀xFx
2. | ∃xGx
3. | Fa   1,∀E
4. | | Ga
5. | | Fa & Ga   3,4,& I
6. | | ∃x(Fx & Gx)   5,∃I
7. | ∃x(Fx & Gx)   2,4-6,∃E

55.4 Derive the indicated conclusion in each of the following argument diagrams.

a. ∃y~Fy
   ∀x(~Fx ⊃ Gx)
   ───────────
   ∃zGz

b. ∃y~Fy
   ∀x(Gx ⊃ Fx)
   ──────────
   ∃y~Gy

c. ∃x(Gx & Hx)
   ∀x((Gx ∨ Kx) ≡ Mx)
   ─────────────────
   ∃xMx

d. ∃x(Gx & Hx)
   ∃y(Hy & Ky)
   ∀x(Gx ⊃ Kx)
   ──────────────
   ∃z(Gz & Hz & Kz)

e. Fb
   ∀x(Gx ⊃ ~Fx)
   ~Gb ⊃ ∃xHx
   ───────────
   ∃xHx

f. ~∃x(Fx & Gx)
   ────────────
   ∀x~(Fx & Gx)

g. ∃x~(Fx & Gx)
   ─────────────
   ∃x(~Fx ∨ ~Gx)

h. (No premises)
   ──────────────────────────────
   ∃x(~Fx ∨ ~ Gx) ≡ ∃x~(Fx & Gx)

i. ∀x~Fx
   ~∃xFx ⊃ P
   ─────────
   P

j. ∀x(~Fx ∨ ~Gx)
   ──────────────
   ~∃x(Fx & Gx)

k. ∀x(~Fx & ~Gx)
   ─────────────
   ~∃xFx

l. $\dfrac{\exists x(Fx \supset Gx)}{\forall x Fx \supset \exists x Gx}$

m. $\dfrac{\forall x(Fx \supset Gx)}{\exists x Fx \supset \exists x Gx}$

n. $\dfrac{\exists x(Fx \,\&\, Ga)}{\exists x Fx \,\&\, Ga}$

o. $\dfrac{\exists x(Fx \,\&\, Gx)}{\exists x Fx \,\&\, \exists x Gx}$

p. $\dfrac{\begin{array}{l}\forall x(Fx \supset Gx)\\ \exists x Fx\\ \forall x(Gx \equiv \sim Hx)\\ \forall x(Hx \lor Mx)\end{array}}{\exists x Mx}$

q. $\dfrac{\begin{array}{l}\sim\forall x(Fx \,\&\, Gx)\\ \forall x(\sim Fx \supset Hx)\\ \forall x(\sim Gx \supset Mx)\end{array}}{\exists x(Hx \lor Mx)}$

55.5 Symbolize the following arguments as accurately as possible using the predicate letters and names indicated with each. Then derive the conclusion diagrams from the premiss diagrams, or else provide either a counterexample or an absurd logically analogous argument to prove the argument invalid. (Note: Any logically analogous argument must at least fit the same argument diagram as the original argument fits.)

a. No industrialists are socialists. All socialists are radicals. Thus, no industrialist is a radical.
(I( ), S( ), R( ); domain: persons)

b. If Fred is going, then everyone will go. Fred will go only if Linda also goes. Linda will go only if everyone is going. So, if someone does not go, neither Fred nor Linda will go.
(G( ), f, l; domain: all those persons the speaker has in mind)

c. If women all have a maternal instinct, then there are such things as instincts in human beings. But if women are like other kinds of animals, they do have a maternal instinct. Of course, women are human beings. Thus, if human beings are like other kinds of animals, they have instincts.
(W( ): ( ) is a woman; M( ): ( ) has a maternal instinct; I( ): ( ) is human instinct; A( ): ( ) is similar to the other animals; H( ): ( ) is a human being; domain: living things)

d. Everyone who jogs regularly has a good lung capacity, and everyone with a good lung capacity is better able to avoid shortness of breath during periods of physical exertion. So, all regular joggers are better equipped to escape shortness of breath while exerting themselves.
(J( ); L( ); B( ); domain: persons)

e. Some of those who were present that night when the police arrived were at the loud party which prompted the arrival of the police. Some of those who were present also were arrested. Hence, some of those who were arrested that night were at the loud party.
(P( ); L( ); A( ); domain: persons)

f. Any loss to your property caused by fire is covered by our policy, except for losses caused by fires you intentionally set. If a loss is caused by a fire you intentionally set, that loss does not come as a surprise to you. The losses you sustained by fire on the seventeenth were certainly a surprise to you. So, those losses are covered by your policy.

(F( ): ( ) is a loss to your property caused by fire; C( ): ( ) is a loss covered by your policy; I( ): ( ) is a loss caused by a fire you intentionally set; S( ); I: your losses on the 17th; domain: all your losses)

g. Everyone who ate at Mack's Diner on the twenty-third got sick; so there is not anyone who ate there on the twenty-third who did not get sick.
(M( ): ( ) ate at Mack's Diner on the twenty-third; S( ); domain: persons)

h. Someone stole my cat last night, and also stole my stereo last week. Anyone who would steal my stereo last week and then also steal my cat last night has a lot of nerve. So, someone has a lot of nerve.
(C( ); S( ); N( ); domain: persons)

i. No intelligent politician would vote for the measure if her voting for it would bring her into disfavor with her constituents. Therefore, if any politician has in fact been brought into disfavor with her constituents, and if all politicians are intelligent, that politician did not vote for the measure.
(I( ); P( ); V( ); D( ): ( ) is brought into disfavor with his constituents; domain: persons)

j. An American man either genuinely supports the concept of equality for women or he is insecure. But there are no American men who genuinely support the idea of equality for women and who nevertheless take the Hollywood macho image as the model for their lives. Therefore, if the only men who model their lives after the Hollywood macho image are American, then all the men who follow that model are insecure.
(A( ): ( ) is American; S( ): ( ) genuinely supports the concept of equality for women; I( ); M( ); domain: men)

55.6 Construct Venn diagrams which prove the validity or invalidity of any of the argument *diagrams* in the preceding exercise to which the Venn technique can be applied. (Many of these argument diagrams are too complex for the Venn technique to handle.) Make sure your results from applying the Venn technique agree with your results in the preceding exercise.

56

Polyadic symbolization

So far, we have worked with formulas of predicate logic which begin with only one quantifier and whose predicate letters are each followed by only one name or variable. This was done in order to keep things simple. But the full power of predicate logic cannot be achieved under such restrictive conditions. The validity of arguments like the following one depends on structure which cannot be revealed under such restrictions:

"Smith" comes later in the telephone directory than "Jones", and "Jones" comes later than "Brown". If one name comes later than some other name in the directory, and that name in turn comes later than a third name, then the first name comes later than the third one. Therefore, "Smith" comes later in the directory than "Brown".

(Try symbolizing the above argument under the restrictive conditions with which we have been working letting the domain be names, and see what a useless diagram you get.)

In this Section, we will take a look at how one might go about symbolizing propositions in a less restrictive way. In the following Section, I will discuss using our rules of inference on the more complex formulas which will result. No new rules of inference will be required. The title of this Section, "Polyadic symbolization", refers to the idea of using predicate letters which are followed by more than one name or variable. Predicate letters followed by just one name or variable on the other hand are called "monadic".

As usual, we begin with the simplest case. Let's take

"Smith" comes later than "Jones" in the telephone directory

for our example. This proposition *relates* two entities to each other, the names "Smith" and "Jones", and that relation becomes crucial in the argument structure above. We fail to capture the relational character of the proposition when, letting 's' stand for the name "Smith", we symbolize the proposition by something like 'Ls', where 'L( )' means "comes later than 'Jones' in the telephone directory".

However, we can fix things up by allowing 'L( )' to change to a *two*-slot predicate letter, followed by *two* blanks rather than one:

L( ) ( )

Of course, what this new 'L' stands for must also change. Let it stand for

( ) comes later in the telephone directory than ( ).

Now we can write

Lsj

to diagram our proposition, letting 'j' name "Jones". 'L( ) ( )' is then officially known as a "two-place" predicate letter.

There is no limit to the number of "places" (or slots) which may follow a predicate letter. For the proposition

Tom followed Sam, Hank's brother, to the store

we could define the predicate letter

F( ) ( ) ( ) ( )

to stand for

( ) followed ( ), the brother of ( ), to ( )

with the resulting propositional diagram

Ftsha

where 'a' names the store, and the other names refer to the obvious people.

Sometimes, it is convenient to define a polyadic predicate letter without keeping the blanks in the same order in which they appear in the English sentence being symbolized. However, then there is no way to keep the order straight when each blank is marked solely by parentheses. I will start using variables to label the blanks, to overcome this problem. Thus,

Fxyzx₁

might be defined as a four-place predicate letter standing for

x, the brother of y, was followed to z by x₁.

Then, our above proposition would be symbolized by

Fshat.

It is obviously important to get the right name in the right slot.

Let's return to our telephone directory argument. We can now symbolize the proposition expressed by the first sentence, without difficulty, as 'Lsj & Ljb'. But the next premiss introduces a need for variables and quantifiers to be used in connection with 'Lxy'. As before, no variable may occur in the diagram of a proposition unless that variable is within the scope of a quantifier. When a predicate letter is followed by two different variables, there will have to be two quantifiers governing it. In the case of the telephone directory argument, the second premiss is clearly about all names, and could be symbolized as follows:

∀x∀y∀z((Lxy & Lyz) ⊃ Lxz)

Carefully note the relative placement of the variables in this formula. 'x' is the "first" name mentioned in the premiss, 'y' is the "second", and 'z' is the "third".

If you were to try to read the above formula literally, it would say something like this:

For all x's, y's, and z's, if x comes after y and y comes after z, then x comes after z in the telephone directory.

Putting all the pieces together, we would get the following diagram for the entire argument:

Lsj & Ljb
∀x∀y∀z((Lxy & Lyz) ⊃ Lxz)
Lsb

(You might try deriving this conclusion from the premisses, using the rules you already have learned. It requires eliminating the three quantifiers one at a time, starting with the outermost one.)

As we go on, keep in mind these principles:

(1) Never immediately follow a quantifier by a variable when there are no other occurrences of that variable within the scope of that quantifier. (Thus, '∀x∀yFxx' is illegal.)

(2) Never allow two quantifiers with overlapping scope to apply to the same variable. (Thus, '∀x(Fx ⊃ ∀xGx)' and '∀x∀x(Fxx)' are both illegal.)

(3) All occurrences of variables must be governed by quantifiers attached to those variables. (Thus, '∀xFxy' is illegal.)

Let's look at some more examples. Suppose in the dead of night you are awakened by a loud noise, and you say "Something hit something". In symbols, we could write

∃x∃yHxy.

It is worthwhile noting that even though two different variables are used here, the above formula would be true even if something hit itself, for in such a case there would still exist something that hit an existing something.

In more complicated cases, you may find it helpful to rephrase the English sentence into a quasi-formula first, and then convert the quasi-formula into a genuine formula. I'll do that with the sentence

Somebody at the party broke something.

The domain must include both people and things; so we will need to make our quasi-formula say something like this:

There is at least one person, x, who was at the party, and there is at least one thing, y, such that x broke y.

Converting to a genuine formula yields this:

∃x(Px & Axp & ∃yBxy)

where 'Px' stands for 'x is a person', 'Axy' stands for 'x was at y', 'p' names the party, and 'Bxy' stands for 'x broke y'.

Alternative symbolizations above, equally correct, would be

∃x∃y(Px & Axp & Bxy)
∃y∃x(Px & Axp & Bxy).

But this is *not* correct:

∃x(Px & Axp) & ∃xBxy

In fact, this last "formula" is illegal nonsense, because the final occurrence of 'x' is not within the scope of '∃x'.

For our next example, we will work on the proposition expressed by "Michael got everything he was supposed to get, except for the bread". Putting this into quasi-formula form, we probably have

For all things x, if Michael was supposed to get x, and x is not the bread, then Michael got x, and if x is the bread, then Michael did not get x even though he was supposed to get x.

In formal symbols, this amounts to

∀x([(Smx & ~Bx) ⊃ Gmx] & [Bx ⊃ (~Gmx & Smx)])

In the preceding formula the part about what is true if x is the bread adds needed information. If we had only written

∀x((Smx & ~Bx) ⊃ Gmx)

it still would have been *possible* that Michael did get the bread.
Presumably,

∀x((Smx & ~Bx) ≡ Gmx)

would have represented an incorrect interpretation of the proposition, since this formula claims, among other things, that if Michael got x, then he was supposed to get x. The proposition expressed by our original sentence did not seem to me to have this implication. Perhaps Michael also got some things he was not *supposed* to get, because he wanted them. We don't want to rule out that possibility.

As you may have noticed, our last correct formula contained only one quantifier even though it also contained some two-place predicate letters. That is perfectly all right since it contained occurrences of only one variable. (Remember that 'm' is a name rather than a variable.)

Polyadic symbolizations may not be checked for correctness with Venn diagrams. But there is another checking technique which can sometimes be helpful. Actually, this technique can also be used for symbolizations involving only one-place predicates, but I waited until now to present it because it is especially helpful in analyzing polyadic formulas. The new technique consists simply of testing a proposed symbolization for truth or falsehood in very small domains, say, one containing just four members, and comparing the results with the truth or falsehood of the English sentence as applied to the same domains. Naturally, in order for the proposed symbolization to count as being correct, it must be true and false in exactly the same domains as the English sentence is.

Let's see how this technique works in some one-place cases first. Symbolize "Everyone came" by 'VxCx' and let the domain be the four people named by 'a', 'b', 'c', and 'd'. In this domain several arrangements of facts are possible. Here are a few of them:

I	II	III	IV
a came	~(a came)	~(a came)	a came
b came	~(b came)	b came	~(b came)
c came	~(c came)	c came	c came
d came	~(d came)	d came	d came

Obviously, the proposition expressed by the original English sentence is true in arrangement I and false in the others listed. If our symbolization is to be counted as correct, then, it too must be true in arrangement I and false in the others. Is it?

You determine the truth value of a universally quantified formula 'Vvf(v)' by considering *all* its "instances, that is, all formulas of the form 'f(n/v)'. In this case, that means you are to consider 'Ca', 'Cb', 'Cc' and 'Cd'. *All* these instances must be true in order for the formula 'Vvf(v)' to be true. Naturally, we find here that 'Ca', 'Cb', 'Cc', and 'Cd' are all true in arrangement I and not all are true in the other arrangements. Consequently, '∀xCx' is true only in arrangement I, as you probably knew all along.

Thus, '∀xCx' appears to be a correct symbolization. Of course, to actually *prove* it is correct, we would have to consider *all possible* arrangements of facts in *all possible* domains of *all possible* sizes. Quite a task! But we have at least verified that '∀xCx' looks correct so far, by working with our limited domains.

Before going on to look at more complicated cases, let's ask how '∃xCx' would have weathered the preceding test. The rule is that a formula of the form '∃vf(**v**)' is true in just those arrangements where *at least one* of the 'f(**n/v**)' is true. Thus, '∃xCx' is true above when at least one of 'Ca', 'Cb', 'Cc', and 'Cd' is true. That is, '∃xCx' is true in I, III, and IV, but not in II. Of course, this implies '∃xCx' is an extremely poor symbolization of the original English we were working on above.

Now, to apply all this in more difficult cases. Students generally become confused in dealing with formulas which begin with a mixture of universal and existential quantifiers. Our new small domain testing procedure should help with that problem.

Consider the proposition expressed by the sentence

Everyone has a boss

which can be symbolized by assuming the set of persons as domains, utilizing 'Bxy' for "x is a boss of y".

Now the question is whether

∀x∃yBxy   or   ∃y∀xBxy   or   ∀y∃xBxy

correctly symbolizes the proposition. To find out, try a four-person domain in various possible arrangements. You might best keep track of the arrangements graphically by letting dots be the elements and arrows be all the relations between them which hold within the arrangements:

I

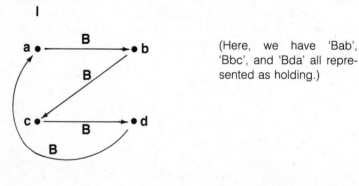

(Here, we have 'Bab', 'Bbc', and 'Bda' all represented as holding.)

II

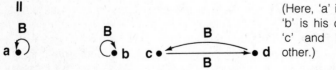

(Here, 'a' is his own boss, 'b' is his own boss, while 'c' and 'd' boss each other.)

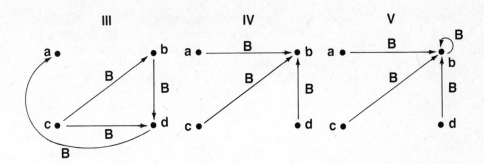

There are, of course, a great many more possible arrangements, but these offer a wide enough variety to distinguish our competing symbolizations for "Everyone has a boss". First, using common sense, let's list the arrangements where this English proposition is true. In arrangements I and II everyone has a boss. Not everyone has a boss in the other arrangements. In III, IV, and V the fellow named 'c' has no boss. Thus, the correct symbolization should be true in arrangements I and II only.

To evaluate '∀x∃yBxy', we noted that first of all it is of the form '∀vf(v)', which means all its instances must be true before it can be true. What do these look like for our 4-member domain? Here they are:

∃yBay   ∃yBby   ∃yBcy   ∃yBdy

In which arrangements are these *all* true? In I, II, and V. That is, '∀x∃yBxy' is true in one too many domains to be the correct symbolization. It essentially says that for each x in the domain, there is at least one y such that x bosses y. That is, each thing in the domain bosses at least one thing. More informally, this says everyone is a boss. It does not say everyone has a boss. But what about '∃y∀xBxy'? Is it different? Since it is of the form '∃vf(v)', it will be true when *one or more* of the 'f(n/v)' is true. These 'f(n/v)' look like this:

∀xBxa   ∀xBxb   ∀xBxc   ∀xBxd

In I, '∀xBxa' is false, since not everyone bosses the person 'a'. Similarly, in I '∀xBxb' is false, '∀xBxc' is false, and '∀xBxd' is false. Thus, *none* of the 'f(n/v)' above are true in I, and as a result we know '∃y∀xBxy' is false in I.

Why does it come out this way? '∃y∀xBxy' is saying there is some one y such that every x bosses *that* y. In other words, to make '∃y∀xBxy' true we need to come up with someone who is bossed by *everyone*. In fact V is such an arrangement. In V, 'b' names someone who is bossed by everyone. In V, '∀xBxb' is true, which means '∃y∀xBxy' is true. V is the *only* one of the above arrangements which makes '∃y∀xBxy' true. (Even IV fails, since 'b' does not boss 'b' in IV.) Thus, '∃y∀xBxy' is not a correct symbolization of "Everyone has a boss". Rather, it would correctly symbolize "Someone is bossed by everyone".

Finally, we can check '∀y∃xBxy'. This will be true when

∃xBxa   ∃xBxb   ∃xBxc   ∃xBxd

are *all* true (because 'Vy∃xBxy' is a universally quantified statement). This happens in I and II only. It fails in III, IV, and V because, for example, '∃xBxc' fails in III, IV, and V. 'Vy∃xBxy' says everyone has a boss, as we wanted. Note that it is true in the correct arrangements for being the desired symbolization.

What then does '∃xVyBxy' say? It says there is an x such that all y's are such that x bosses y. That is, there is someone who has everyone, including himself, to boss. *This* formula would not have been true in *any* of our arrangements I through V.

One thing worth remembering emerges from our discussion of the order of the leading quantifiers: when the leading quantifiers are not all universal or are not all existential, the order in which they appear is crucial to the meaning of the formula.

Polyadic symbolization

56.1 Draw an arrangement of the four-person domain which would make '∃xVyBxy' true.

56.2 Which of the formulas '∃xVyBxy', '∃yVxBxy', 'Vx∃yBxy' and 'Vy∃xBxy' is true in the following arrangements? State how you know.

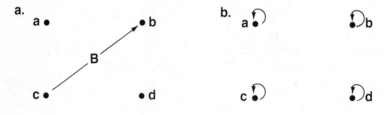

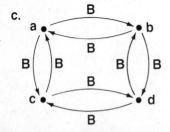 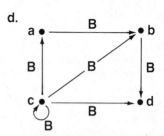

56.3 Symbolize the following propositions, using 'Fxy' for "x has y as a friend". Assume the domain is people.

a. Everyone has at least one friend.

b. Everyone is someone's friend.

    c. Someone is everyone's friend.

    d. Everyone is everyone's friend.

    e. Someone is someone's friend.

    f. No one is anyone's friend.

    g. No one is Sally's friend.

56.4 For each of the propositions in 56.3, construct a three-person domain in which the proposition is true and one in which the proposition is false.

56.5 Symbolize each of the following statements, using the suggested symbols. As you do so, remember that the occurrences of 'x' and 'y' in the definitions of the predicate letters given in the following do not indicate you must use just those variables with those predicate letters. You may use any variable with any predicate letter.

    a. George bought everything. (g; Bxy: x was bought by y; domain: persons, and things George might have bought)

    b. George didn't buy anything. (See exercise a.)

    c. If George didn't buy anything, then he didn't buy everything.

    d. General Corp. is larger than any other corporation. (g; Lxy: x is larger than y; Cxy: x is a corporation other than y; domain: corporations)

    e. Some corporations are larger than General Corp. (See exercise d.)

    f. If General Corp. is larger than any other corporation, then no corporation is larger than General Corp.

    g. Everybody loves a lover. (Lx: x is a lover; Bxy: x is loved by y; domain: persons)

    h. Everybody loves a lover. (Expand the domain to include all living things. Use the preceding symbols, plus 'Px' for "x is a person", and assume the English sentence means every person loves a person who is a lover.)

    i. An ancestor of one person's ancestor is also an ancestor of that person. (Axy: x is an ancestor of y; domain: persons)

    j. Everyone likes one or another fast-moving game. (Px: x is a person; Lxy: x likes y; Fx: x is a fast-moving game; domain: persons and games)

    k. No one likes a cheater. (Lxy: x likes y; Cx: x is a cheater; domain: persons)

    l. No one likes a cheater, but everyone hates a cheater who wins. (Add to the above: Hxy: x hates y; Wx: x wins; domain: persons)

    m. No one knows everything. (Px: x is a person; Kxy: x knows y; Tx: x is the type of thing which might be known; domain: persons and things which might be known)

n. You (that is, any person) can fool some of the people some of the time. (Px: x is a person; Tx: x is a time; Fxyz: x can fool y at z; domain: persons and times)

56.6 Even though the existential quantifier signifies only that there exists *at least* one thing in the domain, it is possible to use the quantifier to say that there exists *exactly* one, or exactly two, or more than one, or more than two, things in the domain. To say there are at least two things, it is not enough to use two variables; thus, '∃xGx & ∃yGy' does *not* say there are at least two G's, because both 'x' and 'y' could end up referring to the same thing. One needs to add to '∃xGx & ∃yGy' the idea that 'x' and 'y' do not refer to the same thing. To add this idea, use an extra predicate letter which says 'x' and 'y' refer to different things. See if you can figure out how to do this in the following exercises. The first one is done for you.

a. There are at least two people in the store. (Ix: x is in the store; Sxy: x and y are the same person; domain: persons)
Answer: ∃x∃y(Ix & Iy & ~Sxy)

b. There are at least three people in the store. (Use the same translation key.)

c. Some people are in the store. (Take seriously here the usual implication in English that this means more than one. Use the same translation key.)

d. There is exactly one person in the store. (The same translation key again.)
Hint: The answer is formed as follows: '∃x(Ix & ∀y(Iy ⊃ ???))'

e. There are exactly two persons in the store.

f. There are several persons in the store.

g. There is more than one person in the store.

h. No more than two persons are in the store.

So, after all, it is possible to use our symbols to accurately represent "some" when it means "more than one"!

57

Polyadic applications

Although no new rules of inference are needed when working with derivations containing polyadic predicate letters or multiple quantifiers, most people have a little trouble seeing at first just how to use the old rules in such cases. The purpose of this Section is to help you apply the old rules to these more complex situations.

   To apply our rules of inference to formulas which begin with multiple quantifiers, there is one simple thing to remember: eliminate the outermost quantifier
first when using a quantifier elimination rule, and when using an introduction
rule, the newly introduced quantifier appears on the extreme left end of the
formula. By comparing the following arrays, you should be able to see how the
abstract formulation of our rules works with multiple quantifiers.

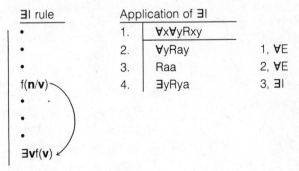

∀E rule   Application of ∀E
                   1. | ∀x∃yFxy
                   2. | ∃yFay   1, ∀E

∀vf(**v**)
f(**n/v**)

Here, '**v**' in the rule stands for 'x' in the derivation, and 'f(**v**)' stands for '∃yFxy'
which is indeed a formula containing an occurrence of '**v**'—that is, 'x'. '**n**' then
stands for 'a' and 'f(**n/v**)' is obtained from 'f(**v**)' by replacing all occurrences of
'**v**' ('x') by occurrences of '**n**' ('a'). Note that 2 is a perfectly well-formed legal
formula, as it should be. In English, the derivation given above might say, for
example, that since everyone has a friend, the fellow named 'a' must have a
friend.

   Here's another slightly more complex example:

∃I rule   Application of ∃I
                   1. | ∀x∀yRxy
                   2. | ∀yRay   1, ∀E
                   3. | Raa   2, ∀E
f(**n/v**)   4. | ∃yRya   3, ∃I

∃vf(**v**)

Several features of the above derivation deserve comment. First, note that the
double universal quantifiers are eliminated in two separate applications of ∀E;
∀E always eliminates only one quantifier at a time. In the first application of ∀E,
'**v**' stood for 'x', while in the second, '**v**' stood for 'y'. Next, note that in the application of ∃I, only one occurrence of 'a' was replaced by 'y'. Given the way the
rule is formulated, you have the option of replacing both of the 'a' 's in line 3 or
only one of them, since either way 'Raa' can be obtained from 4 by substituting
'a' for 'y' in 'Rya' or in 'Ryy' as the case may be. This makes good logical sense.
For instance, if 'Rxy' means 'x is fond of y', 'Raa' means the fellow named 'a' is
fond of himself. Then it would be valid to conclude either that there is someone
who is fond of the fellow named 'a' as we did in the derivation above, or to

conclude that there is someone who is fond of himself, as we would have concluded if we had replaced both occurrences of 'a' by occurrences of 'y' in line 4, obtaining '∃yRyy'.

It would not have been necessary to eliminate both occurrences of "∀" above before applying ∃I, as the following derivation illustrates:

1. | ∀x∀yRxy
2. | ∀yRay   1, ∀E
3. | ∃x∀yFxy   2, ∃I

Of course, here we're getting a conclusion different from that we derived before. By the way, the following "derivation" is *not* correct:

1. | ∀x∀yRxy
2. | ∀yRay   1, ∀E
3. | ∃y∀yRyy   2, ∃I misapplied

Line 3, of course, is not well formed, since it contains two quantifiers with overlapping scope both governing the same variable. But it is instructive to see how ∃I as officially formulated blocks the derivation of line 3. The 'f(**v**)' referred to by ∃I stands for '∀yRyy' above. So, if ∃I were correctly applied in the above "derivation", line 2 would have to be obtainable from line 3 by substituting some name for the variable 'y' in '∀yRyy'. But it is easy to see that there is no way to get '∀yRay' from '∀yRyy' by substituting a name for 'y'. Hence ∃I does not sanction the derivation of line 3 from line 2. This shows that the variable chosen for '**v**' in ∃I must not be one that already occurs in 'f(**n/v**)'.

One final elementary illustration. Suppose there is someone in the group who absolutely everyone in the group likes. It will follow that the fellow named 'a' (assumed to be in the group, which serves as our domain) likes someone. Let's see how the derivation of this conclusion would go. (Let 'Lxy' mean 'x likes y').

1. | ∃x∀yLyx
2. | | ∀yLyb   (Assumption for the purpose of
                                  using ∃E later on. 'b' was chosen
                                  rather than 'a' for our arbitrary
                                  person, because we want to use
                                  'a' later on.)
3. | | Lab   2, ∀E   (Since everyone likes 'b' according
                                  to (2), 'a' likes 'b'.)
4. | | ∃xLax   3, ∃I   (So there is someone 'a' likes)
5. | ∃xLax   1, 2-4, ∃E (Here it is crucial that 'b' was the
                                  arbitrary name, rather than 'a', for
                                  otherwise we would be violating a
                                  restriction on ∃E. Look back at the
                                  derivation's steps to see that 'b'
                                  truly is arbitrary. 'c' or 'd' could just
                                  as well have been used.)

You might be interested to notice that the conclusion we got above does not say that the person named 'a' likes *the same* person as everyone likes. If we want to say *that* we need '∃x(Lax & ∀yLyx)' for our conclusion. But it's easy enough to obtain this longer conclusion:

1.	∃x∀yLyx	
2.	∀yLyb	
3.	Lab	2, ∀E
4.	Lab & ∀yLyb	2, 3, & I
5.	∃x(Lax & ∀yLyx)	4, ∃I
6.	∃x(Lax & ∀yLyx)	1, 2-5, ∃E

Now let's look at an example that's a little more exciting. Here's the argument in English:

There are some examples of countries which share a common border but which have never been at war with each other. However, whenever one country is much richer than another, it always turns out they have been at war at some time or another. Thus, there must be some countries sharing a common border where one of the countries is not much richer than the other.

Letting the domain consist of countries, we can use the translation key

Bxy: x borders y
Wxy: x and y have been at war with each other
Rxy: x is much richer than y

to obtain the argument diagram

∃x∃y(Bxy & ~Wxy)
∀x∀y(Rxy ⊃ Wxy)
∃x∃y(Bxy & ~Rxy)

The derivation then proceeds as follows:

1.	∃x∃y(Bxy & ~Wxy)		
2.	∀x∀y(Rxy ⊃ Wxy)		
3.	∃y(Bay & ~Way)		(Getting set up to use ∃E on 1. Note that only 'x' is affected at this stage.)
4.	Bab & ~Wab		(Getting set up to use ∃E, on 3. A new name must be used here, since 'a' is no longer arbitrary, having appeared in 3. We now have arbitrarily selected two countries to talk about.)
5.	Rab		(We need to show that '~Rab', so we get set up for an application of RA.)
6.	∀y(Ray ⊃ Way)	2, ∀E	(Apply 2 to 'a' and 'b'. Be careful to eliminate only one quantifier at a time.)
7.	Rab ⊃ Wab	6, ∀E	
8.	Wab	5, 7, ⊃ E	
9.	~Wab	4, & E	
10.	~Rab	5-8 & 9, RA	
11.	Bab	4, & E	(Here we are getting set up to get line 12.)
12.	Bab & ~Rab	10, 11, & I	(This is the line we've been working for. It gives us the relations we want between our sample countries.)
13.	∃y(Bay & ~Ray)	12, ∃I	(Introduce one quantifier at a time. Be careful about which slots the variable goes in.)
14.	∃y(Bay & ~Ray)	3, 4-13, ∃E	(Even though 'a' is still present in our formula here, that doesn't matter, because 'b' is the relevant name for this ∃E.)
15.	∃x∃y(Bxy & ~Rxy)	14, ∃I	
16.	∃x∃y(Bxy & ~Rxy)	1, 3-15, ∃E	

In this derivation it is best to wait to do the ∀E's until after getting set up with the temporary assumptions for ∃E. Otherwise, you might not choose the right names when you do the ∀E. Once lines 3 and 4 establish that 'a' and 'b' are going to be the sample countries to be investigated, then you know that 2 will have to be applied to those countries. In fact, it is generally a good rule of strategy to arbitrarily choose individuals for ∃E before doing anything else in a derivation, as soon as you know you are going to need to use ∃E. That way, you can make the rest of your derivation fit those choices.

What remains now for you is to practice.

Polyadic applications

57.1 Derive the given conclusion from the indicated premisses.

a. ∀x∀yFxy
   Fab

b. Fab
   ∃x∃yFxy

c. ∃x∃yFxy
   ∃y∃xFxy

d. ∃y∀xFxy
   ∀x∃yFxy

e. ~∃x∀yFxy
   ∀x∃y~Fxy

f. ~∀x∃yFxy
   ∃x∀y~Fxy

g. ∀x(Fx ⊃ ∀yGy)
   ∀x∀y(Fx ⊃ Gy)

h. ∀x(∃yFy ⊃ Gx)
   ∀x∀y(Fy ⊃ Gx)

i. ∀x∀y((Px & Py) ⊃ Rxy)
   ∃xPx
   ∃xRxx

j. ∃xRax ⊃ ∀xFx
   ∀x∀yRxy
   Fb

k. ∀x(Bx ⊃ ∃yRxy)
   ∀x∀y(Rxy ⊃ Qx)
   ∃xBx
   ∃xQx

57.2 Symbolize the following arguments and derive the conclusion diagram of each from the premiss diagrams, thus proving the arguments to be valid, provided the diagrams are accurate.

a. If one company is a subsidiary of another company, and this second company is in turn a subsidiary of a third company, then the first company is a subsidiary of the third one. Ace Coal is a subsidiary of Energy Products, Inc., a subsidiary of Mammoth Oil Corp. Thus, Ace Coal is a subsidiary of Mammoth Oil.

b. Each person dislikes all those who strongly disapprove of his life-style. But some people, upon reflecting, strongly disapprove of their own life-styles. So some people dislike themselves.

c. Our old insurance policy covered our home against all perils except flooding. Being crushed by a falling tree surely counts as a peril which is not a case of flooding. Our new policy covers our home against everything the old one did. So our new policy covers our home against being crushed by a falling tree.

d. All automobiles pollute the air to some extent. Hence any car owner owns an air polluter.

e. Every time there's a problem in the United States, the people blame the president for it. But if someone blames someone for something, they must believe that person has control over the thing he's being blamed for. The president is a person. So the people think there's some person who has control over all the problems in the United States.

f. Anyone who is a friend of yours is no friend of mine. So, if you have any friends, there's at least one person who's not my friend.

g. Some of the goods confiscated by the police from the house they raided last night were stolen from the Acme Appliance Co. warehouse. However, since everything stolen from that warehouse was owned by Hendricks, some of the confiscated goods are Hendricks'.

h. Any law which has any significant effect hurts someone. Hence, if no one is hurt by the measure passed last year by the City Council, it must be without significant effect. (Assume the premiss that the measure passed last year is a law.)

i. If one event brings about another event, the first event begins before the second one. When one event begins before another one, the events are not identical to one another. Every event is, of course, identical to itself. Thus, no event brings itself about.

j. For each integer (the numbers 1, 2, 3, etc.) there is an integer which comes next and which is larger. No integer is larger than itself. Therefore, there is no integer which is larger than all the integers.

k. No one who thinks for himself will be an unquestioning advocate for any political party. Some people do think for themselves. Hence, no political party has everyone as unquestioning advocates.

l. No company manufactures a vehicle model with all the possible optional equipment currently available on the market. So no matter what model you buy, there will be some currently available options it won't have.

m. Everyone who belongs to the country club thinks he's better than everyone else. (The club has at least two members.) Thus, there are members of the club who think they are better than other members of the club.

58

The limitations of first-order logic

Despite the added power we now have within our grasp to display and work with logical structure, a large number of important limitations to our power still remain. Much of the work done by professional logicians in recent years has been directed toward removing some of these limitations, partly through the development of additional symbolizing devices and added rules of inference. Although I will not attempt to develop symbolic logic any further now, it is still useful to gain perspective by taking an overview of some of the more important limitations of the system you've been learning.

One limitation in predicate logic was inherited from propositional logic, and has not been removed by any feature of predicate logic. This limitation arises because all our symbolic connectives are still truth-functional, while the connectors in English often fail to be truth-functional. Even with the tools of predicate logic, we can't adequately represent the structure of the proposition "Everyone came to the party BECAUSE Fred came". We will still end up using "&" for "BECAUSE", which is of course still only an approximation. Similar problems with "IF. . . THEN", "ONLY IF", "UNLESS", and the like also remain with us.

As you know, some of the other important limitations of propositional logic have been overcome by predicate logic, since we now have quantifiers, predicate letters, and names available as symbolic devices. This allows us to prove valid a great many more arguments. However, some arguments which turn crucially on quantification remain nevertheless out of our symbolic reach, for there are quantifiers in English not adequately representable by "∀" or "∃", such as "almost all", "many", and "few". The validity of an argument can easily depend on the operation of these hard-to-handle quantifiers.

Only a few of the eggs were spoiled

Many of the eggs were not spoiled

is clearly valid, but we couldn't prove it valid in our predicate logic system, because of our inability to represent the relevant quantifiers or to work with them. This particular problem has received very little attention from those who have sought to improve predicate logic.

Some of the other limitations of our predicate logic have received much more attention. One is the limitation to nonempty domains. It seems philosophically objectionable to many people that a logical system require the existence of something. We can, it seems, reason logically about domains which are empty, for when we construct arguments we may not know whether the domain we speak of is empty or not. This might happen, for instance, if we are reasoning about the properties of subatomic particles which we believe might possibly exist, or about the behavior of idealized beings such as perfectly rational persons which we believe do not actually exist. A good logical system, then, ought to be able to cope with such arguments in some way.

Several other limitations of our system which receive considerable attention among logicians are related to devices used in English other than quantifiers and predicates to lend logically important structural features to arguments. There are quite a few such devices, and I can only hope to mention some of them.

One group of such devices consists of various ways we have of referring to things by complex names. So-called *definite descriptions* fit into this category—descriptions beginning with "the", such as "*the* first man to fly an aircraft", "*the* mother of George", and "*the* first person to set foot on *the* eastern shore of *the* largest of *the* Great Lakes". The names in our system of predicate logic have no internal structure, but definite descriptions obviously contain considerable structure, structure which may well be important in an argument. Accordingly, often it will not be sufficient to symbolize a definite description as a simple name.

For instance, the argument

The tall person sitting smugly in the front row is insensitive to the difficulty he's causing for those sitting behind him in seeing the stage. So, obviously, there's at least one tall person who is insensitive to some difficulties he causes for others.

seems valid, but its symbolization is fraught with difficulty. Clearly, it will not be sufficient merely to let 'a' stand for "the tall person sitting smugly in the front row", and then to treat the premiss as being something like 'Da'. We need something more like '∃x(Tx & Px & . . .)' where 'Tx' is 'x is tall', 'Px' is 'x is a person', and so on. This won't be enough, yet, because "the tall person . . . " clearly refers to only one person, while '∃x(Tx & Px & . . .)' merely says there is at least one such person. We need to add something to the symbolization to indicate that there is just one person about whom we are talking. Bertrand Russell proposed adding a clause to our available symbolic techniques which essentially said there is only one tall person sitting in the front row.[1] However, this seems to be too strong a move, since there may be more than one tall person sitting in the front row, and yet we may be referring to only one of them and our reference may be quite clear in the context in which we are arguing. The handling of definite descriptions remains a controversial topic in formal logic, because of problems like these.

Another structure-giving device in English is the adverb, also illustrated by our sample argument about the tall person. The tall person is not merely sitting; he's sitting smugly. Now, it happens that in the above argument this doesn't matter. But in some arguments it does matter:

There's a tall man sitting smugly in the front row. Anyone who's sitting in the front row must be kicked out. So there's someone who must be kicked out.

---

[1] See his "On Denoting", *Mind*. vol. XIV (1905) pp. 479–93, and variously reprinted—for example, in *Logic and Knowledge,* ed. by Robert Charles Marsh (London: George Allen and Unwin, 1956), pp. 41–56. Russell was one of the pioneers in the development of symbolic logic.

This argument cannot be proved valid unless we can find a way to symbolize its first premiss to make it clear that anyone who sits smugly is thereby one who sits.

Try it yourself. It's not obvious how to proceed. Let 'Sxy' be 'x is sitting in y', for instance. Then how do you fit "smugly" in? For practicality, most people wouldn't try—they'd just forget it. But still, there ought to be a way to do it. Sometimes it will be important. So far, no agreement has been reached by logicians about how to handle this problem.

The next troublesome structural device I want to mention is verb tense. We have nothing in our predicate logic symbolism to represent tense. This can be a serious limitation. Because of it, we cannot prove valid the following argument by symbolizing just the given statements, even though it presumably is valid on a normal reading of its premiss:

The game was completed yesterday. So, the game is not underway now.

A logical system which takes tense into account symbolically is called a "tensed logic".

The final structural device in English I want to mention is the use of operators like "necessary", "possibly", "perhaps", "I believe that", "you know that", "he doubts that", and so on. These all have something in common. They, like negation, apply to whole clauses to form new whole clauses. We have no notation to deal with any of them. Thus, we have no way to prove valid the argument

Possibly it will rain tomorrow. Thus, it is not necessarily true that it will not rain tomorrow.

Logical systems dealing with the sub-cluster of these operators related to possibility and necessity are known as *modal logics*.

Aside from the special structure-creating devices I've been discussing, it seems there's another whole area of logical problems our system can't deal with, illustrated by the following argument:

Jill is tall and so is Kathy. Thus, there is something Jill and Kathy have in common.

How should the conclusion be symbolized? Perhaps as '∃x(Hjx & Hkx)' where 'Hxy' means 'x has y'? If so, we will get

Tj & Tk
_____
∃x(Hjx & Hkx)

as the argument diagram. We'd never prove the argument valid that way.

What seems to be needed is the ability to quantify over *predicates* as well as over individuals. Then the conclusion might look something like this:

∃X(Xj & Xk)

where 'X' is a variable ranging over predicates. We might have a new rule of ∃I for predicate quantification to prove the above argument valid:

1. | Tj & Tk
   |_____
2. | ∃X(Xj & Xk)   ? Something like ∃I?

A logical system which allows variables to range over predicates is called a *second-order* logic. Accordingly since our predicate logic does not have such variables, our system is known as *first-order* predicate logic.

What all this shows is that much remains to be done in the development of logical theory, and our system of predicate logic is not detailed enough to handle everything of logical importance in English. Thus, if you symbolize an argument and find that the resulting diagram is invalid, the original argument may nevertheless be *valid*. This is no different from the situation as it existed when we discussed propositional logic. To know that a *diagram* is invalid is in general *not* to know that the original English argument which fits the diagram is also invalid. To prove an English argument invalid, we still have only two techniques: counterexamples and absurd logically analogous arguments. (See Sections 25 and 26.)

The limitations of first-order logic

58.1 Devise a valid argument in English in which the logical relations between "necessarily" and "possibly" are crucial to making the argument valid. (See the example argument regarding modal logic.)

58.2 What kind of logic would be needed to handle the relevant structure of the following argument?

Joe was blond but now he's gray haired. Thus, Joe was blond before he became gray.

58.3 Add five more similar operators to the list given earlier in this Section which included "necessarily" and "I know that".

58.4 Construct another valid argument in English for which one might need second-order logic to get an adequate diagram.

58.5 Sometimes first-order predicate logic is expanded by adding a special predicate symbol, '=', meaning "equals". '=ab' for instance would mean "a equals b", and would often be written in the more familiar-looking form, 'a=b'. Make up a valid rule of inference for this new symbol, a rule which could be added to our system. (A predicate logic with this symbol or an equivalent symbol along with a complete set of rules of inference governing its use is called a "predicate logic with identity".)

58.6 Do you think the argument

Jane is a woman. Hence Jane is a person.

is valid? If not, why not? If so, what problems will arise in symbolizing it to prove it valid?

PART THREE
Applications and Extensions

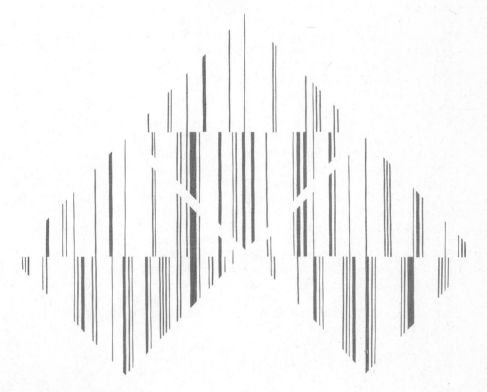

9

Completing an Argument

59

Adding premisses to arguments

Often, an argument as stated by its author is weaker than it needs to be because the author does not include all the premisses. Sometimes this happens when the author believes these premisses are obvious and thus not worth stating, or when the author has argued for the truth of the premisses at an earlier time and now assumes the reader of this argument will remember them. Sometimes it happens because the author is not aware that the premisses are needed in order to make the argument complete.

Whatever the reason for the omission of a needed premiss, a complete and fair evaluation of the argument requires that you supply the missing premisses if you can. Even where the premisses are missing because the author doesn't know he needs them, it is reasonable to try to patch up the argument to see what repairs might work. Doing this will prevent you from being merely picky by attacking an argument as invalid when minor repairs might have saved it.

But there is a catch. *Any* argument, no matter how foolish, can be made into a valid argument through the addition of properly selected premisses. There are three easy ways to do this. (1) Add the conclusion of the argument to its set of premisses, making the argument circular, but trivially valid. (2) Add a premiss which contradicts one of the other premisses, making the argument automatically valid no matter what its conclusion. (3) Add a premiss which states that *if* one or more of the other premisses is true *then* the conclusion is also true. Of these three only the third ever has any interest for us, because the first two tactics show nothing at all about the relation between the original premisses and the conclusion. The third method can be used to make the following invalid argument into a valid one, in a reasonable way:

It is raining.
_____
There are clouds overhead.

Once the reasonable premiss, "If it is raining, then there are clouds overhead", has been added, the argument exhibits one of the most common valid forms. (Even if you wish to complain that the added premiss may not be true in the real world, it is still reasonable to assume that anyone who put forward the original argument was thinking that rain always means clouds above, and adding this premiss still makes the argument valid.)

Since it is *always* possible to make any argument valid by adding premisses, one cannot distinguish an easily repaired invalid argument from a hopelessly muddled argument merely by asking whether it is possible to make the argument valid by the addition of premisses. Accordingly, we will want to ask whether a given invalid argument can be made valid or at least more nearly valid by the addition of certain special, restricted kinds of premisses—premisses which are consistent with the original premisses and which the author of the original argument would have been willing to accept, or premisses consistent with the original premisses which are interesting for some reason.

Naturally, there are no precise methods for deciding whether a given premiss is interesting, or is one which the author of the argument would be willing to add. Often the context in which the argument is presented will help us decide. If we are considering an argument drawn from the last half of a book, we may want to consider adding premisses which the author argued for in the first half, for the context suggests that the author would be willing to use such premisses, and perhaps omitted them from his own statement of the argument merely because he did not wish to repeat himself.

When evaluating the arguments of a particular author, it is reasonable to employ a principle of charity. One seeks to add premisses which help, not hinder, the argument, because one is trying to figure out what may have led the author to say what he said. It would be useless in such an enterprise to add premisses which detract or do no good, unless there is some special reason in the particular case for doing so. (One such special reason might be that you have evidence that shows the author of the original argument actually did believe the premiss you want to add, and that he thought it relevant to the issue.)

To avoid confusion though, keep one thing in mind: every time you add a premiss to an argument, technically, you thereby create a *new* argument. This is true even when the added premiss is something obvious or something the author of the original argument would clearly have approved. The addition of *any* proposition to the existing set of premisses will constitute an enlargement of that set.

This last point is especially important when you're not sure what the added premiss is going to be. You may want to try out various different propositions in the role of missing premiss. In such a case, each time you put a different statement into the argument you are creating a new argument to be evaluated. One such argument might be valid while another might be invalid. The work you have already done on validity should help you decide in each case what premisses might be missing from a weak argument. If you are familiar with standard patterns of inference, you are in a good position to try filling in the missing pieces of an incompletely stated argument, because you will see that certain kinds of pieces will complete the argument and make it fit a valid pattern. Another possible technique to shore up an argument relates to counterexamples: since a counterexample to the original argument displays one particular loophole in the argument, you can improve the argument by adding a premiss which blocks the counterexample, thus filling that particular loophole. Of course, other loopholes and other counterexamples may yet remain.

Let me illustrate these two techniques. Suppose two people are inspecting the scene of an auto accident and one of them is suspicious of the driver's story that the brakes failed. They look for skid marks and find none. The other one of them is more inclined to believe the driver and is encouraged by the failure to find any skid marks. He gives the following incomplete argument:

> If the brakes were working, there would be skid marks. So, the brakes weren't working.

Now you may be thinking that there could have been other reasons why no skid marks were found, such as the possibility that the driver fell asleep at the wheel and never saw the other car and never applied the brakes. But to think along those lines is to question the premiss stated above. That's fine. One of the things to check for in evaluating an argument is whether the premisses of that argument are true. However, that is irrelevant to the validity of the argument.

Looking again at the argument as stated, we see that it is invalid. One could construct a counterexample by supposing that there *are* skid marks and that the brakes were working. To patch up the argument, we could try to block this counterexample by adding a premiss which says that there are *no* skid marks. The new argument, with the added premiss, is no longer vulnerable to the counterexample offered above.

Are there any remaining loopholes? No, because the argument has been made valid by the addition of the above premiss. In fact, you could have seen this by thinking of the form which the argument takes once the added premiss is in place. You could have come up with the added premiss in the first place just by thinking of that form. The original argument had the form

IF A THEN B

NOT A

which is clearly invalid. But by adding a premiss of the form 'NOT B' we make the argument diagram valid, as you may verify for yourself.

Thus, either by blocking counterexamples or by finding a common valid pattern which an argument can be made to fit by adding premisses, one may come up with useful ways to patch up an invalid argument. Even if it is not possible to block all the counterexamples by adding reasonable premisses, if the main ones can be blocked, the argument will at least come closer to being valid.

Adding premisses to arguments

59.1 Why did the speaker in the auto accident situation not state the missing premiss?

59.2 Add a nontrivial premiss to each of the following arguments which will make them valid. Prove the resulting arguments valid.

(a) If John goes to Fred's party, he won't be able to see Linda this Friday, because Fred and Linda don't get along. But if he doesn't see Linda this Friday, she'll be angry with him. Because he likes Linda more than Fred, John has decided not to go to the party. So, Linda will not be mad at John.

(b) We will have to step up production of the new model, or cut back on the number of dealers we are supplying. So, we will cut back on the number of dealers.

(c) The University housing regulations should be revised. After all, they require most students to live in the dormitories unless these students get a waiver from the administration. People should have the right to live where they want.

59.3 Separate the argument contained in the following passage from the background information contained in it, and prove the argument valid once you have added some nontrivial premisses to it.

Because the Social Security System is now paying out more than it takes in, the Social Security Advisory Council decided that Social Security taxes would have to be raised, or else expenses would have to be cut. They chose the latter, and recommended that the cost of medicare be removed from the system, so that medicare would be paid for out of general tax revenues rather than Social Security taxes. This proposal, if adopted, would be the first step in changing the Social Security System into an out-and-out welfare program. Therefore, the Council's recommendation should be dropped.

59.4 The following argument commits one of the fallacies discussed near the beginning of this book. If you first figure out what fallacy the argument commits it will help you to find an interesting premiss to add to the argument. Then add premisses to eliminate the fallacy in the argument and make it valid. You will probably not be able to prove the resulting argument valid unless you are very lucky.

If we abolish the grading system here at the university and give no grades at all, then there will be no way to determine how many courses a student has actually completed satisfactorily. That will mean that prospective employers will adopt their own testing procedures in order to determine whether students have learned anything. But these testing procedures will stress only those things which each particular employer happens to care about, and students will be faced with a wide variety of different tests from a wide variety of prospective employers. The result would be chaos. So, we ought to keep our present grading system at the university.

59.5 Add a validity-ensuring premiss to the following argument:

According to page 123 of the accounting reference manual, assets of this type are to be valued at an amount equal to their purchase price. Accordingly, the correct way to record these assets in our books is to find out what the purchase price was and list them that way.

59.6 Construct counterexamples for each of the following arguments. Then add a premiss or two which are nontrivial to block the counterexamples. Are the resulting arguments valid?

a. If we don't discriminate against women in the job market, they will be encouraged to seek fulfillment in careers. But if women are thusly encouraged, large numbers of them will be unsatisfied. If large numbers of women are going around unsatisfied, that will make for a more unstable society in which all sorts of changes may easily take place. Hence we ought to discriminate against women who are looking for a job.

b. The state will increase its financial support for the university if and only if the priorities of the legislature are shifted in favor of higher education. But if such a shift were to occur, there would be many other projects in the state which could not receive adequate funding, and the people who benefit from those projects would complain both loudly and bitterly. If the state does not increase its financial support for the university, the tuition will have to be raised. So, tuition will have to be raised.

c. Everyone who attended the rally was inspired to act on behalf of the underpriviledged groups in our country. I asked Tom to contribute to the children's food fund, but he refused. So Tom did not attend the rally.

d. Despite the representations made by its promoters, no one who entered the contest won anything. So no one who entered the contest owes any income taxes as a result of entering.

60

Drawing a conclusion from given premisses

In this book you have been confronted almost exclusively with arguments complete with conclusions. Although that may be helpful while learning logical analysis, it is not true to life, for there are many situations in which you need to come up with the conclusion on your own. There is no formula for doing this constructively, but it is one of the most important human mental tasks. It often requires tremendous insight and inventiveness.

Here are just some of the situations in which it would be important to arrive at your own conclusions supported by given premisses:

Suppose you have a set of three beliefs, A, B, and C, about religion, politics, business, or whatever. If these beliefs are related to one another at all, it may be possible to find conclusions which follow from them by means of a valid or a close-to-valid argument. If you knew of these conclusions, you would be aware of some of the things your beliefs commit you to. You might discover then that these beliefs lead to conclusions you do not like, or to conclusions that are interesting for some other reasons.

On the other hand, suppose you are intrigued by a set of claims A, B, and C. You don't know whether to believe these claims or not. Perhaps they constitute a scientific theory or a view about human psychology. One of the things that will help you decide whether these claims are believable will be the set of implications which follow from these claims. If you can discover several important conclusions which are derived from A, B, and C, that will help you to see where these claims lead. If these conclusions disagree with other evidence you already possess independently, then you will see that A, B, and C do not fit well with other things you accept.

That sort of exploration of the implications of a given set of claims constitutes one of the most important critical tools used in evaluating theories. If you can show that theoretical claims A, B, and C together imply D by means of either a valid argument or a close-to-valid argument, and D is something that is known to be false, you have shown either that A, B, and C cannot all be true, or that it is unlikely that they are all true.

Let me illustrate. Suppose you are a health inspector checking into an outbreak of food poisoning among the patrons of Sam's Cafe. You are discussing the case with one of the doctors who treated the people who got sick, and she has a theory about what caused the poisoning. She tells you that all the patients she treated had eaten peas and there didn't seem to be any other dish that all of them had eaten at Sam's that day. On the basis of this evidence, she has tentatively concluded that there was something wrong with the peas.

Let's try to lay out some of the important things everyone probably would agree to in this situation; then, using these things, or some selection drawn from them, we'll see what interesting conclusions might be drawn. Presumably, it is safe to assume at least these three things:

(1) The patients all ate peas at Sam's that day.
(2) There was nothing else the patients all ate at Sam's that day.
(3) If eating peas caused the patients to get sick, then eating peas at Sam's that day also caused anyone else who ate them to get sick.

If you are a health inspector wishing to test the doctor's hypothesis that it was the peas which caused the problem, you might add a statement of that hypothesis to the above set of three assumptions:

(4) Eating the peas caused the patients to get sick.

Then, to see what conclusions this additional premiss allows us to draw from the enlarged premiss set, you just play around with the logical relationships that you can find among all four premisses. You are looking for conclusions which follow validly but which also can be independently checked against the facts. One such conclusion, obtainable from (3) and (4) by *modus ponens,* or horseshoe elimination, or common sense, is

(5) Eating peas at Sam's that day also caused anyone else who ate them to get sick.

There is no mechanical method for finding fruitful conclusions such as this one. If you have studied the rules of inference found earlier in this book, you may be able to notice when some of the rules will help you develop useful conclusions.

Now that you have found conclusion (5), you can go out and test it to see if it fits the actual facts. If after investigation, (5) turns out to be true, then so much the better for the doctor's hypothesis, since the hypothesis leads to a true conclusion. However, if you discover that (5) is false, that would show that something in the premiss set which led to (5) must be amiss. Actually, (5) was obtained from just (3) and (4); so if (5) is false, (3) or (4) or both must be false. (4) is the hypothesis you are testing, and (3) is assumed to be true. Hence, if you continue to accept (3), (4) will have to be given up if (5) turns out false.

The simple maneuver outlined above is well known, and doubtlessly, you have used it before. Nevertheless, people generally do not use this testing strategy nearly as often as it would be profitable to do so. I have found, for example, in teaching various courses at the university that students generally are reluctant to draw out the implications of the views they themselves hold, even though that is one of the very best ways to check the correctness of any set of views.

What do you do after you have obtained a conclusion like (5) above which turns out to be false? That depends on your aims. If you were only trying to show that (4) is incorrect, you might quit. On the other hand, if you are trying to get to the bottom of what happened at Sam's that day, you can't quit yet. Try another hypothesis in place of (4), and see what conclusions follow from it together with (1) through (3). Perhaps you can come up with a reasonable hypothesis which does not have false implications when combined with (1) through (3).

The preceding example shows how you might test the truth of a hypothesis by examining what conclusions it implies. But that is only one of the situations in which it can be useful to draw your own conclusions from a given set of premisses. As I mentioned at the beginning of this Section, you might want to draw some conclusions from some of the things which you believe or know to be true, merely to find out some new truths. This sort of thing is done in mathematics, for example, when someone proves a theorem. In that case, the axioms, definitions, and previously proved theorems of the mathematical theory constitute the set of premisses from which one works. The proof proceeds from those premisses to the conclusion by means of a valid argument. The conclusion then becomes a new known-to-be-true theorem.

This same procedure can be used at *any* time you have a set of propositions and are willing to use them as a premiss set. Quite often, the conclusion you want to derive from a premiss set has to do with what a person *ought to do* in a given situation. The premisses consist of statements about the situation, and the conclusion you want is the true one which tells you the right course of action to take.

Although there is no precise method for coming up with interesting conclusions, I can summarize in a general way the steps you ought to take in making the attempt:

(1)   First, try to become aware of all the relevant premisses. We often have more premisses at our disposal than we can immediately think of. The more relevant premisses you have available, the greater the chances they will together imply something worthwhile.

(2) Next, look for anything which obviously follows from the available premisses. Proceed carefully in a step-by-step fashion, perhaps first of all obtaining some intermediate conclusions, and then using these to obtain your final conclusion. You want conclusions which will help you satisfy your purposes. For example, sometimes you want a conclusion which is independently verifiable, as in the pea-eating example; but on other occasions some other type of conclusion may be of use.

Drawing a conclusion from given premisses

60.1 Background: The university is trying to decide whether to raise the dormitory rates next year. One administrator who prefers to remain anonymous remarks:

We'll have to cut services in the dorms or raise the rates. Of course, if we raise the rates, people will complain. But if we cut services, people will complain, too.

What conclusion is the author of these remarks driving at?

60.2 Derive the diagram of the conclusion you mentioned in answer to 60.1 from the diagrams of the given premisses.

60.3 Background: You're planning a party and you have invited Mike and Joe, even though you don't know either one of them very well. You ask your friend who knows them better whether he thinks they'll come. Your friend tells you:

Mike and Joe won't both come to the party, because they don't get along well with each other.

What can be validly inferred from this statement regarding whether Mike will come? Whether Joe will come? (Try to come up with some interesting conclusions about Mike and Joe's possible attendance.)

60.4 Derive some conclusion diagram from the diagrams of the premiss in the preceding exercise. Translate one or two of these back into English.

60.5 Your roommate, who last semester took the same biology course you are now taking has been watching you study for an upcoming test in the course. Even though you've been studying for hours, you're worried that you will not be prepared to take the test. Your roommate tries to reassure you by saying

Anyone who studies for that course will do well.

This remark is a very incomplete statement of an argument. What conclusion is your roommate trying to get you to believe? What premiss would need to be added to the above remark before one would have a valid argument for that conclusion?

60.6 Derive the desired conclusion diagram from the complete list of premiss diagrams relevant to Exercise 60.5.

60.7 Find an interesting conclusion which you can get validly only by using all the following premises:

Either the witness is out of town or he is in hiding. He could be out of town only if we did not put the roadblocks up in time to stop him from leaving town. But we got the roadblocks up in plenty of time. On the other hand, if he is hiding, he must be staying with a friend.

60.8 Prove in some way that the conclusion you have given above actually does follow validly from the premises, at least in so far as the argument can be accurately symbolized.

60.9 Background: While Republican Richard Nixon was president of the United States, someone broke into the Democratic party headquarters in the Watergate building in Washington, D.C. Three of Nixon's aides, named Haldeman, Erlichman, and Dean were examined in the controversial investigation which followed. There was concern that the president himself may somehow have been involved.

What unstated conclusion is it most likely the author of the following passage wants us to draw from what he says?

If Nixon ordered Haldeman to keep the FBI from investigating the Watergate break-in, Nixon was guilty of conspiracy to obstruct justice. He in fact must have ordered it unless Erlichman did. But Erlichman could have ordered it only if he was in Washington during November, while Dean testified truthfully that Erlichman was in California for that month. Obviously, Erlichman was not present in Washington in November if he was in California.

60.10 Prove in some way that the diagram of the preceding argument, including the conclusion you have added, is valid.

60.11 *The trip to China in May by President Carter's national security adviser . . . was . . . just one more dangerous playing of the "China Card". Briefly, the "China Card" refers to the leverage lately considered available to U.S. diplomats . . . to apply pressure against the Soviet Union by courting . . . China . . . . To play the "China Card" we risk encouraging an already paranoid Russia to perceive itself to be . . . between two fires, faced with the need to fight on two fronts. No one, of course, can predict exactly how the Soviets would react to this situation. But neither can anyone exclude the possibility that they would react much as did Nazi Germany.*[1]

a. What conclusion do you think this author would want you to draw from his premises regarding the advisability of playing the "China Card"? What added premises does one need in order to get close to having a completely stated valid argument for such a conclusion?

[1]Alexander Yanov, in an editorial/comment syndicated by the *Los Angeles Times*, July 16, 1978.

b. Make a list of all the conclusions you think the above passage is intended to support. Do some of these conclusions play the role of intermediate conclusions, to be used in support of others of these conclusions? If so, state which conclusions support which other conclusions.

60.12 What unstated conclusions might be obtained from the following premisses? State any additional premisses needed to arrive at these conclusions. Would the arguments needed be valid, or would they just be close-
to-valid?

It has been known for some time that the use of oral contraceptives increases the risk of death from circulatory diseases, with the amount of
risk increasing as the length of drug use increases. However, pregnancy
also carries with it a risk of death from various causes, such as infection
or hemorrhage. If the risk of death from complications of pregnancy were
equal to the risk of death from the use of oral contraceptives, those who
did not desire children would presumably be better off using the drug,
since its use would prevent the distress arising from unwanted pregnancy. But the risks are in fact not equal. The pill is far more likely to kill
a woman than all the risks of pregnancy combined.

60.13 Suppose premisses A, B, and C validly imply conclusion D. Suppose
premisses E, F, and G validly imply conclusion NOT D. Does that mean
that if A is true, E is false? Does it mean that either A, B, and C are all
true or else E, F, and G are all true? Does it mean that if A, B, and C are
all true, at least one of E, F, and G is false? Does it mean that if E, F, and
G are all true, at least one of A, B, and C is false? Explain each of your
answers.

60.14 Suppose you discover that four of your beliefs when used as a premiss
set imply a contradiction, by means of a valid argument. What does that
say about your four beliefs? What seems to you to be the most reasonable
thing to do about these beliefs in such a case?

60.15 Derive some interesting conclusions from each of the following premiss
sets, utilizing the rules of propositional logic. In each case, use all the
premisses at least once somewhere in your derivation.

a. $P \,\&\, (R \lor Q)$
$\sim R$
$\underline{S \equiv (P \,\&\, Q)}$

b. $R \supset (A \,\&\, \sim B)$
$A \supset (S \supset Q)$
$\underline{S }$

c. $S \lor \sim U$
$\sim S \lor P$
$\underline{U \lor P}$

60.16 Derive some interesting conclusions from each of the following premiss
sets, utilizing the rules of predicate logic. In each case, use all the premisses at least once somewhere in your derivation.

a. M
$\forall x(Px \supset (Qx \,\&\, Rx))$
$\underline{M \supset Pa}$

b. $\exists x(Px \,\&\, Qx)$
$\underline{\forall x(Px \supset Rx)}$

c. $\forall x \forall y(Pxy \supset Pyx)$
$\underline{\forall x \exists y Pxy}$

10

Reasonable
Invalid Arguments

61

Induction and deduction

In the previous two Sections, I occasionally said some peculiar-sounding things about arguments that are "close-to-valid" without being valid. The time has come to explain in detail what it is for an argument to be close-to-valid.

Commonly, when we argue, we want to draw a conclusion only from premisses we already know or believe, but our ignorance limits the number of premisses available to us, and validity is a very high standard for an argument to meet. In the courtroom, evidence is presented and conclusions are drawn, but rarely is the evidence conclusive enough to yield the conclusion *validly*. Similarly, in science data is gathered and theories are constructed but rarely will the data plus other assumptions *validly* lead to the theory as a conclusion.

Yet, in these various situations when we cannot come up with *valid* arguments, we do nevertheless come up with *arguments,* many of which make a great deal of sense even though they are not valid. Many of these arguments seem to "prove" their conclusions true "for all practical purposes", even though there may be some crazy, science fiction-like counterexamples which show them to be invalid. But utilizing concepts presented thus far in this book, all we can say about such arguments is that they are invalid. We will now be turning our attention to invalid arguments, to see what makes some of them more well-reasoned than others.

I should warn you that the area of logic which we now are entering is not nearly so well developed as is the logic of validity. In fact much of what I am about to say is controversial, and the terminology which I use is not standard. Nevertheless, if something useful is said about techniques for evaluating and discriminating among invalid arguments, that should prove valuable to you.

Amidst all the controversy surrounding these topics, one point of terminology has become fairly standard: the theory regarding what constitutes a valid argument is known as *deductive* logic, while the developing theory which we are about to look at, regarding the logical discrimination between invalid arguments, is known as *inductive* logic. (However, even here there is a small problem. Many authors talk as though the techniques of inductive logic apply only to *some* invalid arguments, arguments of a special kind which they call "inductive arguments". I will not make such a restriction, nor will I attempt to set up any special category of inductive arguments. It seems easiest merely to think of inductive logic as applicable to all invalid arguments.)

Invalid arguments naturally come in abundant variety. Unfortunately, logicians don't have a theory that can handle all of them, or even most of them, in any detail. Consequently, the usual approach consists of picking out a few special types of invalid argument and trying to say something about how one might go about evaluating arguments of those types; I'll do that in the following three Sections. In this Section I will discuss the evaluation of all invalid arguments.

In order to conduct inductive analysis, we need to know more precisely what we are looking for. What exactly is it for an argument to be "close-to-valid", or "well reasoned"? We need a precise definition of a basic evaluative concept to be used throughout inductive logic, a concept which will play a role in inductive logic analogous to that played by the concept of validity in deductive logic. It will turn out that there are *two* such central concepts in inductive logic.

The first such concept is the notion of a *tight* argument. *An argument is tight if and only if it is invalid but the truth of the conclusion is highly likely, given the truth of the premises.* If the conclusion has only a fifty-fifty chance of being true, given the premises, then the argument is not tight because the conclusion is not sufficiently likely on the basis of the given premises. In determining whether a particular argument satisfies the definition of "tight argument", one must consider only what is included in the argument itself, without bringing to bear any additional background information or beliefs one might have concerning the subject matter under discussion. In this regard, the evaluation of arguments with respect to tightness parallels their evaluation with respect to validity, for in the latter case the argument was considered valid only when it contained within itself sufficient support for its conclusion to render the conclusion certain given the premises.

Accordingly, the following argument is *not* tight:

That man is standing there with a smoking gun in his hand.
He recently fired the gun.

There is no probable connection presented by the argument between a gun's being fired and smoke. The argument may seem tight, but only because you are assuming some extra things about smoking guns not stated in the argument. As far as this argument itself goes, smoke could come from guns all the time, as from chimneys, without any shooting. In order to make the above argument tight it would be necessary to add a great many premises.

Note that it is not necessary that the premises of a tight argument be true. In fact, the truth of the premises is irrelevant to the tightness of an argument, in just the same way it was irrelevant to the validity of the argument. When determining whether a given argument is tight, you only ask whether the conclusion would be probable given the truth of the premises. Naturally, if the premises are false, then the tightness of the argument will say nothing about the truth of the conclusion. Here too, the situation is like that which exists for valid arguments, as the validity of an argument says nothing about the truth of the conclusion when one or more of the premisses is false.

Tight arguments, in the specific sense defined here, do not occur frequently in real life. The only examples that come readily to mind are those which involve explicit reference to random processes, such as the following:

P1. This fair coin has been tossed fairly 1,000 times, and has landed showing just one of its sides each time.

P2. One of the coin's two sides is "heads" and the other is "tails".

P3. The distribution of heads and tails which results from the repeated fair tossing of a fair coin which lands each time showing just one of its two sides is random.

C.   No more than 600 heads showed up in the 1,000 landings of the coin.

Although the logical relationships in this argument are complex, and a full analysis of them would require a long development of the mathematical theory of probability, I think it is fairly clear that the probability of C's being true, given P1 through P3, is high. If so, this argument fits our definition of a tight argument.

Although not many arguments found in real life are tight, a few are, and, more importantly, we can use the notion of a tight argument in order to build the second central notion in our inductive logic, namely, the notion of an argument which is *tight relative to background assumptions*. (For ease of reference in the future, I will abbreviate the name of this notion by the acronym "TABA", standing for "tight argument relative to background assumptions".) There are a great many arguments in real life which are TABA's.

Many arguments which strike us as being reasonable even though they are invalid also fail to be tight; but these arguments would become tight if a great many assumptions were added to them—premisses which specify the relevant portions of our commonly held beliefs. Once those assumptions are made explicit by being added to the premisses of the argument, the argument will become valid or tight. But we don't take the trouble in everyday affairs to try to actually spell out all these missing premisses. It would take too long, and would often be impossible since we don't even know what all of our assumptions are in many cases. *An argument which is not tight, but which would become tight once all the relevant background assumptions were added to its premisses is known as a TABA.*

Examples of TABA's are easy to come by. Here's one we saw before:
It's raining outside; therefore it's cloudy overhead.

This argument is clearly invalid as it stands, for after all it is possible that rain might fall in the absence of clouds. We might imagine that we are living in a world where rain very frequently condenses out of the clear air and clouds are almost unheard of. In such a world, the truth of the premisses would not make the conclusion even likely to be true. Therefore, I think the argument fails to be tight or valid.

However, to show that the argument is a TABA, I need to show that its conclusion is probable given the premisses—once enough background assumptions are added to its premiss set. What added premisses would be natural here? Presumably, some like these:

(1) We are not in possession of any information to indicate that this particular rain storm is unusual with respect to the condition of the overhead
sky.

(2) Almost always, when it rains outside, it's cloudy overhead.

Once premisses like these are added to the original premiss which says
that it is actually raining outside, the conclusion does become probable relative
to the set of premisses.

It will usually not be very clear in any given case of a TABA just what premisses can be added to make the argument tight. However, it is important to note
that in order to count as a TABA, only certain kinds of premisses may be added
to an argument to make it tight. The only ones that may be added are those
which can be reasonably described as *relevant background assumptions* about
the topic being discussed in the argument.

Nevertheless, herein lies one important reason why people disagree with
one another so much: we do not all share the same background assumptions.
Thus, when I hear an argument and evaluate it, I may think it is a TABA while
you may not, because we may not be thinking of the same background assumptions. For example, some audiences would readily consider this argument
to be a TABA:

Last week, the *National Enquirer* reported that several "top psychics" predicted that California would endure a major earthquake during the next
year.[1]

There will be a major earthquake next year in California.

Obviously, the argument is invalid, and not tight as it stands, but the question
is, What assumptions can count as relevant background here? Those who find
the argument persuasive presumably would want to add at least these:

(1) The *National Enquirer* is a reputable, careful paper which generally
carries accurate reports.

(2) Those labeled "top psychics" by the *National Enquirer* have the ability
to foretell the future with a fair degree of accuracy.

On the other hand, I myself would deny both (1) and (2), since I believe the
National Enquirer is not especially accurate in its reporting and I do not believe
that so-called psychics can foretell the future by any special ability nor that they
can obtain a high degree of accuracy. Without (1) and (2), it seems there will
not be any other reasonable premisses to add to the argument to make it tight.
Thus, I do not find the argument persuasive.

What can logic do about such disagreements? It is crucial to emphasize
that with respect to determining whether an argument is a TABA, the premisses
which one may add to the argument to make it tight must be actual background
assumptions being made by real people who have to deal with the original incomplete argument. Of course, all arguments are either tight or valid once the
right premisses have been added, but the argument should count as a TABA

---

[1]The *National Enquirer* is a nationally circulated newspaper, sold primarily in drug stores and
groceries, featuring articles on psychics, medicine, the private lives of famous people, and business
advice of a popular nature.

only when the added premises which make it tight are ones which can reasonably be said to be assumed by real people within the context of the argument. This puts a limit on what premises can be added to an argument in order to determine whether it is a TABA. Thus, with respect to the *National Enquirer* argument above, there are such relevant background assumptions, and thus the argument is a TABA. My personal disagreement with the argument turns out not to be so much a disagreement with its internal logical structure as it is a disagreement with the background assumptions.

To help settle disputes about the *actual truth* of the conclusions of arguments like the preceding one, we make explicit the background assumptions and then if there is disagreement about their truth, we construct *new* arguments about their truth. Accordingly, if someone really wanted to persuade me, on the basis of what the *National Enquirer's* psychics say, that there will be earthquakes in California next year they will have to persuade me of the truth of the necessary background assumptions, and that will take new arguments.

So far, we have seen that some arguments are valid as they stand, while others aren't. Those that aren't can sometimes be made to be valid by the reasonable addition of premises, as in Section 59. Those that can't be made valid in that way are subject to evaluation by the techniques of inductive logic. The first key concept of inductive logic is the idea of a tight argument; when an argument is tight that does not mean its conclusion is true, but it does mean that its conclusion is *probably* true given its premises.

Some arguments that are not tight as they stand can be made tight by the addition of premises which are relevant and, within the context of the argument, can reasonably be called background assumptions. These are called TABA's. An argument may be a TABA and have a false conclusion, or false premiss, or false background assumptions. In discussing validity, we used the special term "sound" to refer to valid arguments which had *true* premises. It would be useful to have a special term for a *tight* argument with true premises; so, I will call such argument *strong*. And when we have a TABA, that argument is *strong* relative to its background assumptions if all its premises and background assumptions are true.

You need one additional piece of information before the basic structure of inductive logic will be in place: probability comes in *degrees,* and is not merely a simple "yes" or "no" matter. Thus, when we talk about an argument's tightness, that is also something that comes in degrees: some arguments are tighter than others. Normally, when you say that something is probable, you mean that it is fairly highly probable; I will follow the same practice in labeling arguments as "tight". Only those arguments whose conclusions are fairly highly probable, given their premises, will be called *tight*. Of course, this is all somewhat vague, since the exact point at which a conclusion ceases to be fairly highly probable and becomes not quite highly probable isn't specified. No one knows how to measure the exact degree of probability associated with the conclusions of all the different kinds of arguments which come up in real life, but the mathematical theory of probability and statistics can do the job for certain special cases. For most arguments, we have to be content with being vague. We do not have the time to get involved with the details of the mathematical theory of probability here, but I recommend it to you as a subject in logic worth pursuing. If an argument is a TABA, or a strong argument, does that mean it's a "good" argument? Well, that depends on what you mean by "good". Rather than asking how "good" a particular argument is, or how "logical" it is, usually it is more fruitful to ask a more specific question, such as, "Is the argument valid?" "Is it tight?" "Is it tight relative to some reasonable set of background assumptions?" "If so, are these assumptions true?" "Is the argument valid, once some reasonable premisses are added?" "If so, are these premisses true?" And so on. There are many different ways in which arguments can be "good".

Note, though, that we continue the practice begun at the very beginning of the book, separating the questions concerning the truth of the premisses from the questions concerning the relationship between premisses and conclusion. An argument which is tight may still have all blatantly false premisses. Similarly, a TABA may be a TABA and yet require background assumptions which are ridiculous. There are always then two main questions in argument evaluation: Are the premisses true? Are the premisses related to the conclusion in such a way as to make the conclusion true or probable if the premisses are true? In general logic by itself cannot answer the first of these two questions, although it can help to answer it when coupled with experience or other data. The second of these two questions is the province of logic.

As a practical matter, when you are trying to determine whether a given argument is tight, you are going to be doing something very much like constructing counterexamples. Of course, if you are dealing with an invalid argument (as you always will be when you are using inductive techniques properly), there will be counterexamples. But now you are not going to be satisfied merely to know there are counterexamples. You are going to want to try to think of various types of counterexamples, and various types of story in which the conclusion may not actually be false, but only unlikely to be true. The reason: you are trying to judge how likely it is that the conclusion would be false if the premisses are true.

Accordingly, in looking at the earlier argument regarding the earthquakes as it was originally stated (before the addition of background assumptions) I can say that "psychics" are all frauds or deluded people who have no particular ability to predict the future, and that there have been enough recent quakes in California to relieve the pressures there, making it unlikely there will be a major quake next year. With all this detail built into my counterexample, it no longer matters whether there will in fact be an earthquake in California next year in my story. What matters is just that it is *unlikely* within the story world I've created. Since I have included the premiss of the original argument in my story and yet its conclusion is unlikely in the light of the premisses, that shows the original argument, as stated, fails to be tight.

A somewhat different counterexample would have worked just as well. I could have imagined that there are a few special people, the true psychics, who can predict the future by means of special powers they possess, but that the *National Enquirer* does not interview these special people in order to write their sensationalized stories. Instead, according to my counterexample, the *National Enquirer* just makes up their stories, according to what they think will sell newspapers. Perhaps the so-called psychics they supposedly quote do not even exist. Suppose also that there will probably not be any major earthquakes in California next year. This story presents one more way in which the original argument fails to be tight, for this story explains another way in which the premisses could be true without thereby ensuring that the conclusion is likely.

Naturally, the background assumptions which I mentioned earlier in connection with this argument (those stating the *National Enquirer* is reputable and the psychics are able to predict things) could be added to the argument by those who like the argument in order to block my examples. With those premisses added to the argument it becomes much more difficult, if not impossible, to think of a coherent example which would show that the resulting argument is not tight. I leave it to you to think about that. Perhaps, then, the argument which results from adding those background assumptions is in fact tight, so that we can say the *original* argument is a TABA.

The following chart summarizes the relationships between the various basic evaluative concepts now available for your use:

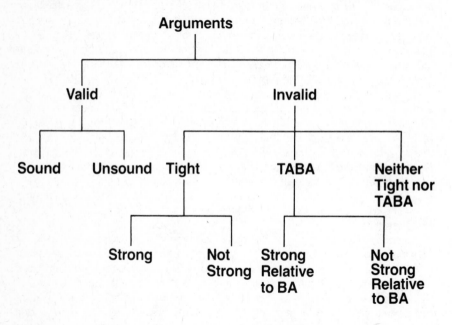

Induction and deduction

61.1 Which of the following statements are correct?
   a. All valid arguments are tight.
   b. All strong arguments are tight.
   c. All tight arguments are strong.
   d. An argument which is strong relative to a set of background assumptions is always tight relative to those same background assumptions.
   e. Some tight arguments are valid.
   f. Some sound arguments are strong.
   g. A tight argument does not have any counterexamples.
   h. A strong argument does not have any counterexamples.

i. The concepts of validity and soundness belong to deductive logic, while the concepts of tightness and strength belong to inductive logic.
j. An argument with all true premisses is either sound or strong.
k. An argument with all true premisses is either tight or valid.
l. If an argument has a false conclusion, then it cannot be tight.
m. If an argument has a false premiss, then it cannot be strong.
n. If an argument has a false conclusion, then it cannot be strong.
o. All strong arguments have true conclusions.
p. All strong arguments have true premisses.

61.2 Prove that the following argument is not tight.

Mary is pulling weeds out of the lawn, in the direct sun, even though it's a hot, humid day with temperatures in the 90's.

Mary is perspiring.

61.3 Is the argument of 61.2 a TABA? Defend your answer in detail.

61.4 Is the following argument tight? Is it a TABA? Defend your answers in detail.

My friends Jack, Mike, and Terry all decided to go to State U. last year after they graduated from high school, and all three of them have told me how much they like it there. Jack is majoring in history, Mike in psychology, and Terry in business. Jack is an outgoing kind of person who likes to have a good time, but the other two are more quiet, serious types. So, I guess most of the freshmen at State U. are fairly well satisfied there.

61.5 If the conclusion of the above argument were changed so that it read

I guess all the freshmen at State U. are fairly well satisfied there

would the above argument become tighter or would it become less tight? What if the conclusion were changed instead to read

I guess most of the freshmen at State U who are like my friends in most respects also would be quite satisfied with the place?

61.6 The next few questions pertain to the following passage:

What justifies the law in punishing people? Some have held that the only possible justification lies in the possibility of reforming the wrongdoer through the process, while others have argued that the sole purpose and justification of legal punishment stems from its deterrent effect on society at large. The primary difficulty with the second of these theories comes because this theory of punishment justifies the occasional punishment of an innocent person. Someone believed by potential bank robbers to have committed a bank robbery should be punished, according to the deterrence theory, whether or not he actually robbed the bank, in order to deter the potential robbers from making the attempt themselves.

a. Assuming that the first two sentences of the passage are of an introductory nature, and that the last two sentences state an argument, what is

the stated conclusion of that argument?
b. Is that argument valid? Tight? A TABA? Sound? Strong? Defend your answers.

61.7 The questions which follow pertain to this passage:

It cannot be denied that a person who has to nurse his or her spouse through the final agonizing and terrible months of a prolonged, painful, and incurable illness may be so deeply disturbed by the process that they come away from it ruined mentally or physically. So, in such a case, the nursing spouse would have been better off if they had been permitted to have a lethal dose of a sedative administered to their ailing partner, with his or her consent of course, to end the process before it took its toll.

a. Why is this argument not tight?
b. What premises might reasonably be added to it in the effort to make it tight?
c. Do you think it likely that *reasonable* premises could be added to this argument to make it valid? Why, or why not?

61.8 Answer the same set of questions as in Exercise 61.7, but with respect to the following argument:

I got a "C" in that course rather than the "B" I deserved, because the instructor didn't like me. I know he didn't like me ever since I gave him a hard time about the way he graded that first test. Sure, maybe the answers he chose were right, but the questions were so picky, and half the stuff we studied wasn't even on the test. And then there was the time I proved in class that he was wrong about how to do one of the problems. Man, was he embarrassed! (Hint: assume the first sentence expresses the intended conclusion.)

62

Arguments by similarity

Since there aren't any general techniques applicable to the evaluation of all the differing kinds of invalid arguments, I will turn to talk about some of the common types of such arguments in an attempt to leave you with some clues and methods for their evaluation. In this Section, we will look at one of the most simple and most common types—those arguments which are built upon similarity.

There may be a number of different ways in which similarities are employed in argumentation, but the ways I wish to look at share one common theme: in these arguments, one draws a conclusion about some things by claiming they are similar to other things which are described in the premises. One might draw a conclusion about tomorrow's weather by comparing it to the weather experienced on other days in the past which were preceded by today's conditions. Or a conclusion about the mating habits of a newly discovered species by comparison to those of similar species already known.

Clearly such arguments play an important role in our everyday lives. In particular, we "size up" people and situations very quickly as we encounter them, often using a reasoning process which compares them to similar, previously encountered people or situations. This reasoning fits the format described above. The sort of "image" a person projects (that person's appearance and public behavior) leads people to conclude that he is comparable to others, or to certain idealized types. It is as though we said to ourselves upon meeting someone,

> This person is similar to the dynamic executive type in respects A, B, and C.
> Dynamic executive type persons behave in ways D, E, and F.
>
> Therefore, this person will behave in ways D, E, and F.

Please note that I am only pointing out the extent to which we use this type of reasoning; it remains to be seen whether it is logically respectable.

Although a great many arguments by similarity like the preceding one contain conclusions about *individual* things or people, I believe the most fundamental type of this argument contains instead a conclusion about a whole *class* or set of individuals. In such arguments, a few individual items are described in the premises; the argument then proceeds to conclude that the members of some larger set of individuals, including those described in the premises, are similar to the members described in the premises.

In fact, it seems the above argument may be seen as an abbreviated version of the following argument chain:

> This person is similar to the dynamic, executive type person in respects A, B, and C. (That is, the dynamic, executive type has characteristics A, B, and C, and so does this person.) Dynamic, executive type persons behave in ways D, E, and F.)
>
> Anyone with characteristics A, B, and C will behave in ways D, E, and F.
>
> This person will behave in ways D, E, and F. (From the first premiss and the intermediate conclusion.)

I will attempt to represent all arguments by similarity that have conclusions about an individual thing as a chain like the above.

Accordingly, the fundamental outline of an argument by similarity will contain a universal conclusion—a conclusion about all the members of a certain class. Then, conclusions about particular members of that class may be drawn in additional steps, as above. (Note that in the above example the ultimate conclusion follows *validily* from the first premiss and the intermediate conclusion. Thus, in this case it is the first part of the argument which is invalid and which requires inductive evaluation. This is typical of such arguments.)

Here, then, is the general outline of one basic type of argument by similarity:

> P1 outline: Individuals a, b, c, d, etc. (or the members of set S) all have the characteristics F, G, H, etc.

P2 outline:	These same individuals differ from one another in respect to other characteristics R, S, T, etc., in various ways:
C1 outline:	Any individual with (some of) the characteristics F, G, H, etc., also has the other characteristics listed in P1.

This basic type of argument by similarity will be known as the *categorical* type, as contrasted with the *statistical* type, to be discussed later in this Section. In categorical arguments by similarity, premisses of the form P2 are optional. It is also optional to list all the individuals in P1—often some general description of these is used instead, such as "everyone I've ever met" or "all the examined cases" or "dynamic executive type persons".

As we saw earlier, arguments by similarity sometimes suppress conclusion C1 in order to move directly to a conclusion about an individual. The above outline may be supplemented to show what typically happens in such a case when the full argument is spelled out. Under C1 above, add the following material:

P3 outline:	Individual w (not included in the list given in P1) has (some of) the characteristics F, G, H, etc.
P4 outline:	(Often omitted) Individual w lacks some of the characteristics F, G, H, etc.
C2 outline:	Individual w has all the remaining characteristics on the list F, G, H, etc.

However, even when the argument as stated omits any explicit statement of the type C1, you will mentally insert a conclusion of that type in order to evaluate the argument, for the conclusion of type C2 depends on C1. Our argument about dynamic persons can be seen to fit this outline.

Although this argument as originally stated did not include anything fitting under P2 or P4, it can readily be expanded to include such material, to illustrate the use of premisses of these types in a realistic way. (Imagine the following argument presented in a meeting called for the purpose of making a hiring decision.)

> Dynamic, executive type persons are outgoing, well organized, and articulate. Such people take charge of a situation, keep calm under pressure, and remain gracious in defeat, no matter whether they be young or old, tall or short, male or female. The candidate we just interviewed for the position certainly was outgoing and articulate. So, I think she would keep calm under the pressures of the job, despite the fact that she appeared to be somewhat less than well organized.

Fitting this argument to the standard outline is instructive:

P1.	Dynamic, executive type persons are outgoing, well organized, and articulate. They also take charge of a situation, keep calm under pressure, and remain gracious in defeat.
P2.	Such persons are sometimes young, sometimes old, sometimes tall, sometimes short, sometimes male, and sometimes female.
C1.	Anyone who is outgoing, well organized, and articulate will take

charge of a situation, keep calm under pressure, and remain gracious in defeat. (This conclusion remains unstated in the original argument.)

P3. The candidate for the position is outgoing and articulate.

P4. She appeared not well organized.

C2. She will keep calm under pressure.

Note that we now have premises P2 and P4 in our argument, and that in this example, unlike the earlier version, the move from C1 to C2 via P3 and P4 is *invalid,* and thus subject to the evaulation techniques of inductive logic.

Thus far, I have been talking about the structure of arguments by similarity without mentioning how one might go about *evaluating* such arguments.

Let's turn now to evaluation. A brief glance at the outline of the basic argument structure should show that if an argument fits the structure that does not guarantee it will be either valid or tight. In fact, if one were looking merely at the structure without considering concrete examples of its application, I believe one would wonder why anyone would ever be persuaded by any argument of this sort. The lone fact that *some* known individuals are F, G, and H surely seems to do little to prove *all* individuals which are F and G are also H; and adding the fact that the known individuals also differ from one another in some other ways seems irrelevant. Nevertheless, we often find concrete instances of this type of argument to be quite reasonable. Are we just deluded?

As you may have guessed, if we are not deluded in the normal case, it is because background assumptions are being made. I believe we find particular arguments by similarity persuasive only when we see them as strong relative to background assumptions. (Actually, there are some special cases in which what I've said fails to hold, because the argument in question has more going for it than usual. For instance, if one were willing to add to the standard argument outlined above the background assumption that the individuals mentioned in P1 constitute the entire domain, then C1 follows *validly.* I will ignore such atypical cases.)

Accordingly, I suspect that anyone who finds the argument about our job candidate, given above, to be a reasonable argument not only accepts its explicit premises but also makes some background assumptions which lie hidden from view and which are going to be very difficult to state precisely. Perhaps those assumptions would include at least these:

B1. Being outgoing, articulate, and well-organized are personality traits.

B2. How a person behaves depends in large measure on the person's personality traits.

B3. The traits mentioned in B1 serve as reliable indicators of a type of personality that tends to exhibit itself in behavior which includes keeping calm under pressure, remaining gracious in defeat, and taking charge of a situation.

B4. Age, body type, and sex are factors which sometimes influence behavior.

The addition of these background assumptions to the original premiss set would make it much clearer why someone might find the argument persuasive. Although needing the addition of yet more background assumptions before it would become tight, the original argument does become considerably tighter with the addition of the preceding assumptions. We can now see why P2 might be thought to help the argument, for its function now appears to be to show that age, body type, and sex are irrelevant in connecting the personality traits mentioned in P1 with the behavior predicted for the candidate and for all others with the same personality traits. The main function, however, of the background assumptions lies in their showing why the similarity described in the argument might be likely to mean something, why it works. For this function, B1 through B3 are useful, for they suggest the kind of connection between personality and behavior which is required for the original premisses to indicate anything about the behavior of the job candidate.

Our job candidate argument is fairly typical of these arguments. Such arguments will seem "logical" only when it appears the various characteristics on which the similarity rests are likely to have something to do with each other. To see this more clearly, compare the job candidate argument, P1 through C1, with the following argument which fits exactly the same format, but which seems to me to be hopeless because the relevant characteristics seem unrelated:

P1. The five automobiles which have passed by my station during the last four minutes have all been dark in color, with automatic transmissions, and bucket seats. They also have been clean, occupied by at least two people each, and they have had at least 30,000 miles on them.

P2. Some of them had white walls, while others did not; some had the windows open while others had all windows closed; some needed gas, while others had full tanks.

C1. Any car which is dark in color, with an automatic transmission, and bucket seats will be clean, occupied by at least two people, and have at least 30,000 miles on it.

This tale has a simple moral: in order for an argument by similarity to be a TABA, or tight, there must be some reasonable way to establish or state a connection between the various characteristics mentioned in the premisses and in the conclusion. Normally, this connection will have to be added by stating a background assumption. If such a background assumption cannot reasonably be described, and there is no connection already explicitly made in the argument as stated, the argument will not be a TABA (nor will it be tight or valid). If such background assumptions or explicit connections are considered false, the argument will not seem strong—strong relative to background assumptions, or sound.

Although the most important consideration in evaluating arguments by similarity will probably be the matter of connectedness just discussed, there are a few other evaluative considerations also worth mentioning:

1. Referring back to the basic outline, you can see that P1 contains a list of characteristics, F, G, H, etc., which the argument will have to try to relate to one another in some way. A subset of these characteristics is used in C1 to specify the set of individuals to which the conclusion applies. The longer the list

of characteristics included in this *subset,* the easier it will be to make the argument tight as a general rule. Why? A longer list of characteristics used to define the set will generally make the set more narrowly and specifically defined, with the result that the individuals described in the premises have a better chance of fairly representing the set defined in the conclusion. If we had concluded in our job candidate argument that anyone who is *outgoing* will keep calm, be gracious, and take charge, our argument would have been more difficult to make tight than it was when we concluded instead that anyone who is *outgoing, well organized,* and *articulate* will keep calm, be gracious, and take charge. (However, a similarity based on 1,000 loosely related characteristics will never be as reliable as one based on two or three closely related ones. As usual, the overriding consideration lies in the question whether there are reasonable background assumptions available to relate the relevant characteristics.)

2. The *number* of individuals covered by P1 may be important. If this number is too small relative to the set described in the conclusion C1, the argument will commit the fallacy of hasty generalization, unless it escapes this fallacy through the addition of background assumptions which imply that the small set is likely to be representative of the large set. Especially if we are working in a scientific context where there are known laws relating characteristics F, G, H, etc., such assumptions may be available. These laws may be assumed as background with the result that data about a very few individuals may yield reliable information about a large class of cases. From an examination of the mating practices of just one or two members of a particular species, we may be able to reliably conclude things about the practices of all the members of that species, by employing an argument by similarity made tight through the addition of a number of assumptions drawn from biological science. However, lacking such background assumptions, one generally needs a relatively large sample in P1 if one is to reliably draw conclusions like C1. In terms of our standard evaluative terminology, this means that an argument is more likely to be tight or a TABA if the set of individuals in P1 is large, relative to the set about which the conclusion C1 makes its claims.

3. Our arguments become more likely to be tight or TABA's when they include premises of the type P2 which describe a large number of relevant differences between the individuals referenced by P1. Here again, the overriding consideration will be how much importance can be attached to these differences through the addition of background assumptions. However, if these differences are found to be relevant, they help tighten the argument, by making it more likely that the individuals described in P1 are widely representative of the class described in the conclusion C1. Recall that one can commit the fallacy of hasty generalization not only by choosing too small a sample on which to build, but also by choosing that sample too narrowly, excluding some types of individuals about which one wishes to draw a conclusion. A premiss of type P2 which specifies that the individuals in the sample diverge widely in *relevant* ways will make less likely the commission of this second form of the hasty generalization fallacy. In our last version of the argument about the job candidate, the differences mentioned in P2 were made to appear relevant to the issue because it was reasonable to make the background assumption that said age, sex, and body type might affect how one behaves with respect to taking charge, keeping calm, or being gracious. But in our automobile example, the differences men-

tioned in P2 didn't seem to help much, since these differences seem irrelevant to the type of characteristics mentioned in the conclusion.

4. If an argument contains a conclusion of type C2 about a specific individual not mentioned in P1, and a premiss of type P4 which describes how that individual differs from all the individuals in P1, then that argument generally becomes less likely to be a TABA the more characteristics are mentioned in P4. The characteristics mentioned in P4 come from the list in P1; the list in P1 constitutes the basis for the argument. If the individual w (in P4) lacks many of these characteristics, there isn't much reason left for concluding that he does after all have some of the remaining characteristics from the list. Of course, here again, the negative force of this consideration may be eliminated if background assumptions are available which show the characteristics mentioned in P4 to be irrelevant or unimportant. If we have a background assumption telling us that being outgoing is the *only* personality trait relevant to keeping calm under stress, then it wouldn't matter to the tightness of our argument if we had a P4 which mentioned that the job prospect is not articulate or well organized. Normally, however, the kinds of things which are brought up in real arguments in premisses of type P4 are things which might well be relevant and important; in such cases these things constitute obstacles to the argument's being classified a TABA.

The preceding evaluative considerations do not yield an automatic procedure for generating verdicts about the tightness or strength of arguments by similarity. However, these considerations represent useful things to look for.

We have been discussing *categorical* arguments by similarity. Such arguments also take another form, the *statistical* form. Statistical similarity arguments have a structure similar to that of the categorical arguments, and their evaluation is also similar in many ways. However, it is worth while to look at the details a bit, given the importance of these arguments in today's society.

In outline, the statistical argument by similarity looks like this:

P1 outline: X% of the individuals a, b, c, d, etc., which have characteristics F, G, H, etc., also have characteristics M, N, P, etc.

P2 outline: These same individuals differ from one another in respect to other characteristics R, S, T, etc. in various ways:

C1 outline: Approximately X% of the entire set of all individuals which have characteristics F, G, H, etc., also have characteristics M, N, P, etc.

As with the categorical type of argument, one may add onto the above outline to obtain a chain of arguments arriving at an additional conclusion, C2, about the chances that some particular individual, w, has characteristics M, N, P, etc., where w is stated to have some or all of the characteristics F, G, H, etc. in premisses P3 and P4.

Here again, the standard argument which fits this outline will not be tight or valid. In fact, it may be highly *improbable* that the set of individuals described in P1 represents accurately the entire set described in C1. For example, perhaps 75% of the individuals living west of the Mississippi River in 1860 died before reaching the age of 45. These individuals differed from one another in a great many respects, as to age, sex, occupation, educational level, and exact location. But it would be sheer silliness to conclude from such information that about 75% of all the people living west of the Mississippi die before the age of 45. In order to make such an argument tight, it would be necessary to add reasonable background assumptions which would have the effect of insuring that the sample in P1 is likely to be representative of the entire population in C1. Presumably, no such background assumptions are available for the particular argument just described; consequently, it is not even a TABA.

On the other hand an argument fitting this outline will be tight, or possibly even valid, if the premises specify that the list of individuals in P1 was randomly chosen with respect to characteristics F, G, H, etc., and the list is large. In a random sampling, every member of the total population under investigation has an equal chance of being selected for investigation. The results of the investigation would then be placed in the premises of the argument. A random sample of things that are F's, G's, H's, etc., which is sufficiently large will have a very good chance of being approximately representative of the whole population of things that are F, G, H, etc., and for that reason the conclusion of such an argument may well be probable relative to the premises.

Randomness and representativeness are often confused. A random sample might not be representative; a representative sample need not be random. A representative sample is one which has relevant characteristics similar to those of the whole population; in other words, its characteristics *represent* those of the whole population. On the other hand, a *random* sample is a sample whose members were chosen in a random way, which has the result that every member of the population has an equal chance of being included in the sample. Accordingly, a random sample of voters is a sample chosen in such a way that each voter has an equal chance of being included. A representative sample would be a sample which, no matter how it was chosen, would be similar to the whole set of voters with respect to whatever features we were interested in. The connection between randomness and representativeness is this: if I choose a very small random sample from a rather large population, the chances that my sample will be representative are not very good, and even if I choose a large random sample from the population, the chances of my sample's being *exactly* representative are very poor. But if one obtains a large enough random sample from a given population, the chances are high that it is *approximately* representative, thus making it possible to utilize the results of the sampling in a tight argument about the whole population.

Statistical analogical arguments are used in industry when testing mass-produced items which are too numerous to each be examined for defects (such as nuts and bolts). Instead of checking each item for defects, a random sample of the product may be tested and found, say, to be 95% free of defects. If the sample is large enough (perhaps 500 to 1500 pieces) and truly random, then it will be possible to conclude *validly* that *probably* about 95% of the whole population of items are free of defects since we can employ the premiss that a large randomly selected sample is likely to be representative. Another way of putting this: the conclusion that approximately 95% of the items are free of defects will be obtainable by means of a *tight* argument.

(The theory of mathematical statistics is designed to make all this more precise. That theory can answer questions such as, "Exactly how probable does the conclusion become when the sample consists of exactly 1,000 randomly chosen items?" Or, "Exactly what do we mean when we say 'approximately' 95%?")

This same procedure is employed in public opinion polling. A known population may be divided into subgroups, and within each subgroup a random sampling procedure picks out certain individuals to be interviewed. The results of the interviews become the premisses of the argument, and at first the conclusions will be conclusions about the particular subgroups. These conclusions are then combined to yield a conclusion about the whole population. (A simpler procedure consists of merely sampling randomly from the whole population, without dividing it into subgroups. The subgrouping works only if something about the population and its subgroups is known in advance of the polling. If subgrouping is used, the procedure is called *stratified* random sampling.) Political polls which use these techniques often are reported in the newspapers, and businesses use these techniques in doing market surveys to discover the need for products or services. Again, the mathematical theory of statistics is required to obtain exact results.

When the results of polls are published, it is often overlooked that sampling never yields exact information about the whole population, except in the special case where the whole population becomes the "sample". For instance, when a poll of 1,500 voters (which is a large enough sample) shows that 54% of them prefer candidate A over candidate B, it is tempting to put forward the following argument:

P1.   Out of 1,500 voters recently polled, 54% prefer A over B.

P2.   These voters were randomly selected from among the total set of 1,000,000 voters in this county.

C1.   54% of the voters in the county prefer A over B.

Naturally, everyone recognizes that the conclusion is not *certain*, but it is easy to think this argument is at least *tight* (that is, that the conclusion is probable given the premisses). However, that temptation should be resisted. The above conclusion is *not* even probable relative to the premisses. The appropriate *probable* conclusion contains a slight but all-important modification: "*Approximately* 54% of the voters in this county prefer A over B." (Or: "Between 50% and 58% of the voters in this county prefer A over B".)

Keep this in mind when you read in the newspaper that a poll shows one candidate in the lead by a few percentage points. The one which leads in the poll may not be in the lead within the whole population, if the lead in the poll is sufficiently narrow. When reading the standard Gallup or Harris survey reports in the papers, to be safe you should include perhaps a 6% total spread—3% in either direction of the reported figure—if you want to tightly draw a conclusion about the whole population.

There are also some famous difficulties associated with obtaining a truly random sample of certain kinds of populations. In some cases it is impossible, as with the voter. In order to predict election outcomes (assuming the outcomes are honestly and accurately determined by the votes cast) one wants a random sample of voters who will actually vote, and one wants to find out how the members of this sample will vote. Neither is possible. The poll must be conducted in advance of the vote, and there is no completely certain way to predict who will actually vote on election day. Nor is there any way to ensure that those sampled will tell the truth about their preferences, although professional pollsters have techniques which apparently encourage truth-telling. But even if the truth is obtained, that truth will be a truth about the voter's attitude at the time he is polled, not at the time he votes. Hence, the best one can do with a poll about a forthcoming election is to conclude that among those *likely* to vote, *approximately* X% prefer candidate A. Even then, your conclusion will at best be probable and never certain.

I have spent so much time talking about random samples in connection with statistical arguments that you are in danger of being misled into thinking that all such arguments refer to random samples. Don't forget that the general outline of this type of argument does not require that the sample be random. If the argument is to be tight, though, the sample will have to be *likely to be representative,* and the argument will have to contain enough premises to insure this. Premises which report on sufficiently large random samples constitute one, but only one, way to accomplish this. Other ways are not so well understood, but clearly do exist.

If a premiss somewhere in the argument states that F, G, and H are related in some special way to M, N, and P, then it might well become likely that something of the form C1 would be likely to be true, and the argument would be tight. This need not involve any reference to randomness; instead, there could be some reference to a causal connection between the relevant characteristics. Here is an example containing some statistics I made up to illustrate the point:

> 37% of those people we studied who smoked at least one pack of cigarettes per day for at least ten years up to the times of their deaths, and who died of natural causes, died of lung cancer. These people, numbering in the thousands, differed from one another in every other relevant respect, such as sex, age, general state of health, residence location, diet, and the like. None of these other factors made any difference to the rate at which such people died of lung cancer. We may conclude that about 37% of all such people die of lung cancer.

The key to the tightness of this argument does not lie in any claim about randomness, but rather in the claim that all other relevant factors had been ruled out. (I have not the slightest idea how anyone could know that such a claim was true, but that is a different issue.)

If an argument fitting the statistical pattern does not appeal to randomness, then it will require some kind of assumption about relevance or connection between the various characteristics, in the same way that categorical arguments did, in order to be tight. This will be the most important of the background assumptions if it is not explicitly stated in the argument. In this respect, the statistical and categorical forms of the argument are alike. The statistical form can escape this requirement for tightness only if it appeals to randomness.

Moreover, the other factors relevant to the evaluation of a categorical argument by similarity remain applicable in the statistical case. It would be a good idea to review those factors now, before trying the exercises.

Arguments by similarity

62.1   Assuming that the following pairs of arguments are presented in the same context, with the same background assumptions, which one of each pair is tighter or more easily made into a TABA? Explain your answer by appealing to the criteria for evaluating arguments by similarity. Assume the background assumptions are irrelevant in your comparison.

(I) We have noted that Senators Highpoint and McDougal, both liberal Republicans from large industrial states, favor the compromise bill. From this evidence we conclude that the liberal Republicans will support the bill despite their resentment over the way they have been pressured.

(II) We have noted that Senators Highpoint and McDougal, both liberal Republicans from large industrial states, favor the compromise bill. From this evidence we conclude that the liberal Republicans representing large industrial states will support the bill despite their resentment over the way they have been pressured.

(I) The primary mining product of the two mountain states, Colorado and Idaho, is zinc. Both these states and the mountain state of Wyoming have similar geological features, although Colorado and Wyoming are located in the Rockies, while Idaho is not. Therefore, the primary mining product of Wyoming probably is zinc.

(II) The primary mining product of the two mountain states, Colorado and Wyoming, is zinc. Both these states and the mountain state of Idaho have similar geological features, although Colorado and Wyoming are located in the Rockies, while Idaho is not. Therefore, the primary mining product of Idaho probably is zinc.

(I) The twelve members of our club who I have spoken to about this issue all agree that it should be put to a vote in a regular meeting. Among the twelve, there were old members and new members, influential members and members no one pays attention to, members from all parts of town, and from varied economic backgrounds. So it is reasonable to conclude that everyone in the club thinks we ought to vote on the matter in a regular meeting.

(II) The twelve members of our club who I have spoken to about this issue all agree that it should be put to a vote in a regular meeting. Among the twelve, there were old members and new members, influential members and members no one pays attention to, and members from all parts of town. So it is reasonable to conclude that everyone in the club thinks we ought to vote on the matter in a regular meeting.

62.2   Unfortunately, many people have a strong tendency to make arguments similar to the following highly emotional one. Your task here is to look beyond the rhetoric and the highly charged subject matter to evaluate the logical force of an argument like this. In fact, one of the advantages of studying logic ought to be the development of a skill that will enable you to examine arguments like this one from an objective point of view, no matter what your own personal background might be, and no matter how offensive you might find the claims being made.

Two businessmen with whom I served on the community beautification committee were pushy and arrogant, hardly letting me express any of my ideas. They both ignored me, probably because I'm a woman. Now I hear I've got to serve with another businessman on next year's committee. He's going to be pushy and arrogant too.

a. If this argument were put into the standard format discussed in this Section, there is another conclusion, not explicitly stated above, which must be thought of as part of the argument. What is that conclusion?
b. Obviously, this argument is not tight as it stands. What background assumptions would be needed in order to make it tight? Are these assumptions likely to be true of our real world?

62.3   The following questions pertain to this argument pair:

(I) Senator Smith and the president are both Democrats who have spent most of their adult lives holding public office. They are both thoroughly committed to the preservation of the current system of government in the United States. Thus, all Democrats who have similarly spent their lives as officeholders are similarly committed.

(II) Senator Smith and the president are both Democrats who have spent most of their adult lives holding public office. They are both thoroughly committed to the preservation of the current system of government in the United States. Thus, all who have similarly spent their lives as officeholders are Democrats committed to the preservation of the current system.

a. In order to fit these arguments to the basic similarity argument outline, the characteristics F, G, H, etc., which appear in that outline must be identified. Make a list of all the characteristics mentioned in the above arguments which should be thought of as corresponding to a letter from the outline, and state as a part of your list which letter from the outline you wish to use to stand for each characteristic.
b. Given your lettering of various characteristics, state in outline form, using the letters instead of the characteristics, the conclusion of argument (I) and then state the conclusion of argument (II).
c. Which of the criteria for evaluating similarity arguments applies directly to the comparison of arguments (I) and (II) for tightness? What does the application of that criterion yield in the way of a comparison between these two arguments? Explain.

62.4   Suppose you wish to conduct a random survey of the adult residents of a certain poor neighborhood to determine their attitudes about the city government. Which of the following techniques would be best for selecting the people to be interviewed? Explain why you chose the technique you did.

a. Obtain a special telephone directory of the city, arranged by street address rather than by name of subscriber, and pick out all the telephones listed in the neighborhood you are interested in. Then pick every fifth one for interviewing, if it is listed as belonging to a residence. Interview the person listed as the telephone subscriber for those numbers selected.

b. Pick a street in the neighborhood at random and interview all the residents who live on that street in the block closest to the center of the neighborhood.

c. Obtain a list of each separate dwelling unit in the neighborhood, listing each home, each apartment, each rented room, and so on. Randomly select a large enough number of these, and then interview one adult resident from each selected dwelling unit. In case more than one adult lives in a given dwelling unit which has been selected, flip a coin or roll a die to determine which person will be the one interviewed.

62.5   Here are some premisses:

In a recent survey of campus dormitory residents, 65% of those responding thought that the dormitory managers did not exercise enough control over the residents, and 30% thought the managers exercised too much control.

Which of the following conclusions is the most probable, given that the survey was random and covered a significant proportion of all residents?

a. Between 60% and 70% of all the dormitory residents think the dormitory managers do not exercise enough control over the residents.

b. Between 62% and 68% of all the dormitory residents think the dormitory managers do not exercise enough control over the residents.

c. 65% of all the dormitory residents think the dormitory managers do not exercise enough control over the residents, and 30% think the managers exercise too much control.

62.6   Which of the following two statements can go wrong more easily, in the situation described in the previous exercise?

a. Between 60% and 70% of all the dormitory residents think the dormitory managers do not exercise enough control over the residents.

b. Between 60% and 70% of all the dormitory residents think the dormitory managers do not exercise enough control over the residents and between 25% and 35% of the residents think the managers exercise too much control over the residents.

Therefore, in reference to the preceding argument, would a. or b. be the more probable conclusion?

62.7   How could a random sample fail to be representative? Give an example. How could a representative sample fail to be random? Give an example.

62.8 Given that a random sample, even a large random sample, can fail to be representative, how does randomness help to yield reliable conclusions in statistical arguments by similarity?

62.9 The following argument exemplifies a special problem which sometimes arises in drawing conclusions from polling the public. Identify the problem, explaining how it can make the conclusion unlikely despite the truth of the premisses.

In a recent nationwide survey of 2,500 adults, only 3% of the respondents stated that they were either homosexuals or lesbians. The other 97% claimed to be heterosexual. This survey was conducted in a suitably random manner by professional pollsters and included persons from all walks of life. This sample size is sufficient to yield reliable results. Thus, we may conclude that the heterosexual population comprises about 97% of the adults in the country.

62.10 The following questions pertain to this argument:

During 1979 thirty-seven patients suffering from humititis were treated with weekly small doses of x-ray radiation on the affected areas for eight consecutive weeks. Thirty-two of these, or 86%, reported receiving considerable relief within this time span; two patients, or 5%, reported no change in their condition; and three, or 8%, got worse. Since humititis left untreated grows rapidly worse, the two patients whose condition stabilized during treatment may be considered as having been affected by the treatment. Thus, thirty-four out of thirty-seven patients (92% of the total) found relief to some extent during the course of treatment. These patients were of various ages and in varying degrees of general health, with both advanced and early cases of humititis. No correlation between the success of the treatment and any other factor in the patients' histories was noted. Indications are, then, that humititis in all its stages will respond favorably to the x-ray regimen in approximately 92% of all cases.

a. Is the sample on which the conclusion rests here a random one?
b. Is it likely to be a representative one? What background assumptions would one have to make in order to say "yes"? (There should be quite a few. Give some thought to this.)

62.11 Not all arguments by similarity are laid out in standard form. Here are some excerpts from an editorial which appeared on January 4, 1971, in the Louisville *Courier-Journal,* on the subject of the SST (Supersonic Transport)—an airplane which, as it turned out, was never built. (The SST was to have been an especially fast-flying commercial passenger airplane, costing millions of dollars to design and develop, and billed by its supporters as the airplane of the future. Supporters of the plane wanted the federal government to pay in part for its development.)

When they have exhausted all other arguments for the supersonic transport plane, backers of the SST fall back on the contention that thousands of people will be thrown out of work if the plane isn't built . . . .But . . . following this line, a good case could be made for indefinite production of aircraft carriers and submarines, whether or not they are needed, since curtailment of production will displace thousands of skilled, highly-paid workers. It could be similarly argued that the government should have underwritten the Edsel . . . .[1] *The criterion by which such things must be judged is not whether or not they create jobs, but whether or not they are needed. The SST is not needed.*

a. The conclusion of this argument is not stated explicitly. What is the conclusion, in your estimation?
b. Rewrite the argument, putting it into standard form for a categorical argument by similarity, letting the individuals in the argument be the various projects mentioned in the editorial. Add missing premisses that clearly belong to the argument. Include the conclusion you gave in answer to part (a).
c. What additional background assumptions are relevant here?

63

Causal arguments

When we assert that certain characteristics are related to one another, as we often do when arguing by similarity, the relation is often a causal one—we believe that things with one set of characteristics tend to cause things with another set. But how do we come to *know* what causes what? Presumably, arguments can help to reveal these relationships.

In this Section we will look at arguments with *conclusions* of the general form 'X causes Y'. Naturally, such arguments have wide application, not only in connection with supplying background assumptions for similarity arguments, but also in connection with many other activities.

Not all arguments leading to conclusions about causal connections are invalid. However, the commonly occurring valid ones presuppose that you already know about some particular causes and their effects so that you can use that knowledge in your premisses. But, ultimately, our knowledge about causes and effects seems based on arguments which are not valid—arguments which do not have any premisses that tell you one thing causes another. Those are the arguments we are concerned with in this Section. (The valid arguments, on the other hand, have already in effect been dealt with earlier, in Part II, along with all the other valid arguments.)

[1]The Edsel, for those who have not heard, was an automobile line attempted by the Ford Motor Company a few years ago. It was a notorious flop—notorious enough that the writers of the editorial given above could assume that their public would recognize reference to it as a reference to a product which the public definitely did not believe it needed.

To analyze and evaluate arguments which claim to show one thing causes another, we need a fairly clear idea of what it means to say one thing is a cause and another its effect. I will not attempt a full analysis of these notions here, instead, I will merely try to clarify a few points which are troublesome in argument evaluation.

We often talk about "the" cause of a particular explosion, death, or day of nice weather. Our talk is misleading, because it suggests that "the" cause is the only relevant factor in bringing about the situation. In every interesting case I can think of, the thing which is most naturally labeled "the" cause is only one factor among many which all cooperated to bring about the effect. Yet we pick on only one factor, and bless it with the title of "cause", ignoring all the other things which could have been called part of the cause. When we say the short was the cause of the explosion, we ignore the fact that the explosion would never have happened if the gas had not been in the tank, if the power had not been turned on, if the gas had not been of an explosive nature, if the gas had not been of a sufficient concentration, if the temperature had not been right, and so on.

These remarks are not intended as a recommendation that we change our ways; rather, I am trying to point out something about how we use the idea of "cause". If we aren't specific about such things, it may lead to unnecessary disputes later. Jones, the owner of the exploded plant, may claim "the cause" of the explosion was Smith's turning on the power, while Smith, wishing to avoid the blame, may say instead that "the cause" of the explosion was the short in the wire. Clearly, a dispute of this nature has more to do with blaming than it does with causation.

What does it mean, then, to say that something is "the" cause? It appears many things may be "the" cause. Why single out one of them? I believe this happens because we tend to think of the factors which lead to the effect event as being made up of the general background conditions which we more or less take for granted as a part of the way the world was prior to the event plus the one crucial fact which we wish to emphasize or draw attention to, which we label "the" cause. We may not even be aware of all the relevant general background conditions.

For us, then, these background conditions all blend together into a mass of undifferentiated facts which we do not think of as crucial to the explosion. Nevertheless, the fact of the matter is that the explosion would not have happened without them, and someone who wished to think of the short as part of the general background could well settle on one of these other conditions as the crucial one to draw attention to by labelling it "the" cause.

Among the conditions which generally get included in the general background are the conditions and events which helped to bring about some of the other factors which were more closely and directly connected to the effect. If Jones, the plant owner, continues to blame Smith for causing the explosion because Smith turned on the power, Smith may resort to calling attention to one of these special background conditions, namely, that his foreman, Robinson, told him to turn on the power. Thus, Smith may say it was Robinson's order which caused the explosion. Probably, Robinson's order would normally be included in that general haze of background conditions which never get considered for the role of "the" cause, but now attention is shifted and poor Robinson gets all the heat.

If the truth of the matter is that all these things are just "part of the cause", is it wrong to single out one of them as "the" cause? No, I don't think so, provided that we keep in mind what we are doing. In fact, it would seem entirely out of line to insist we must explicitly include *all* the background conditions as "the cause" of a given effect.

Within the maze of background conditions and causal factors, the one or two factors which may be most reasonably singled out to be labeled "the cause" are those which our practical purposes on that occasion highlight. If our purpose is the fixing of legal responsibility, "the cause" of the accident may be the factory owner's negligence, while if our purpose is to improve electrical systems so they are less susceptible to failure even when maintained by negligent people, "the cause" may be insufficiently insulated wires. We isolate one or two factors which helped to bring about an event—factors which we wish to blame someone for, or use as clues about the future. We call such factors "the cause" of the things we wish to explain.

How does all this messy business tie in with the argument evaluation? (1) When you are confronted with two arguments, each claiming to prove a different thing caused X, it is possible both are correct, and that the only difference between them is that they assume different parts of the total story as background conditions. One argument proclaims that Smith's turning on the power is "the" cause of the explosion, while the other labels the short as the cause. There is no need to decide between them. (2) A causal argument cannot be evaluated unless one has at least some vague idea which conditions are being taken for granted as part of the general background. Since the general background is vast, no argument ever mentions all of it, nor can one think of it all. But causes work only against a background set of conditions, and there is no way one can tell if it is even reasonable to think of X as the cause of Y without some idea of background conditions. These conditions can play the same role as that played by background assumptions in similarity arguments—they can be hidden premises which must be included in the argument if it is to be tight.

It is possible to avoid some of the special problems created by the label, "the cause", by using a less committal phrase in its place. For instance, one could say that Smith's turning on the power was a "causal factor" in bringing about the explosion, or that it was "causally connected" with the explosion. Careful writers sometimes phrase things that way. This does not eliminate the role of the background conditions; it makes it more obvious that they had a role to play.

One final word of caution before we look at more causal arguments: an argument may be a causal argument without ever using the word "cause" or any of its derivatives. Many alternative locutions indicate causation, such as: "bring about", "make", "create", "inhibit", "develop", "enhance", "extinguish", and on and on. Any verb which implies that one thing makes something else happen will do the job. Be aware of the wide range of things which can be thought of as causes: thoughts, people, animals, events, facts, omissions, beliefs, and many other sorts of things.

Probably the most basic invalid argument pattern for the conclusion that X causes Y is the "method of concomitant variation". The idea behind this argument strategy is straightforward enough: when two conditions, X and Y, tend to vary together from time to time or place to place it seems likely that X and Y are causally connected somehow.

For example, let X be the occurrence of inflation in the economy, and let Y be the occurrence of a deficit in the federal government's budget, requiring the government to borrow money in order to operate. (Many people believe there is some kind of connection between these two occurrences.) According to the method of concomitant variation, in order to argue for the conclusion that X and Y are causally connected, we should attempt to determine whether X and Y have in the past tended to vary together. Two basic patterns of concomitant variation are possible: (1) Inflation tends to be worse when the budget deficits are greater. (2) Inflation tends to be worse when the budget deficits are smaller. Type (1) concomitant variation is termed "direct", while type (2) is called "indirect". If we determine that either type of concomitant variation has held in the past, we report that as a premiss, and from that premiss, we conclude that inflation and government deficits are causally connected. At least in its barest form, that is how the argument would go. Both inflation and deficits are capable of coming in various degrees; that makes it possible to have very nice correlations between the one and the other, where the greater degrees of the one are associated with the greater degrees of the other. A report of such a tidy correlation would be a helpful premiss in establishing the causal connection.

On the other hand, we could keep things very simple by ignoring the various degrees in which these phenomena occur. The only perfect concomitant variation which is then possible consists of either one of the two following arrangements: (1) Inflation occurs whenever deficits occur. (2) Inflation occurs whenever deficits do not occur.

The four different kinds of concomitant variation discussed above are illustrated in the following chart:

	Inflation	Deficit	Inflation	Deficit	Inflation	Deficit	Inflation	Deficit
Year 1	high	high	high	low	yes	yes	yes	no
Year 2	low	low	low	high	no	no	no	yes
Year 3	medium	medium	medium	medium	no	no	no	yes
Year 4	low	low	low	high	yes	yes	yes	no
Year 5	high	medium	high	medium	no	no	no	yes
	Direct variation		Inverse variation		Direct variation		Inverse variation	

Note that the correlation does not have to be perfect in order to count as a significant one. In Year 5, above, there were slight aberrations from the patterns established in the earlier years, without a complete breakdown in the pattern.[1]

To use one of these sets of data in an argument of the simplest form, one would merely say something like this:

---

[1]The second two types of correlation are traditionally said to be used, not in the method of concomitant variation, but in another method called the joint method of agreement and difference. I see this as a minor terminological matter, and prefer the unification of theory which is possible by viewing the latter two kinds of correlation as being a special case of concomitant variation. The tradition to which I refer apparently started with Francis Bacon, in his *Novum Organum* (1620), and was popularized by the English philosopher John Stuart Mill in 1843.

During the year 1975 through 1980 the following rates of inflation and amounts of government budget deficit occurred: (Give data.) It is apparent that inflation is always higher when the deficit is higher. Thus, we may conclude that the two are causally connected.

Although an argument like this is too rudimentary to be tight as it stands, it forms the kernel of a good idea about how to argue for a causal connection. We'll worry about fixing it up later.

There's another application of the method. Sometimes it is only possible to look at the cases in which X did occur, rather than looking at both the cases where it did and did not. Generally, this is not as effective in finding causes of X, but nevertheless, it is a possible strategy. Using this strategy above would mean that we would only look at years in which inflation did occur, and if we then found that in all or almost all those years there was also a federal deficit we might conclude that the two were causally connected. (This strategy is sometimes called the method of agreement.)

Still another strategy would be to look at only one year in which inflation occurred and several other years in which it did not, and see what was different about the year of inflation. If we discovered that the year of inflation was the only year in which there was a deficit, we might conclude a causal connection once again. (This method is sometimes called the method of difference.) Since all these strategies are fundamentally alike, you can think of them as types of the method of concomitant variation.

You might wonder why anyone would use either of the last two strategies since they seem so much weaker. Sometimes there is no choice; sometimes it doesn't seem to matter. We don't have too much trouble figuring out what holds the pages in the binding in a paperback book, even though we don't generally have a sample of paperbacks made both with and without glue for us to test. Here the "method of agreement" is being used. Or, if we do after all come across one paperback in which the pages are all loose and falling out, and we discover that it alone, amongst all the paperbacks we have seen, has little or no glue in its binding, we have no difficulty concluding that the lack of glue is the cause of the loose pages. Here we have used the "method of difference". So, there are some occasions on which these special types of the method of concomitant variation are useful.

Although the method of concomitant variation does allow for causes and effects to come in varying degrees, it nevertheless treats both causes and effects as units which have no parts. This constitutes a limitation on the method, a limitation which the other primary method tries to overcome. This second method for establishing one thing as causally connected with another is traditionally known as the method of residues. According to this strategy, one might have to break up an effect into parts and claim one part is causally connected to one thing while another part is connected to something else. The connections between the *parts* and their several causes are determined still by the method of concomitant variation. Accordingly, we might want to see if we can correlate a *portion* of the inflation in the economy with the federal budget deficit and another portion with something else, such as shortages of raw materials for key manufactured goods. Obviously, in order to be successful in such an endeavor, one would have to look very carefully at just how much inflation could be accounted for by each factor, and to do this would require very careful research into just how much inflation occurred with just how much deficit in the past, and just how much occurred with certain specific degrees of shortages.

Arguments which follow the methods we have already discussed, employing no other premisses besides those required by the methods themselves, will not be tight. In fact, one can get absolutely ludicrous results by blindly using these methods, especially if one tries to conclude that one has found "the cause" of a particular phenomenon using the methods alone. For example, by observation we may become quite certain this premiss is true: "Whenever the shadows are shortest, the sun is highest in the sky." Employing the method of concomitant variation, we might conclude that: "The shortening of the shadows causes the sun to rise in the sky." This shows why it is far safer to avoid conclusions of the form X causes Y, and go instead for conclusions which merely assert a causal connection. Doing so here yields the conclusion that "The shortening of the shadows is *causally connected* with the sun's being high in the sky", which is not ludicrous, and in fact is true.

However, even if we stick to the latter version of the conclusion, our methods can lead us seriously astray, because our methods do not rule out coincidences. Coincidences create chance correlations which do not indicate any real causal connections. For example, it could happen that every time I have washed my car during the month of June, it has rained the next day. Using our methods, of course, it appears we may conclude that my car washing is causally connected with the coming of the rain. But most people would agree that such a conclusion is more of a joke than a serious causal hypothesis.

The point is not merely that our methods can lead us to draw *false* conclusions from true premisses. *Any* invalid argument has that feature. Rather, the point is that the conclusions we get by these methods may not even be made *probable* by the premisses. The difference between good and bad arguing lies in the background assumptions which can reasonably be added to these arguments, background assumptions which say things like these:

Shadows are caused by an object blocking light.

The only possible causes of inflation are Federal spending policies, shortages of raw materials, shortages of labor, ...

Such assumptions must be added to our arguments before they will be tight, for without them there is simply too much chance of coincidence. These assumptions tell us something about what is likely to be accidental and what is not. If they tell us enough, we might end up with a tight argument.

It's not easy to spell out all the background assumptions which lie behind typical causal arguments. Many of them we take for granted so continually that we are not even aware of them. For example, when trying to figure out why someone acted the way they did, we just automatically assume that people's actions are determined (caused) by what they want, by their attitudes, by what they see as their options, and by what they see as the facts. That is, we have a set of possible causes already in mind in a given case, and our arguments about the particular cause of a particular behavior will center solely on which combination of the possible causes is the correct one. So, even if that behavior also happens to be correlated with the price of pizza in Peoria, we will not take that into account at all. Because in most cases it is very difficult to say exactly what the background assumptions are for a given causal argument, most of these arguments will be at best *suspected* TABA's; hardly any will be tight or easily capable of being made tight.

Although most causal arguments like those we have been discussing will fail to be clear TABA's, and are certainly not tight or valid, there is often not much which could be done to improve them without introducing premisses which no one could know to be true, such as a premiss which listed all the possible causes for a given event, or introducing a huge array of background assumptions that would take three years to write down. A person who learns the intimate details of organic chemistry would be hard pressed to list all the assumptions he makes in giving a causal explanation of one particular reaction. Consequently, it may well be impossible to demand of each argument that how to make it tight be clear before we call it reasonable. One may have to settle for having just a general idea of how it might be done.

Aside from the general problem about ruling out coincidence, there are some other difficulties with the methods of concomitant variation and residues which deserve particular mention since these difficulties must all be avoided by a tight argument for a causal connection. These are realistic difficulties which you must face quite often in your struggle to find the truth about how the world really works.

The first of these difficulties is the problem of multiple effects: a single cause may, under varying circumstances, bring about vastly different effects. Suppose a federal budget deficit causes inflation, except in years of general business decline. Then, if we naively collect our data about inflation and deficits, we may be faced with a bewildering pattern which seems to show inflation is not correlated with deficits at all, and thus will not be causally connected to deficit spending by our method of concomitant variation. To avoid this problem, we have to know to look for the more complicated pattern, and then use the method of concomitant variation. This will result in correlating deficits in good business years with inflation.

The second problem is the difficulty of moving from a conclusion which asserts a causal connection between X and Y to a conclusion which asserts that X causes Y. For example, refer back to the shortening of the shadow argument given earlier, which illustrated the difficulty of deciding from concomitant variation which is the cause and which is the effect. Obviously, one needs added premisses to help one decide. But the difficulty goes even further. If X and Y are causally connected, it does *not* follow that one of them causes the other. True, in our shadow example it turned out that one of the two did after all cause the other, but consider this example: a rapidly falling air pressure reading on a barometer is generally quite well correlated with a change in the weather toward more unstable or stormy conditions. From this, by the method of concomitant variation we conclude that a falling barometer and stormy weather are causally connected. So far so good. But which one causes the other? The correct answer is, apparently, neither. The falling barometer reading *and* the change toward stormy weather are *both* caused by a third factor not yet mentioned—a lowering of the atmospheric pressure. In other words, when X and Y are causally connected, it may be because both X and Y are caused by Z.

The third problem arises because the same effect may be brought about by differing causes. Suppose inflation may be brought about by various different things, such as federal deficit spending, high wage demands by labor, and scarcity of raw materials. If we then try our method of concomitant variation to look for a correlation between inflation and other items, we may not find any simple correlation. Sometimes inflation occurs when there is a deficit, and sometimes when there is not. Sometimes it occurs when there is a shortage of raw materials and sometimes when there is not (because one of the other causes is at work). The best way to take care of this problem is to use concomitant variation on the *combination* of causes. Thus, one can find that after all inflation is correlated with federal deficits *or* the occurrence of high wage demands by labor *or*. . . . But, of course, one would have to know to *look for* such a complicated correlation before one would find it.

When someone ignores all the problems and quickly jumps to the conclusion that X causes Y merely on the grounds that occurrences of X and Y are correlated, some authors of logic texts label the argument as fallacious, in just the same way that other arguments were labeled fallacious early on in this book. A good name for this fallacy would be *the fallacy of questionable causation.* You may now add this fallacy to the list discussed earlier. Since the avoidance of the problems with causal argument generally requires the extensive use of background assumptions, it would be fair to describe a case of the fallacy as an invalid argument about causation which cannot be made tighter through the addition of any reasonable set of background assumptions.

I have been talking about correlations, patterns, and various kinds of repetitions. How does any of that apply to the analysis of the cause of a *single* event, such as a particular fire or explosion? A single event cannot form a pattern or correlation. It just occurs. Because it does not make a pattern, the only way to argue for causal connections between a single event and anything else, using the methods outlined in this Section, is to classify the single event, and then to look for correlations between members of that class and members of other classes of events or situations. That is, we give a causal argument about the cause of a particular explosion by subsuming the particular explosion under the general heading of explosions, or possibly under a somewhat more specific heading, such as *explosions of methane gas,* or *explosions of volatile gases.* Then, we apply our methods to the whole class rather than to just the particular event in question. Doing so may yield the conclusion that explosions of this type may be caused by sparks, excessive heat, and so on. This general conclusion may then be used in connection with added premises about this particular case to yield the result that this explosion was caused by a spark, since none of the other possible causes was present. One might then ascertain by a separate application of the method of concomitant variation that shorted electrical wires spark under certain conditions. This will complete the story about the particular explosion, for it was the shorted wires that caused the spark. The general point, then, is that the only way to use our methods to explain the causes of a particular event is to classify that event (perhaps in some fairly complicated way) and thus allow that whole class of events to be investigated for correlations with other events or situations. Creative classification is the essence of effective causal argumentation.

Causal arguments

63.1 Suppose you are in charge of deciding how to get people to try a new automobile polish manufactured by your company. The product is of exceptionally high quality—better than any of its competitors—but it faces the problem of getting people to try it for the first time. In order to decide what marketing strategy to use, you gather data about what has happened in the past when various strategies were tried with new products in the automotive supply industry. You restrict your data to products which performed at least as well as their competitors, and which were manufactured by established companies like your own with networks of distributors who are able to make the product available. (You make these restrictions because you sensibly believe in some background assumptions which say that inferior products or products of limited availability might not do well even if the strategy adopted for getting people to try the product was satisfactory.) Assume that all the relevant factors are listed here in the summary of the data. From that data, make an argument, using our methods, for a causal connection between a particular strategy or combination of strategies and success in getting the public to try a new automotive supply product. The table of data records whether the strategy was successful, and which strategies were used. The strategies include these: mailing out small free samples, advertising on television, on the radio, in newspapers, in automobile specialty magazines, in general purpose magazines, and finally, setting a price lower than the competition.

Product	Success?	Samples?	TV?	Newspapers?	Radio?	Auto magazine?	General magazine?	Price?
A	yes	yes	no	yes	no	yes	no	yes
B	yes	yes	no	yes	no	yes	yes	yes
C	no	yes	yes	yes	no	yes	yes	no
D	no	no	yes	yes	no	yes	no	no
E	no	no	no	yes	yes	yes	no	no
F	yes	yes	yes	yes	yes	yes	yes	yes
G	no	no	yes	no	no	yes	no	yes
H	no	yes	no	no	no	yes	no	yes

63.2 This table summarizes another different set of data regarding the same situation. This time, degrees to which the factor listed was present are indicated, on a scale of 1 to 10; 10 indicates the highest degree. Note, though, that a rating of 10 does not necessarily mean that factor was *influential* to a high degree; it only means that factor was *present* to a high degree.

Product	Success?	Samples?	Tv?	Newspapers?	Radio?	Auto magazine?	General magazine?	Price?
A	10	0	8	9	0	10	2	5
B	8	0	8	2	0	9	2	5
C	3	5	4	5	0	4	0	3
D	3	5	2	8	0	4	3	0
E	2	0	0	0	1	2	5	0
F	8	10	10	0	5	3	10	0
G	7	10	10	1	5	3	0	0
H	5	0	6	9	0	4	0	3

a. Which single column of strategies correlates the best with success?
b. Is that a direct correlation, or an inverse one? If it were inverse, what would that mean with regard to the effectiveness of that strategy?
c. Since even the best correlation between success and a single strategy is not very good, that suggests more than one factor may be important. Can part of the success of the successful products be accounted for by one strategy and part by another? (Refer back to the method of residues.) Which combination of two strategies correlates well with success?
d. Suppose someone claimed that product E failed to gain much success because its manufacturer did not send out free samples. What would you reply?
e. Suppose someone said that product G would have done even better if its manufacturer had used radio advertising more extensively. What would you say?
f. Suppose someone claimed that the data indicates that lowering prices below the competition does not seem to have any effect, and that therefore, in order to increase our profits, we should raise the price of our new product, without fear that it will cut sales. Would you agree that the data support all of these claims? Some of them?
g. Make a list of as many background assumptions as you can think of which are likely to operate in the arguments put forward by those who believe the data in the above table is reliable for revealing the causal connections between success and the strategies listed. (There should be quite a few of them.)

63.3 The next set of questions concerns the following passage:

From the most ancient times, in primitive societies, the men have been assigned to hunt while the women function as gatherers. This division of labor between the sexes lies at the root of male domination over women in modern society, for the early division of labor established a tradition of male dominance which never broke down even when hunting and gathering largely ceased. Although the sex roles, hunting and gathering, were designed to protect the women and their suckling children from the dangers of the countryside, the hunters obtained dominance from their role, in proportion to its importance to the culture. Among the San people, who rely only little on hunted meat, the degree of male dominance over women is slight, while in the traditional Eskimo society which depends almost exclusively on meat hunted by the men, women count for very little. Among the Hadza people of Tanzania, the men contribute only an insignificant proportion of the community's diet through their hunting, while the women are responsible for almost everything through their gathering. Hadza women enjoy approximately the same social status as the men.[1]

a. What is the ultimate conclusion of this passage?
b. What is the conclusion concerning cause and effect which the data about the San, Eskimo, and Hadza supports?
c. What is the connection between these conclusions? (Write out the steps connecting them.)
d. What background assumptions are prominent in the argument from the San, Eskimo, and Hadza peoples? (Hint: Think of possible objections to this argument, and then supply background assumptions to block these objections. It is likely that such assumptions are being made, if your objections are fairly obvious ones.)
e. How could this argument be made stronger by the addition of further data? (Explain what kinds of data would help.)
f. What kinds of discoveries would weaken this argument if they were added to our background knowledge?
g. This passage was drawn from an article which includes more premises than those mentioned above. In that article, the author explains how it is that hunting gives men the dominant role when hunting is important to the society. The explanation is that the returning hunter, laden with precious meat, has a time of triumph and adulation when he ceremonially distributes the meat to the group, while the more continual and less dramatic successes of the women in gathering go unnoticed. This information does not fit into any of the patterns for causal argument we have discussed. Yet adding this information to the argument in the passage makes that argument stronger. Why?
h. Suppose someone claimed that the hunter role was irrelevant to causing male dominance in cultures where hunting is important. Could such a person still claim (with reason) that the hunter role was *causally connected* with male dominance in those cultures? Should the author of the above passage have modified his conclusion to assert causal connection rather than straight causation between the hunter role and male dominance? (Give an example of how the hunter role could be causally connected with the fact of male dominance without being its cause.)

---

[1]Summarized from Boyce Rensberger, "Our Sexual Origins," Special Edition of *Science Digest,* Winter 1979, p. 114.

64

Arguments to the best explanation

One of the reasons we try to discover causes is that we want to *explain* things about the world. Humans seem to hunger after explanations almost as strongly as food. Whether it be for practical purposes, to control nature or other people, or merely for the sake of curiosity, explanation of one sort or another engages much of our time and energy.

But what exactly is explanation? The term is used so widely that it applies to just about any descriptive discourse which clarifies something for someone, or is intended to do so. Since people can become puzzled about almost anything, and any descriptive discourse designed to remove puzzles can be called an explanation, we are not being told very much when we are told that something is an explanation. There are not merely many explanations given by people every day, but these fall into many categories, which may be quite different from one another, operating according to quite diverse rules. To mention just two examples, compare the type of explanation given by way of a chemical formula with the type given by describing motives for an action. The formula tells of what the thing being explained is composed; the description of motives certainly does not tell what the action it explains is made of!

Because explanations come in such wide variety, I need to limit our range of concern. I will follow a hint given by the ancient Greek philosopher, Aristotle, who also happened to be interested in this topic. (In fact much of his philosophy is centered on this concept.) He suggests that we can think of explanation, insofar as we might want to investigate the topic, as always being an answer to the question, "Why?" Not everything which normally counts as an explanation will fit this conception, but this conception will be broad enough to give us a sufficient range of things to talk about. According to this conception of explanation, if someone is telling me how the chess pieces were arranged on the board at a particular stage of an interesting match, he might in common parlance be "explaining" how the pieces were arranged, but his description would not count as an explanation under our Aristotelian conception unless it somehow was directed toward answering some kind of "Why" question.

We have already noted that there are many different kinds of explanation. Corresponding to many of these will be different types of "Why" questions. When we ask "Why?" we do not always want the same kind of answer. When I ask the teacher "Why did I fail that test?" I may not be looking for the same sort of thing as when I ask "Why do we have to have a test?" In the first case I might well be looking for something like a cause of my failure. In the second, I am looking for some kind of justification, and not a cause. We argue a great deal about explanations of all sorts, and it seems appropriate in a book like this one to talk at least a little about how you can tell when the arguments for or against an explanation are good ones. Since the best of these arguments are most often invalid, the discussion of their evaluation belongs to this chapter.

There are two main types of arguments about explanations that I will discuss here. The first consists of invalid arguments for a conclusion that looks like
this:

X is the best explanation for Y.

The second consists of invalid arguments for a slightly more complex conclusion
that looks like this:

X must be true because X is the best explanation for Y and Y is true.

(One gives this kind of argument when one argues that the girl down the hall
must be jealous of Joan because that is the best explanation for her going
around criticizing Joan to everyone else.)

To some extent we have already discussed such arguments in the last Section—particularly those of the first type—for many explanations are causal, and
in that Section effective arguments for a conclusion of the form 'X causes Y'
were discussed. Of course, there are also many other kinds of explanation
which we have not yet talked about. But I will not go through each type of
explanation separately. Instead, I will attempt to give some general principles
regarding what will count as good arguments for conclusions of each of the
two basic types listed above, no matter what kind of explanation is involved. It
will then turn out that the material discussed in the previous Section will be a
specific spelling out of these principles as they apply to causal arguments.

In preparing to evaluate an argument of either basic kind outlined above,
one needs to know how far the conclusion is intended to go. Does it claim that
some preferred explanation is the best of all explanations of any kind? Or only
that it is the best of its own kind? Generally, the context of the argument will
make this clear. Probably most everyday controversies over which explanation
is better are concerned with the comparison of explanations all of the same
kind, but when one engages in certain more theoretical or philosophical enterprises there may be disputes as well over explanations of varying kinds.

Suppose that John committed a murder, and we want to "explain" his act.
One possible approach would have us understand John's act as stemming from
motives, attitudes, and goals once we see what John believed about the facts
of his situation. We might be told that John wanted to eliminate a source of
tremendous frustration from his life by committing the murder, that he thought
he could get away with it and thus not have to suffer much, and thus that he
thought that he would be happier if he committed the murder. On the other
hand, another approach would speak of the *causes* of John's act. This approach might make mention of some of his attitudes as predisposing him to act
in violent ways, but it would be likely to also mention things about the particular
details of the circumstances immediately surrounding the murder—how John
felt, how the victim looked at him, how the rage mounted, how particular words
triggered an emotional reaction, etc. These two explanations are quite different.
Is one better than the other? This is the sort of general philosophical question
mentioned above. Each explanation may be right, and the best *of its own kind,*
but the comparison between the two may be very difficult (and interesting).

Presumably, the same tests which determine the best explanation of any
given kind might also be used to determine which of two explanations belonging

to different kinds is better. But since such cross-kind comparisons occur with less frequency in everyday life than same-kind comparisons, and the cross-kind comparisons are usually more difficult, I will concentrate my attention on same-kind comparisons. However, you could try to apply the tests of a good explanation to the murder case described above, as an interesting exercise.

There seem to be three basic tests to see which one of two explanations is better. Thus, the very best explanation is that explanation which satisfies these tests better than all other explanations. (There is no guarantee that there is only one "best" explanation, since two explanations could be tied for first place.) An argument designed to establish that one explanation is the best needs to give good evidence for accepting that its explanation best satisfies these tests. The explanation must

(1) fit the facts
(2) fit the assumed background theory
(3) be simple, complete, and intellectually satisfying.

The best explanation must fit the facts best. Here I am referring to particular facts about particular things. A good explanation should not only "fit" the fact being explained, but also all other particular facts. Let's suppose you have gone to an instructor to ask for a justification of a grade you received on an essay you wrote for that instructor's course. The explanation you receive is quite short: "In my professional judgement, your work merits a 'D'." Does this explanation fit all the facts? Well, if your paper received a grade of "D", it does fit at least that fact. But what about other facts? On the surface, it seems that the explanation "fits" them, too, since the explanation offered is very limited, and thus seems not to be involved with any other facts besides the one grade you received.

Although it may *seem* that the proposed explanation for your grade does fit all the facts, it will be worthwhile to see exactly how one tells whether a given explanation really does.

In order to decide whether an explanation fits the facts, first describe a real fact which the explanation ought to fit if it is correct. Then, see whether the truth of the proposed explanation would in any way tend to show that your description is false. Since you know your description is not false, any tendency on the part of the explanation to show that the description is false would count as a mark against the explanation.

More formally, the test can be represented as an argument with the conclusion that your description of the fact is false. The main premiss of this argument will be the proposed explanation which you wish to test. You will also need an auxiliary premiss which describes the general background conditions in which the proposed explanation operates. In outline form, the argument will look like this:

P.   The proposed explanation.
P.   Description of the general background conditions.
C.   The claim that some given factual description is false.

You ought to be able to see that if this argument is tight or valid, the proposed explanation fails to fit the facts, since the conclusion of this argument fails to fit the facts, and the second premise of this argument is definitely true, leaving only the proposed explanation in the first premiss as the culprit.

Applying this technique to the paper-grading example, we could start by building our conclusion around a description of any known fact. Perhaps because you spent three days of hard work writing the essay, doing the best job you could, you felt quite confident when you handed in your essay that it was a good paper which would receive a grade of "B" at least. Does the instructor's explanation fit *these* facts? To test it, see if anything about the explanation strongly suggests or proves that any of these facts are not the case. That is, try a conclusion of the following sort in the above argument form:

I did not spend much time working on the essay.

Or, again, maybe a conclusion like this:

I did not think the paper was a good one when I handed it in.

Is there anything about the proposed explanation which tends to show that these conclusions are true? If so, there is something about the proposed explanation which does not fit the facts as you know them, and thus there is something about the proposed explanation which is faulty. However, it is fairly clear that nothing about the proposed explanation tends to show that you believed the paper was a bad one when you turned it in, unless you are in a position to add as a reasonable premiss something to the effect that you almost always judge your essays in advance to be of the same quality which your instructors' professional judgement indicates. *If* you *are* in a position to add such a premiss, then the following argument, relative to this assumption, looks like a TABA. The proposed explanation fails to fit the fact (which we are assuming for our example) that you felt confident of the high quality of your essay when you handed it in.

P1. In the professional judgement of my instructor, the essay merited a grade of "D". ·

P2. I turned in an essay to my instructor for grading. It was turned in on time, in the proper form, and it was on the assigned topic.

C.   I thought the paper was a bad one when I handed it in.

Similarly, *if* you know that you almost always write an essay judged to be of high quality whenever you spend very much time writing it, *and* you turn it in on time, on the appropriate topic, then you are in a position to use this knowledge in the argument, and the result will be that the explanation given by your instructor for your low grade on this particular essay will not fit the fact that you spent three days of hard work writing it. That is, the following argument under these circumstances will be a TABA, with true background assumptions:

P1. (As above.)
P2. (As above.)

C.   I did not spend very much time working on the essay.

We have seen now that *under certain circumstances,* it *is possible* for the proposed explanation not to fit all the known facts where they include such things as those mentioned above. On the other hand, if you are *not* in a position to know that you almost always realize in advance when your essays will be judged to be poor, or you are *not* the sort who almost always writes an essay judged to be good when you work at it, then you will *not* be able to claim that the proposed explanation fails to fit these facts on the basis of the above arguments. It will then be entirely possible that you could have worked hard, spent considerable time, and felt confident, even though your essay would be judged to be of low quality.

This process of checking the proposed justificatory explanation against the proper facts can continue so long as you can think of facts relevant to the issue which the explanation ought to fit. In each case, the procedure is the same: find a fact, then try to see if the proposed explanation tends to contradict that fact by seeing whether a reasonable TABA or a valid or strong argument for a conclusion contradicting your fact can be obtained by using the proposed explanation.

Let's suppose that you are an average student who often has some difficulty writing essays, and who is often not quite certain of what grading standards will be employed in grading essays. In such a case, the proposed explanation for your grade does seem to fit all the facts we've discussed so far in connection with this case because the preceding two arguments under present conditions will not be TABA's with believable assumptions. Nevertheless, there is surely something unsatisfying about the instructor's explanation for the grade you received. This shows how important it is to continue looking for *all* the possibly relevant facts before deciding the explanation fits the facts.

Which facts, if any, does the instructor's explanation not fit? We've checked out some of the possibly relevant facts already, but we have not checked the explanation against another kind of relevant fact, namely, the fact regarding whether your grade was justified. Suppose the fact is that your grade was *not* actually deserved. Then, taking that lack of justification as a fact, see whether the proposed explanation contradicts it (in the loose sense of contradiction described above). That is, test the argument

P1.   (As before.)
P2.   (As before.)
_____
C.   The paper deserved a grade of "D".

Clearly, this argument can easily be thought to be a TABA with true background assumptions. The relevant background assumption which would be reasonable to make here in order to convert the argument into a tight one would be something like this:

My instructor's professional judgement is generally reliable in the evaluation of student work turned in to him for grading.

Thus, there is a conflict of sorts between the fact that the grade was not justified and the instructor's explanation for your grade, *if* the background assumption is correct.

This shows that no matter what explanation your instructor gives, so long as you continue to insist the grade was not justified, the explanation will continue to contradict one important *assumed* fact (the general reliability of the instructor), and you will presumably continue to reject the explanation, as one might expect. However, if a detailed and sensible justificatory explanation is offered by the instructor, that explanation will fit all the facts other than the one you hold onto, namely, that the grade was not justified. That explanation will then be the best, according to the present test, and your continuing to hold onto the belief that the grade was not justified will begin to appear untenable. It would be appropriate for your instructor, or someone else, to offer arguments to the effect that your continuing to hold your belief could itself be explained best by psychological factors which prevented you from seeing the plain truth.

To summarize our discussion of the first test of a good explanation, I can say merely that the best explanation best fits all the real facts, but in actual practice we have to test proposed explanations against what we *think* are the facts. We will have to try to decide when an explanation contradicts a supposed fact whether that is the fault of the explanation or the fault of the supposition. When we *know* that something is a fact, we can criticize any explanation which fails to fit that fact. But when we only *believe* something to be a fact, and a proposed explanation fails to fit that belief, we must be prepared to allow for the possibility that it is our belief rather than the explanation which is wrong.

The best explanation fits the assumed background theory best. This next test of a good explanation in many ways resembles the first. There are only two new elements in this test, replacing those used in the first: (a) In this test, we appeal to theory rather than facts. (b) Also in this test we allow appeal to obvious assumptions rather than restricting ourselves to what is claimed as actual fact. I will discuss each of these differences in turn.

The distinction between theory and fact cannot be made here with any great precision. Nevertheless, most people can sense some important difference between the general principles we use to organize our understanding of the many facts about particular things on the one hand and those facts themselves on the other. By using the word "theory" I mean to refer to general, organizing, descriptive, principles of all sorts. I do not mean to imply that these principles are in any way dubious or unproved. (Sometimes people use the word "theory" to make such an implication, as when they say things like, "That's only a theory.") Thus, according to my use of the word, the laws of physics are theories about the ways physical things work, the religions of the world are theories about the meaning of life, the way to live properly, and the general aims and origins of the universe, communism is a theory about how best to organize society, and various more particular and specialized theories abound throughout our lives. One can have a theory about what kinds of teaching techniques work best in certain kinds of situations, about what sorts of candidate will appeal most to certain kinds of voters, and so on. Facts, on the other hand, are to be seen as being more particular and local. If I have a belief, not yet proved, about how a particular auto accident happened, in ordinary speech, but not in this book, I could be said to have a theory about the accident. Here, I will not call that a theory, because it is too specific to one time and place, but rather I will call it a belief about the facts, a belief not yet known to be true.

Why make this fuzzy distinction between fact and theory? Because of the second way in which the second test for the best explanation differs from the first test for the best explanation. According to the first test, the best explanation fits the *actual facts* best. But according to the second test for the best explanation, such an explanation fits the *assumed* background *theory* best. So there is a difference between the appeal to the *actual* facts and the *assumed* theory.

Why not just say that the best explanation should fit the really correct true theory best? Because it is not at all clear that there is some kind of objectively correct, actually true theory for everything. Theories, as we are using the term, are general organizing principles. Facts are much more particular local affairs. It seems there may well be many ways to organize such particular local facts by means of differing theories, and that each such theory may be equally correct. To give a very simple illustration, if you had to construct a "theory" describing how a deck of playing cards is organized, you might describe the cards as being divided into suits, or as being divided into face cards and nonface cards, and each could be correct. Thinking of such descriptions as little theories, you might now imagine how similar kinds of alternative descriptions could be constructed for other kinds of complex arrangements. Such alternative descriptions, using alternative concepts, can grow into genuine alternative theories about the nature of the world, or human action, or life. It is not at all clear that one theory about a given range of phenomena could be the only one that could be called "correct".

But there is more. Even if there were such a thing as the single correct theory for a given subject matter, that might well not make any difference for what counts as the best explanation for something within the range of the theory. I am claiming that an explanation in order to be a good explanation should fit the *assumed* theory, even if that theory is *not* the correct one. Why? Because an explanation makes no sense without background theory, and the only background theory available is the one that is being *assumed* in the context in which the explanation is put forward.

Thus, I claim that there are two reasons why the best explanation is properly judged by reference to whatever relevant background theory is assumed in the context, not by reference to the actually true and correct theory. The first reason is that I am not at all convinced there is such a thing as *the* actually true and correct theory. The second is that explanation makes sense only against the background of the assumed theory.

Let me say a bit more about the need for assumed theory in the background. Remember that an explanation is an attempt to answer a "Why" question. These "Why" questions, it has been claimed, fall into various categories, or kinds. Some of them ask for the cause, others for the justification, for instance. But one cannot understand a causal explanation without some kind of theory about what a cause is and how causes relate to effects. That is, causal explanation is not possible unless present in the context in which the explanation is given, there is some kind of theory about causation in general. Now, suppose that it makes sense somehow to say that the assumed general theory about causation is wrong. I don't see how that would make any difference to judging the various competing causal explanations for a given particular event. They would still be causal explanations, and their force would be a function of the causal theory that in fact was wrong. In other words, explanation takes place within the framework set up by the background theory assumed by the people in the context of explanation, and without that framework, no explanation is possible.

Let's return to the paper-grading example for an illustration. What kind of assumed background theory is relevant here? Although it would take a long time to spell it all out, I can mention some parts of the assumed background theory which will give you the idea of the sort of thing that would count:

Instructors grade papers for the purpose of evaluating them.

The evaluation of student work is part of the overall process of education, contributing to that process by showing students where their strengths and weaknesses lie.

In order for the evaluation of student work to serve its overall purpose in education, the evaluation must be done in such a way that it shows the students where their strengths and weaknesses lie.

In order to serve its purpose in education, evaluation of student work must be based on genuine and not on imaginary strengths and weaknesses.

In evaluating student work, professionally trained instructors should use professional judgement to pick out these strengths and weaknesses.

It is against this kind of background that the request for justification of the grade given to the essay makes sense. Without such a background, no justification is possible, for there would be nothing to justify, since there would not be a grading system at all. If a small child writes the letter "D" on your essay with a red pencil, that would not constitute a grade. Any good explanation must fit the assumed background theory as well as is possible.

Does the instructor's explanation of your grade—our earlier example—fit the background theory relevant to grading? If the grade was merely marked on the paper without comment, and the only justification given by the instructor was the one brief statement that in his professional judgement the paper deserved that grade, then it would seem that the explanation offered by the instructor would not justify the grade in a way that would tie in with the kind of background theory mentioned above. That theory requires grading to be justified by reference to its value in showing students where their strengths and weaknesses lie, not by reference merely to the instructor's judgement. At least that is what the background theory as sketched above seems to be getting at. So, also by reference to this second criterion of a good explanation, the brief justification offered by the instructor fails to be a good explanation. A better explanation would tell the student where the paper failed to be successful and where it was successful.

Before we leave our discussion of background theory and its role in explanation, however, there is one very important modification which must be made. The preceding discussion makes it seem as though the background theory never can be questioned within the process of explaining. However, that's not so.

Background theories are in the process of changing fairly frequently. Various social forces continually mold and shape theories accepted by a given group of persons or a single individual at a given time. Also, what we may call "logical forces" play a role in changing those theories, as can well happen when we attempt to give explanations which fail. Described in the abstract, the change in background theory under pressure from "logical forces" may happen like this: someone assuming background theory *T,* which covers a certain range of phenomena (such as a theory about weather, or about motives), is confronted with a specific case within that range of phenomena that needs explanation. (For instance, a particular storm, or a particular act.) He attempts to construct the explanation, but discovers that no explanation of the particular case can be constructed using theory *T* as background—at least no explanation which meets the other two criteria of a good explanation can be constructed using *T*. He then sees that some modification of *T* appears to be necessary.

Note that I am not saying it is sometimes possible to construct an exemplary explanation without appeal to background theory. Rather, I am saying that sometimes it is not possible to give a good explanation of something until the background theory has been changed, and that the very attempt to explain something may result in the changing of the background theory. Once that theory has been changed, of course, the test of a good explanation will be to see if the explanation fits the new theory.

The best explanation must be the simplest, most complete, and intellectually most satisfying explanation. This third and final test of a good explanation is the least precise. It is not at all clear how one tells in every case which of two possible explanations fits this criterion best. But in many cases, one can clearly tell. If I explain my friend's smile as being the product of his desire to fool me into thinking he is still my friend when in fact he has turned against me and has been plotting with everyone else I know to cause me embarrassment and shame, then my explanation is quite clearly not as simple as an explanation which holds that he smiled because he was displaying friendliness.

On some occasions, simplicity fights with satisfaction. The instructor's brief explanation of your grade was *simple*—it is hard to imagine a simpler explanation—but it is not intellectually satisfying. An intellectually satisfying explanation would pinpoint with skilled precision exactly what features of your essay were deficient. Thus, in utilizing this test of a good explanation, one must somehow make a balanced judgement, weighing simplicity, completeness, and satisfaction whatever way seems appropriate.

I realize all this is vague and incomplete. But it is clear that some such test is important. Without this test, an extremely cumbersome, overly complex, tortured, *ad hoc,* incomplete explanation that happened to fit all the facts and the background theory would count as being just as good as an explanation of the same thing which was simple and satisfying. Our paper-grading example was an example of an incomplete explanation. One can, with imagination, come up with outrageous examples of overly complex ad hoc explanations. One could say, for instance, that most people stop when the light at the intersection turns red because they are afraid that the red light would injure them if they proceed. The injury occurs only to people who notice the light (and that is why inadvertent red light-running does no harm) and only to people who are afraid of the injury. (Which is why calloused red light-runners escape). Although one would be hard pressed to explain why normally afraid people are not hurt by the light when forced to go through the intersection in an emergency, with enough ad hoc complications added to the theory, one might be able to make it fit the facts. But such an explanation seems not to be the simplest or the most satisfying one of the phenomenon in question.

Not only is this test for the best explanation vague, but also it appears to have nothing to do with the *correctness* of the explanations which pass it. Why should simplicity or intellectual satisfaction have anything to do with being correct? And if they have nothing to do with being correct, why should they be used as a test of a good explanation at all?

It would take an entire book to deal with these questions adequately, but I suspect that one might begin by saying that the purpose of requiring a good explanation to be as simple and intellectually satisfying as possible is merely to make it easier to work with. We have trouble working with explanations that are complex. So if we have a choice we will choose to work with the easier, simpler explanation if it satisfies our taste. We also find it easier to deal with explanations that strike us as satisfying, partly because we may remember them better and partly because we find it more personally rewarding to work with them. In other words, I am citing psychological reasons why simplicity and satisfaction are important. I am not trying to prove that the simpler or more satisfying explanation has a better chance of being correct.

This completes our discussion of the tests of a good explanation. Since there are three independent tests, the best explanation will be the one that satisfies all three the best. I have no idea how to weight these tests so as to be able to compare two explanations which satisfy differing tests to differing extents—for example, one explanation which satisfies the first test very well but does not satisfy the second test fully, and another explanation which does not satisfy the first test too well but which satisfies the second one perfectly. Without such comparisons, our discussions of the tests for the best explanation remains incomplete; nevertheless, in many cases these tests do work and in all cases they at least tell you what to look for.

At last we are ready to apply all this to the original task of argument evaluation. We got into our discussion of what constitutes the best explanation because we wanted to be able to evaluate arguments of two different kinds: (1) Arguments for conclusions of the form, 'X is the best explanation for Y', and (2) arguments for conclusions of the form 'X must be true because X is the best explanation for Y and Y is true'. Naturally, these two kinds of arguments are closely related—one is merely a variation of the other. In both kinds of arguments, the key to evaluation will be the same—to ask whether X really has been shown by the argument to be the best explanation for Y. That's where our tests for the best explanation come in.

There are a great many ways one might go about arguing that X is the best explanation for Y. One might give a list of possible explanations, and then proceed to eliminate all but one of them, using various arguments for each. Thus, one might encounter something like this:

> There are only three possible explanations for the cause of the accident. One is that the brakes failed; another that the driver dozed off; and the third that the driver was thinking about something else at the time. We can rule out the first because examination of the wrecked vehicle showed no malfunction in the braking system. The second is unlikely because the driver had been driving for only a few minutes before the mishap, and he had gotten plenty of sleep the night before. Only the third fits all we know about the situation.

Here, we are not told why there are only three possible explanations of the causal sort. But leaving aside the question whether the first premiss above is true, we can see here that our tests of a good explanation are at work. The first two proposed possible explanations are ruled out on the grounds of not fitting simply and satisfyingly with the known facts. (Given a little reasonable background theory, for instance, one can make a fairly good case for the conclusion that the driver was tired and had been driving for a substantial period of time if one accepts the second explanation proposed above; thus, we might say that that explanation does not fit the facts too well.)

In each case, when you are asked to evaluate an argument and decide that one explanation is the best (of its type), you can respond by considering whether that argument does indeed show or tend to show that the preferred explanation satisfies our three tests better than all its competitors. That is the relevance of our three tests to argument evaluation.

Arguments for the conclusion that 'X must have occurred because X is the best explanation for Y and Y occurred' are, of course, merely extensions of the first type of argument. In fact, it is best to see such arguments as consisting of two stages: in stage I, it is argued (or assumed) that X is indeed the best explanation for Y. Then, in stage II, it is argued (or assumed) that Y did occur, and the final conclusion is drawn, namely, the conclusion that X must have occurred. So the ultimate conclusion in such a case is merely that X did occur.

I point out this type of argument to make one simple point: arguments in favor of a certain explanation need not stop with the conclusion that X is the best explanation. Often, there will be some further aim, some further conclusion. In any case, it is worthwhile noting that once one has established that X is the best explanation for Y, then one has grounds for conclusions beyond that, conclusions such as the one now under consideration.

(One minor point: Some kinds of explanations do not describe events or anything else that could reasonably be thought to be the kind of thing that occurs; in such cases, one cannot sensibly conclude that X *occurred* once one knows that Y occurred. For example, a justifying explanation for someone's death may be offered in terms of God's will, but it will not be proper to conclude that since someone died and the best explanation for the death is that it was God's will, therefore God's will must have "occurred". Rather, in a case such as this, the proper conclusion will be that it must have been God's will—that is, that X is *true*.)

Before we leave the topic of explanation, I think it important that you realize just how limited our conclusions about explanations will be. The very best explanations fit all the facts, fit the assumed background theory, and are the simplest, most complete, and most intellectually satisfying. But in practice we can't tell if a proposed explanation fits all the facts; we can only tell if it fits the facts we know or believe in. And a proposed explanation may fit the assumed background theory even though that theory itself is very poor and will soon be discovered to be fatally flawed. Moreover, simplicity and satisfaction seem to have nothing or little to do with the correctness of the explanation. Thus, what counts in practice as the best explanation shifts and slips, depending on our knowledge of the facts, our acceptance of background theory, and our conceptions of simplicity, completeness, and satisfaction. At best, explanations can be fairly well grounded for a time. But over the years, as background theory changes and more facts become known or forgotten, what counts as the best in explanation is very likely to change. Yet, there seems to be little we can do about this. It is not as though new criteria of explanation would work better. Nor can we be expected to know all the facts. We are thus inevitably left in the position of seeking the best explanation in terms of the criteria mentioned here (if I am right), knowing that this means today's best may be tomorrow's throwaway. Does this mean, though, that you can reasonably give any explanation you wish for anything? Not at all! To ignore the presently best-established background theories, to ignore the facts that you *do* know about, to concoct inelegant, ad hoc, incomplete explanations, is to give up on using your rational faculties.

Arguments to the best explanation

64.1 Is the following argument valid? Explain. Is this argument tight?

Bob changed his mind about Christianity and now attends services regularly. The best explanation for this change is that his roommate and his girlfriend presented the Christian religion to him in a new way and got him to think seriously about it for the first time. Therefore, we can be sure it was his girlfriend and roommate's influence which played this role.

64.2 In order to establish that Bob's roommate and his girlfriend played the role attributed to them, the preceding argument appeals to the premiss that this is the best explanation of Bob's changed behavior. Make a list of some bits and pieces of evidence which might be relevant toward establishing the truth of that premiss; include at least five items on your list. For each item on your list, state whether that item is a factual item or a bit of background theory. If all the items on your list turn out to be fact items, add some reasonable background theory items which would be relevant here. (You don't have to personally believe the theory you mention.)

64.3 Assume that you are confronted with a situation in which you have performed a laboratory experiment in chemistry as assigned for your course, using the directions supplied by the instructor in your lab manual, but you did not get the expected results. Let's say that the chemical with which you were working was supposed to change color, according to the manual, but when you did the experiment, no change occurred. List three possible reasonable explanations for what happened.

64.4 Here are some facts, possibly relevant to the preceding explanations, which you may now assume are known: (a) The same set of directions has been used in this course for years. (b) Your friends in the lab got the color change as expected, using the same supplies you used. (c) Your equipment was clean when you started doing the experiment. Take each of your explanations from 64.3 and pair it with each of the facts, for each pair constructing the kind of argument which you are to use in deciding whether the explanation fits the fact. After writing out all nine of these arguments, talk about each one of them, stating whether it shows the proposed explanation to fight with the relevant fact.

64.5 Choose one of the preceding explanations which fights with one of the facts and modify that explanation so that it no longer fights with that fact under the presumed circumstances. Has your explanation become less simple or less intellectually satisfying in the process of modification?

64.6 Describe one or two explanations which fit all the facts described in 64.3 and 64.4 but which are decidedly less simple and less satisfying than any of those proposed in answer to 64.3.

64.7 Scot Morris, a former editor of *Psychology Today,* has given us a report on a classroom demonstration he and one of his colleagues interested in "ESP" (extrasensory perception) conducted in a psychology course at Southern Illinois University.[1] The students in the class chose a card from an ordinary deck of playing cards which they were allowed to examine, and having chosen the card, they informed the instructors of their choice. Dr. Morris' colleague, Steve Werk, then proposed to attempt sending the identity of the card to a friend of his, located in a distant office, by telepathic communication. After concentrating in front of the class for a while, he asked the students to choose three of their classmates to call the friend on the telephone to see if he got the message from Prof. Werk. Prof. Werk gave the students the telephone number and told them to ask for the man by name and to tell him that they are checking on the results of the ESP experiment just conducted. The students did as they were told, and soon returned to report that the man had gotten the identity of the card correct.
   a. Evaluate the following argument for one possible explanation of what happened in this demonstration: "The man in the distant office was able to get the card's identity correct, but not because of ESP. No one has ESP. Some kind of trick was involved." (As a part of your evaluation, identify any background theory being appealed to, and isolate facts which are known and which the explanation agrees with or fights with, and say something about the simplicity, completeness, and satisfaction for the explanation.)
   b. Is the following argument better than the last one? About the same? How do you tell? The man at the other end of the line was able to identify the card because he received a telepathic message from Prof. Werk. No other explanation is possible.
   c. What elements of background theory are assumed by the second explanation (in (b)) but denied by the first one (in (a))?
   d. When one explanation of something assumes a particular piece of background theory and another explanation of that thing denies that piece of background theory, then the piece of theory becomes itself an issue. The way to think about this is to include the disputed theory within the explanation that uses it, and then evaluate the whole explanation, including the piece of disputed theory. Of course, that greatly enlarges the dispute. In our example, it now becomes an issue whether the explanation which includes ESP as a piece of assumed theory is better than one which excludes ESP. All sorts of evidence would be relevant to such a dispute—evidence not presented in the classroom demonstration. Some of this evidence involves all sorts of facts about what other people have been able or not able to do in other tests of ESP. Suppose, just for the sake of argument, that all these other facts can be adequately explained without appeal to the reality of ESP, but that these same facts can also be explained adequately by means of an appeal to the reality of ESP.

[1]In the *Skeptical Inquirer,* Spring 1980, pp. 18–21.

Which type of explanation would then be simpler and more intellectually satisfying—the ESP type, or the non-ESP type? Why?

e. Suppose we alter the argument given in (a) above by describing what the trick was, and by eliminating reference to the general claim that there is no such thing as real ESP. For example, suppose we say that the man on the telephone was able to identify the card because he had the room bugged, and thus overheard the choice made by the students. Is this a good argument, given what you know about the situation? Why not? Would the explanation it offers be simple and intellectually satisfying? Why doesn't this seem to matter at this point?

f. Prof. Morris and Prof. Werk in fact had arranged this demonstration to illustrate to their students the value of careful scientific investigation and the importance of healthy skepticism about published reports of demonstrations regarding ESP. The room was not bugged, and the deck of cards really was quite ordinary. The students chosen were not party to the trick. Nevertheless, the students in the class had been tricked. Come up with one or two possible explanations of how the trick was done. (The story as told above contains all the relevant facts you need to know to find a possible explanation.)

g. If someone, after being told how the trick was accomplished, still insisted that ESP was used, then how would one evaluate their ESP explanation?

11

Arguments About
What We Ought To Do

65

Varieties of valuation

In this chapter, we are going to discuss how to evaluate arguments about what we ought to do, about what course of action we ought to take. Such arguments might be presented concerning relatively trivial matters, as whether one ought to wear the blue coat on a forthcoming date, or important matters of life and death, as whether a particular individual ought to be executed for his crimes. Arguments about "oughts" and "ought-nots" constitute one of the most complex, and yet most interesting and important, areas of applied logic.

The word "ought" has several senses, not all of which concern us here. I want to talk about the kind of "ought" which expresses evaluations of things as being good or bad in some respect. Of course, you are very familiar with this sort of "ought". It is used when someone says "You ought to say you're sorry" or "You ought to buy that car". But "ought" does not always express an evaluation of this type. One can say, "Given the look of the sky, I'd say it ought to start raining any minute now". This "ought" expresses what might be called a "natural expectation" on the part of someone, and does not evaluate anything as being good or bad. We will not be doing anything in this chapter to help clarify or evaluate arguments about the nonevaluative "ought".

The "ought's" which concern us, the evaluative "ought's", have been the subject for bitter intense debate for as long as there has been human history. These "ought's" include the "ought's" of moral judgement, the "ought's" of political policy, the "ought's" of business decision-making, and so on, through the lists of our most important concerns, concerns about life and how to live it. Despite their importance, and perhaps because of it, these "ought's" engender continuing and seemingly unresoluable disagreements. Especially in the moral and political realms, there seems to be no clear way one could even attempt to settle once and for all the correctness of some particular "ought" claim. Such claims do not seem, at least on the surface, to be subject to empirical investigation in the way that claims in science about gravity or psychoses are. It does not seem, at least on the surface, that one could go out and do any kind of scientific experiment to find out what is moral—what from the moral point of view one ought to do.

Since there does not appear to be any clear-cut empirical procedure for discovering what one *truly* ought to do, morally speaking, it has been urged by some philosophers that moral judgements are in fact neither true nor false—in other words, that sentences such as "You ought to give back the typewriter you borrowed last month" do not express propositions.[1] If these views about moral "ought" sentences are correct, then our logic does not apply to them, for our logic has to do with evidence-giving, or support-giving for *propositions*—for truths or falsehoods. (Such philosophers have sometimes thought that moral judgements merely emit emotions toward certain types of actions, rather than expressing propositions. With this view of moral judgements, to say "Torture of animals is immoral" is nothing more than saying in a very disgusted tone, "Animal torture! Yuk!", or to say "You shouldn't steal that book" is just the same as saying "Your stealing that book! Yuk!")

Even if these philosophers were right about *moral* "ought" sentences, that would not show that other kinds of "ought" sentences are similarly lacking in truth value. If someone claims that in order to maximize profit we *ought* to do X, there seem to be some clear tests of whether he is right, and probably no one would claim that such an "ought" sentence fails to express a proposition. Thus, even if those philosophers who deny that moral "ought" sentences express propositions were right, there would be quite enough "ought" sentences from realms other than morality for us to talk about here, where we restrict ourselves to treating propositions.

Moreover, even within the *moral* realm of speech, we can argue for the appropriateness of a given moral judgement, in much the same ways we argue for the truth of other kinds of "ought" sentences, or for the truth of other kinds of sentences not containing "ought" at all. To support a moral "ought", we cite evidence, point out the consequences of our acts, and appeal to general principles, such as we do when we say "Look what would happen if everyone behaved in the way you want to behave". These procedures make it look as though logic should apply also to such arguments. After all, some moral arguments make much more sense than others, and this seems to be a matter of logic. If someone says "You ought to treat people of all races with equal respect because my mother says so", that does not seem to be nearly as good an argument for its conclusion as the argument, "You ought to treat people of all races with equal respect because it would be unjust to do otherwise".

We shall want to be able to look at the logic of the arguments for moral claims, using the standard techniques of argument analysis. An added reason for doing this is to be found in the similarity between arguments about moral "ought's" and arguments about nonmoral "ought's"—"ought's" such as the one found in the business example above. It will be good to have a *general* account of how arguments involving "ought" are to be treated, regardless of whether the "ought" is a moral one or not, just so long as it is an "ought" of the evaluative kind.

---

[1]For example, see A. J. Ayer, *Language, Truth and Logic* (London: Dover, 1946).

There are some special problems about "ought" arguments requiring special treatment before one can become adept at the logical analysis of such arguments. The first of these problems is that varying *kinds* of evaluations, made from differing perspectives, are possible, and result in differing kinds of "ought" statements. The "ought" judgements based solely on practical business considerations will be of a different kind from moral "ought" judgements, even when these judgements agree on what course of action ought to be followed.

As soon as I say that "ought" judgements, expressed in "ought" statements like "You ought to give that back", or "You ought to play your ace now" come from different perspectives, utilizing differing kinds of criteria or standards, you are quite likely to think that I am merely pointing out that different people have different "values", so that different people will make different "value judgements", resulting in differing "ought" statements. Accordingly, you might well suppose I am saying that a person coming from one cultural background will say, for example, "You ought to carry the baby to full term and then give it up for adoption", while another person from a different background will say, on the contrary, "You ought to have an abortion".

But, that is not what I am saying.

Of course, I recognize that people differ from one another with respect to what they value, and that those differences may result in differences of opinion regarding what one ought to do in certain circumstances. But *that* is not the relevant point here. Rather, I am saying that two people might say exactly the same words—for example, "You ought to pay the fine"—but *not* be expressing the same proposition. That will happen if the two people are using the word "ought" in two different ways, appealing to different sets of standards for judging behavior.

The crucial thing here is to determine what sorts of reasons lie behind any particular "ought" statement. If those reasons consist entirely of considerations belonging to morality, then we have a moral "ought", while if the reasons pertain only to what course of action would best further one's business interests, then we have a business interest "ought". There are quite a few different kinds of reasons one might have to back up an "ought" statement, to show why one ought to do so such and so, and each different kind of reason will result in a different kind of "ought", even when these "ought" statements are expressed in exactly the same words.

Sometimes, you can tell what sort of "ought" statement is being made by the way it is worded. For instance, someone might say, "According to the law, you ought to pay the fine." The leading phrase here reveals that the reasons being used here are legal ones, or perhaps legal-moral ones. But often there will be no verbal clue in the "ought" statement itself which will tell you what kind of "ought" you are dealing with. In such cases, context may help you decide, but in many cases you won't be able to tell without questioning the speaker.

The reasons supporting an "ought" statement can be grounded in moral, legal, or business considerations, as well as other concerns. Moreover, these broad categories of concern can be broken into subcategories, as is revealed by the following case: suppose an actress cheats and ruthlessly steps on people to get a leading role in a second-rate film. Someone assessing the situation might say "She ought not to be playing that part". That might be a statement about the filmmaker's business interests, based on the idea that the actress is not suited to the role and was chosen for it only because of some shady dealings. Or it might be a statement about the actress' own business interests, based on the idea that the role is not good enough to warrant the future trouble she will be likely to encounter from having treated others badly, others who will perhaps some day be in a position to strike back. Or, again, it might be a statement about morality, based on the idea that she does not deserve the role. It could even be a legal "ought" statement, based on the idea that there are provisions in the contract between actors, actresses, and producers which have been violated by her cheating.

My purpose will be to clarify all this sufficiently to allow us at least a beginning in the process of evaluating arguments about what we ought to do.

As a first step, let's consider how the ideas we have been discussing above relate to logical analysis and evaluation. One relationship is fairly obvious: if someone is trying to prove a moral "ought" statement, the premisses should appeal to moral considerations, and not to considerations irrelevant to the moral issue. Or, if someone is trying to prove a legal "ought" statement, the only premisses which could be used are ones which relate to legal truths. *The logically appropriate way to establish the truth of a certain sort of "ought" statement must always be to appeal to considerations of the sort relevant to that type of "ought".* Thus, if I am trying to prove that the actress ought not, *morally speaking,* have the role, it would be irrelevant for me to point out that the producer of the film will be likely to lose money on the film if he has this actress play the part, (unless I can connect this loss with morality).

There is also another general logical point which can be made once one recognizes that there are different kinds of "ought" statements. The point is that *sometimes* there is no real disagreement between "ought" sentences that *appear* to be contradictory. Someone may say, "She ought to be playing that role", while someone else may say, "She ought not to play that role", and yet there *might* be no disagreement between these two "ought" judgements! For, the first sentence may express the thought that she is well suited to the role in virtue of her ability, and is in fact better suited than anyone else, while the second sentence may express the thought that her playing the role is immoral. There is no true disagreement between such thoughts. They could well both be true.

It is tempting to continue by asserting that the word "ought" is multiply ambiguous—"ought" has many different meanings. According to this strategy, when, on the basis of moral considerations someone says that the actress ought not have the role, the word "ought" means something different from what it means when someone says she ought not have the role on the basis of business considerations or legal considerations. Now it is correct to suppose that such claims of ambiguity will solve some of the problems we have been discussing. We could say that the string of words, "She ought not play that role," could express many different propositions because the word "ought" could have many different meanings in that string of words.

Although this approach is tempting, I would urge you to avoid it. Why? Because if you take the approach just outlined, you will be stuck with an almost infinite number of different senses for the one word, "ought". There are too many different sets of considerations one might use to determine what one ought to do.

It is already apparent that at least one sense of "ought" would pertain to business considerations. But more than one business "ought" would be needed, since there are different *kinds* of business considerations against which one can measure acts.

Let's say, for example, that I am trying to decide whether to go ahead and have a piece of merchandise shipped to a customer from my company's warehouse before I have a signed order from that customer. I know the customer well as being one of my company's best customers, and they need this merchandise to be shipped right away. But written company policy says that I am not to have anything shipped without a signed order from the customer. In describing this situation, there is one clear way in which I can say without doubt that I *ought not* have the merchandise shipped without the signed order, for the company policy is clear. Talking this way, I might say that although I *ought not* do it, I will go ahead and do it anyway to keep my customer happy. On the other hand, in another mood, I can equally legitimately say that despite the company policy, I really *ought* to go ahead and have the merchandise shipped, because that will accord with the overall aims of the company to keep its customers satisfied. Note how using one set of business considerations leads me to say perfectly easily that I ought not to do it, while appeal to the other set of considerations yields the opposite result! This is exactly what the ambiguity theory is supposed to take care of; that is why we will need to say there are two different senses of "ought" at work here, *if* we are going to use the ambiguity theory to solve this kind of puzzle.

I conclude from considerations like these that the ambiguity theory will do its work only if we have an *enormous* number of different senses of "ought". For that reason, it is awkward to use the ambiguity theory, since it is not very simple. (See Section 64 about the desirability of sticking to simple explanations.)

Fortunately, I think there is a better way. I suggest that we deal with "ought" in much the same way we might want to deal with words like "short". A short person is not the same height as a short bush or a short piece of wire or a short baby. Yet we do not want to say that there are different *senses* of "short", one for people, one for bushes, one for wires, and one for babies. We can explain the differences in the ways we apply the designation, "short", to people and to bushes by saying that the word "short" is applied to different classes of things *according to standards appropriate to those classes*. The standards for shortness for people are different from the standards for shortness in bushes, wires, and babies. But we don't need to say that "short" changes its sense or meaning every time a different set of standards becomes relevant. Rather, "short" always means roughly the same thing, namely, "of less height or length *than the relevant standard*".

How can we use this idea to deal with "ought"? We can say that "ought", in its evaluative uses, always means roughly the same thing—"required by relevant standards". As those standards change, differing "ought" judgements about one and the same act may be generated, but "ought" means the same thing each time. There may be more than one set of relevant standards of conduct pertaining to a given act, as in our business example above. Each set will generate its own "ought". Accordingly, for now we can view "ought" sentences of the form

Person P ought to do X

as being abbreviations for the more fully expressed sentences of the form

Person P ought, with respect to standards or reasons R, to do X.

In the next Section this view about "ought" sentences will be modified, but for now we will adopt it. Of course, similar remarks may be made about "ought not" statements. Using this pattern of analysis, we may look at the business examples already presented.

(1)   You ought not to have the merchandise shipped

is short for

(2)   With respect to the stated company policy, you ought not to have the merchandise shipped.

But

(3)   You ought to have the merchandise shipped

is short for a different sentence:

(4)   With respect to what best accords with overall company aims, you ought to have the merchandise shipped.

Something important happens once we make these moves. The seeming incompatibility between (1) and (3) disappears! Once (1) is expanded to (2) and (3) is expanded to (4), we see that there is actually no incompatibility between (1) and (3) at all, because (2) and (4) could easily both be true. Since (2) says the same thing (1) says, and (4) says the same thing (3) says, there is no actual logical incompatibility between (1) and (3). Thus, it is possible for "P ought to X" and "P ought not to X" *both* to be true!

Moreover, if the written company policy says explicitly that certain matters are left up to the discretion of the manager, we get another startling result. Suppose the policy says that it is up to the manager to decide whether a tardy employee should have his pay docked or should stay late to make up the work, but one of these two penalties must be assessed. Then, *from the point of view of the policy,*

(5)   It is not true that the manager ought to dock Fred's pay

when Fred is tardy. However, it may well be that Fred will be consistently late to work if he is allowed to stay late rather than being docked, and the manager knows that this will not be in the best interests of the company. Thus, *from the point of view of the overall best interests of the company,* it is true that

(6)   The manager ought to dock Fred's pay

when Fred is tardy. This shows that two sentences of the forms

P ought to do X   and   It is not true that P ought to do X

can both be true! Once (5) and (6) are expanded according to the method outlined earlier, the seeming contradiction between them will disappear.

In technical terminology, what I am saying about such pairs of seemingly contradictory "ought" statements is this: when the set of standards relevant to the statement "You ought to do X" is different from the set of standards relevant to the statement "You ought not to do X", or "It is not true that you ought to do X", then both statements *may* be true, because *the proposition expressed by* "You ought to do X" *varies with the set of relevant standards.* The proposition varies because "You ought to do X" is really an abbreviated way of saying something more complicated that makes reference to the relevant standards, as was already explained. Thus, we are going to have to know something about what kind of standards are relevant to the particular "ought" in the conclusions of arguments before we can tell whether the premisses are of the right sort to prove the truth of the conclusion, at least normally.

Does this mean you can just make up any set of standards you want and then go around saying "You ought to do X", where the "X" is filled in with your own favorite acts? I suppose so. You *could* do that, but why would anyone take arbitrary or selfish standards seriously? You can make up rules of a game and then say that, according to those rules, other people ought to do such and so. But this will have a point only if others are playing the game. Moreover, if you claim that your standards are *moral* ones, creating new moral obligations for other people or for yourself, what justifies you in making *that* claim? I have certainly not shown that important standards of conduct such as moral standards can be made up arbitrarily or created out of thin air.

You will be more aware of the various possible kinds of standards or reasons relevant to asserting that someone ought to do some particular thing if you have an idea of what kinds of standards or reasons are typical. Evaluative standards or reasons for ascribing obligations to people generally fall into the following categories:[1] moral, legal, political, religious, intellectual, aesthetic, customary or etiquette-bound, practical.

Moral "ought" statements are quite familiar. A statement falls into this category when the standards have to do with what types of behavior are just, righteous, or morally proper, and what types are unjust, evil, or morally improper. I will refer to such statements as moral "ought's", even though it is the whole statement and not just the word "ought" which is moral in nature.

The legal "ought" is sometimes confused with the moral one. However, there is a clear distinction between the two. One's legal obligations are normally spelled out in some sort of formal code of laws, and once this has been done, one ought, according to the law, do X when the law says that one ought to do X, no matter how immoral or otherwise objectionable the law might be. It is thus entirely possible that according to morality one ought to do X when according to the law one ought not to do X, or vice versa. The political "ought" is distinct from the legal one. The legal "ought" measures acts with reference to the laws. The political "ought" measures them against what is expected in the political system as it exists in actual practice. In some societies, there may not even be a legal system as such, but every society has some kind of political system by which its decisions are made. Thus, one might say, "You ought to listen to the tribal elders", even where there is no law which requires it. This "ought" might well be a political one, expandable into the following: "According to the way

[1]This list comes almost entirely from Paul W. Taylor, *Normative Discourse* (Englewood Cliffs, N.J.: Prentice-Hall, 1961), Chapter 12.

things are done in our society, to make decisions about the direction in which we are going to go, you ought to listen to the tribal elders".

Religious "ought's" can be quite significant. One might say, "You ought to thank God for your many blessings", but one certainly wouldn't mean that thanking God was morally or legally required, nor that it was politically obligatory. Rather, of course, one would be saying that according to the standards of conduct within the religion, thankfulness is required.

"Ought's" coming from intellectual considerations are easily overlooked. For example, we might have discovered that you hold several beliefs that are mutually inconsistent, and that you refuse to give up any of those beliefs, even though there is no doubt about their inconsistency. Perhaps you just don't want to give up the security of those beliefs. You might believe that a person is trustworthy only if they keep all their important promises to you, and that your girlfriend or boyfriend is trustworthy. Yet you might also believe that your friend promised to go out only with you, and that is an important promise. In addition, you might believe that another friend is completely honest with you about important matters, and that friend has told you that your girl or boyfriend went out with someone else. These beliefs are, naturally, inconsistent. If you continue to hold onto all of them, we might say to you, "You ought to get your head straightened out" or "You ought not continue to believe all these things". Here, we are appealing to standards for intelligent dealing with facts and rational investigation of the truth. This is the intellectual "ought".

The aesthetic "ought" occurs when the standards or reasons relevant to the obligation have to do with beauty. Using this kind of "ought", we might say, "You ought to move that piece of furniture over to the corner", or we might talk about how a piece of music ought to be played, or how a poem ought to be organized.

"Ought's" arising from custom or etiquette should be familiar. It is this sort of "ought" which sometimes occurs when one says something like "You ought to say 'Thank you' when someone gives you something". But watch out. The line between moral obligation and the obligation imposed on us by custom is a very fine one, and it may be difficult to tell in a particular case what type of obligation is being asserted. A breach of custom may be looked down upon by others, but a breach of morality is seen as being *evil.*

Finally, there is the vast array of other practical "ought's". The examples from business given earlier fit into this catchall category. A basketball player "ought not" tackle another player, according to the rules of basketball. One "ought not" try to fix an electrical switch without turning off the current. One "ought" to keep his checkbook balanced. And on and on. Naturally, within this vast category of considerations, there will be important subcategories, such as the category of business factors. But I will not try to make a list of all of them.

Of course, in any given situation when someone says that you ought to do X, not all of the above kinds of "ought's" will be relevant—not all of them could sensibly be used at that time. If you are wondering whether to make a certain chess move, and someone says that you ought to do it, presumably there would be no religious "ought" involved. On the other hand, in most situations there are several different kinds of "ought's" that might be used with good sense, and it will take some determination on your part to discover for sure exactly what kind of "ought" is being thrown at you. In many cases, you will never find out, because you won't have enough background information about the speaker to know. Moreover, on many occasions the "ought's" that are presented with abandon by speakers of varying interests and backgrounds will not be pure examples of any one of the types listed above; *many "ought's" are mixtures of types,* where the backup standards for the "ought" come from more than one of the categories described above.

While we are about the business of clarifying "ought" statements, there are one or two related matters to be considered. One of these becomes obvious almost immediately once it's mentioned: There are several words and phrases which may be used to express the same ideas as "ought" expresses. Here is a list of comparable ways of expressing "ought" thoughts:

You ought to do X	You are required to do X
You should do X	You are obliged to do X
You must do X	It is necessary for you to do X
It is your duty to do X	It is incumbent upon you to do X
	It is your responsibility to do X

Similarly, there is a list for "ought not" statements. Notice the placement of the word "not" in some of these expressions. For instance, "It is not your duty to do X" does not say the same thing as "It is your duty not to do X", and only the latter of these expressions is similar to "You ought not to do X".

You ought not to do X	You are required to not do X
You should not do X	You are obliged to not do X
You must not do X	It is necessary for you to not do X
It is your duty to not do X	It is incumbent upon you to not do X
You are forbidden to do X	It is your responsibility to not do X
Doing X is prohibited to you	You are not allowed to do X
You may not do X	

When confronted with some statement of one of these forms, you must clarify what kind of standards or reasons are lying in the background of the statement before you know what statement is being made.

This same point holds also for certain more complex forms of expression involving "ought" or its equivalents. Consider statements like these:

If you are going to drive there yet tonight, you ought to get going.

If you want to win, you ought to put Smith on the first string.

If you want to be just, you ought to give the part to Karen.

If you are going to obey the law, you ought to pay the fine.

Such statements are known as hypothetical "ought" statements, or conditional "ought" statements. I'm sure the reason for such a label is obvious. The points made earlier about types of "ought's" still are applicable. Of course, as you might guess from the above examples, it is often fairly obvious what sort of "ought" the speaker has in mind once one looks at the antecedent.

You might be wondering whether these hypothetical "ought" statements really say anything different from the nonhypothetical ones we were discussing earlier. (The nonhypothetical "ought" statements are called "categorical" statements.) For instance, does the statement, "You ought to give the part to Karen", when understood morally, mean the same as "If you want to be just, you ought to give the part to Karen"? Strictly speaking, I don't think so. I admit that in ordinary conversation the two statements would properly be thought completely similar in expressive power or meaning. But the second statement very strictly understood seems to me to talk only about my obligations under the circumstance that I happen to want to be just, and it does not seem to say whether I have any obligations when I don't happen to want to be just. The first sentence, on the other hand, clearly states that I have the obligation, period. It does not place any condition or limitation on my obligation in the way the second sentence, taken strictly literally, does. So, from the very strict point of view, the sentences do not express the same proposition. Thus, to be clear, one should use conditional obligation sentences only under circumstances where one really does mean the obligation to be conditional.

Varieties of valuation

65.1 From among the following sentences, pick out all those which express either categorical or hypothetical "ought's", and identify each as to whether it is hypothetical or categorical.
   a. If you don't stop doing that, you'll drive me crazy.
   b. It's your responsibility to take care of that problem.
   c. If you want to have ice cream for supper, it's okay with me.
   d. If you want to go to the movies, you have to finish your supper first.
   e. You have to pay your brother back for all the help he gave you.
   f. Might makes right.
   g. You must turn out the lights and lock the door before you leave the office when you are the last to be there at night.
   h. Whenever the temperature reaches the critical point, you are to turn off the valve.
   i. Whenever I see him, I always say something nice.
   j. Eat your cake.
   k. It's your duty to clean up this mess.

65.2 This exercise is designed to help you distinguish "ought not" from "not ought".
   a. When someone says "You don't have to do that", are they saying the same thing as is said by "You ought not do that", or by "It's not the case that you ought to do that"? Is it the same as "That's something you ought not do"? Is it the same as "You have no obligation to do that"? How about "You have an obligation not to do that"?
   b. Make up some additional ways of saying that you don't have to do that.
   c. Make up some additional ways of saying that you ought not to do that.
   d. When it is true that you ought not to do that, does it follow logically that you are not obligated to do that? (Assume that the type of "ought" in both these obligation sentences is the same.) When it is true that you are not obligated to do something, does it follow logically that you are obligated not to do it? Explain.

65.3 Make up your own examples of a moral "ought" statement, a legal "ought" statement, a religious "ought" statement, a political "ought" statement, an aesthetic "ought" statement, and an intellectual "ought" statement. With each example say a bit about the context you are assuming when you claim that these are appropriate examples—say enough to make clear why you think each example is indeed an example of the appropriate category.

65.4 Describe two situations which seem fairly realistic in which there are two seemingly conflicting "ought" statements made—two conflicting statements for each of the situations. Then explain how the seeming conflict in each case dissolves once one specifies what type of "ought" is involved in each statement. Rephrase each "ought" statement so as to bring out what type of "ought" it involves.

65.5 Rephrase each of the following "ought" statements so as to make explicit what type of "ought" it expresses. (There is not just one right way to do this.)
   a. You ought to attend Harvard.
   b. You ought not knock down old ladies.
   c. If you are going to drive that car to town, you ought to have a driver's license.
   d. If you are going to drive that car to town, you ought to check to see how much gas is in it before you leave.

66

Relative vs. absolute "ought's" and prima facie vs. deep disagreement

---

In the previous Section we saw how two sentences of the forms "You ought to do X" and "You don't have to do X" might not express propositions which contradict one another. For instance, the first sentence might mean "From the legal point of view, you ought to do X" while the second sentence might mean "From the religious point of view you don't have any obligation to do X". I find it suggestive to refer to this sort of analysis of meaning as *dissolving* apparent disagreement between the original two "ought" sentences. The disagreement was only apparent, not real, and once the sentences were analyzed, the apparent disagreement disappeared, or dissolved. Disagreements which might dissolve like that are prima facie disagreements. This name literally means "disagreement at first view". Prima facie disagreements about what one ought to do are disagreements expressed in sentences which have not been analyzed to state from what point of view the evaluation is being done; so that one cannot yet tell whether once the point of view is made explicit the disagreement might be dissolved.

A disagreement is *merely* prima facie when it is a prima facie disagreement that *would be* dissolved once the analysis described previously has been done. Probably the last Section left you with the mistaken impression that I think *all* prima facie disagreements may be dissolved. It is time now to correct that impression, time now to say clearly that many disagreements about "ought's" are genuine ones which do not disappear once the relevant point of view has been made explicit.

For instance, one judge, using the body of relevant law, may conclude that the defendant ought to be set free, while another judge, using the same body of law as his reference or standard, will conclude the opposite. We enlarge these conclusions in the manner of the previous Section, to make the standards explicit, and we get "From the point of view of the law, he ought to go to the penitentiary" and "From the point of view of the law, he ought to be set free". Between these, there seems to be a contradiction that will not go away—that will not dissolve. (Strictly, to get the contradiction we ought to make the second judge say "The law does not require he be sent to the penitentiary.") Disagreements that do not dissolve are more than prima facie disagreements, more than merely apparent disagreements. I will call such disagreements *deep*.

Do not be misled by the word "deep". Deep disagreements may exist about *trivial* matters. I may say to my wife, "We ought to buy another bottle of dish soap" and she may disagree. I may be speaking from the economical or practical point of view and so may she. A deep disagreement is merely one that does not dissolve after applying the analysis described in the previous Section.

Many people I have spoken to about matters such as these tend toward some kind of relativistic point of view about "ought" statements, a point of view which would make genuinely deep disagreements about "ought's" impossible. The sort of relativistic view I have in mind runs something like this: "Whenever two people (or societies) differ about what one ought to do, the two people (or societies) are each using their own standards by which to judge, and the *only* reason for the disagreement is that the two people (or societies) have different standards". I hear this kind of relativism being expressed particularly about moral "ought's". But note that this kind of relativism rules out the possibility of deep disagreements about "ought's", because this kind of relativism says essentially that all disagreements about what we ought to do dissolve once the relevant standards are brought into the picture.

Once *this kind* of relativism is brought out into the open, it seems clear to me that it is wrong. The previous example of the judges should be fairly convincing, for they did not have a merely prima facie disagreement. Yet, you may wonder just how it is that deep disagreements can arise.

How can two people using the *same* standards of evaluation come up with opposing conclusions about what should be done in a given situation? A moment's thought will reveal, I think, that in nearly all cases, one essential step in establishing specific "ought's" is first coming up with a view of the factual circumstances surrounding the situation being evaluated. Before deciding whether the defendant ought to go to prison, the judge tries to determine the facts about what the defendant did. Before deciding whether I ought to try driving home for Christmas, I try to determine the facts about the road conditions, the weather expected, the condition of my car, and so on. In other words, a premiss, in the usual kind of argument in favor of a specific "ought", will describe the factual situation to which the "ought" applies.

Clearly, there is room for disagreement between two people about what the facts are. Two people using the same standards of evaluation may come up with different "ought" judgements as a result. Perhaps the first judge thinks the defendant knew that the checks he wrote had no funds to back them up in the bank, while the second judge thinks the defendant did not know. Because of such differences, the two judges, using exactly the same standards of evaluation, namely the relevant laws, make different judgements about what ought to be done with the defendant. This, then, is one way in which deep disagreements about what ought to be done may arise.

Not all instances of such disagreement about the facts are as straightforward as described above. For instance, two people—let's say senators—may differ about whether the federal government ought to enter into an arms limitation treaty with the Soviet Union. One of the bases for the disagreement may be disagreement about some fairly complex facts regarding the likely effects of certain aspects of the treaty. Or the disagreement may arise from the two senators' different perceptions of Soviet trustworthiness.

Disagreement about the facts is not the only source of deep disagreement. If we think of the reasoning which leads to an "ought" statement as following something of the following form, we will see why:

Premisses describing the factual situation.
Premisses describing the relevant evaluative standards.

The "ought" judgement about what one should do in the situation.

The second source of deep disagreement is simply this: not everyone reasons well. Arguments fitting the above format may be quite logical or quite illogical. When two people agree about the factual situation and about the evaluative standards which they will apply to that situation, they may still arrive at differing conclusions because one of them reasons well and the other one reasons badly, or because both reason badly.

Perhaps you have borrowed someone's car and have gotten into an accident with it. You and the person you borrowed the car from want to know whether the car owner's insurance company is obligated to pay for the damages. You want to know whether the company *ought* to pay. Both of you read the policy, which contains the set of relevant standards, and you come up with opposite conclusions about whether the company ought to pay. You do not disagree about the relevant facts. You do not disagree about the relevant standards. (Let me emphasize that you are not concerned about whether it is "fair" for the company to pay, but only about whether the company is obligated to pay under the policy.) You do disagree deeply about what the company ought to do. This disagreement may well have arisen because one or both of you have not reasoned correctly about what the policy obligates the company to do. The disagreement is a deep one because it cannot be resolved by making the evaluative standards explicit. Those standards already are explicit in the argument, for the policy contains them and both of you are using the same ones.

The relativist who denies the possibility of deep disagreements about "ought's" has forgotten that people are not always true to their own evaluative standards—people do not always apply those standards correctly—and when they fail to stick to the actual logical consequences of their standards, people using exactly similar standards may end up disagreeing even when they also agree completely about the facts of the situation.

There is at least one more way in which deep disagreement about "ought's" may arise. The evaluative standards may not happen to yield just one evaluation when applied to a given situation, even when they are applied correctly. This can occur in at least two ways: (1) The evaluative standards may be too vague to yield any one clear-cut "ought" in the situation.   (2) The evaluative standards may be internally inconsistent. In the first case, the standards are open to inter- pretation by those who apply them, and different people may interpret the stan- dards differently. This kind of problem arises within formal evaluative systems like the law, written company policy, rules for playing certain kinds of organized games, and the like. In the second case, where the standards are internally inconsistent, one can get differing "ought" conclusions from the standards merely by emphasizing differing aspects or parts of the standards. Each part of the standards may be perfectly clear and not at all vague, but the parts don't work well together.

We can modify our insurance example to illustrate the point. Suppose the policy says that the insurance covers the damages when the car was borrowed "for a limited time, for a specific purpose, with the consent of the car owner". Suppose also that you borrowed the car "for a few months", for the purpose of giving yourself a way of "getting around campus". Does that count as a "limited time" and a "specific purpose"? Some might say "yes", others "no". As a result, some would conclude the company ought to pay; others would conclude the company is not obliged to pay. This seems to be a case where the relevant evaluative standards are vague enough that reasonable persons could reason- ably disagree about the obligations of the company.

The second sort of problem about standards which do not yield just one "ought" conclusion may also be illustrated by the insurance policy story, if we make a few modifications. Suppose, for instance, that on page four of the policy we find the statement, "This policy insures only the named insured and mem- bers of his/her immediate family residing with the named insured in the same household". That seems clearly enough to indicate that people who borrow the car from the named insured and who are not members of the immediate family are simply not covered. But then, let's suppose, on page seven in the policy document we find another statement—something to the effect that persons who borrow the car with the permission of the insured for a limited time and for a specific purpose are also insured. This part of the policy yields a different con- clusion about whether the borrower is covered. Presumably, the policy writers have made a mistake; the policy is inconsistent.

We have thus seen that there are at least three different ways in which deep disagreement about "ought's" may arise: (1) There may be disagreement about the facts to which the evaluative standards need to be applied.   (2) There may be incorrect reasoning used to get from the facts and standards to the conclusion about what ought to be done.   (3) The standards themselves may leave open what ought to be done in the given situation, either by being vague, or by being inconsistent.

As things stand now, you are in a position to clarify "ought" sentences by making explicit what standards are being used to arrive at the "ought". This clarity is a necessary first step in evaluating "ought" arguments, but as I have been arguing above, achievement of this clarity will not by itself eliminate the need for evaluation of arguments about what we ought to do, since deep dis- agreements are still possible.

However, we are not as yet in a position to clarify all evaluative "ought" sentences. In fact, the most important ones are still missing from our analysis, and it is to these that we now turn our attention. The most important "ought's" are not the propositions which say, "According to morality, you ought to do X" or "According to the law, you ought to do Y", both of which make reference to some special set of standards. Rather, the most important "ought" propositions are the ones that say

All things considered, you ought to do X.

First, let me explain why this is the most important kind of "ought" proposition.

When company policy says you ought to tell prospective customers they are getting something for free but morality says you ought not to tell them that because it isn't true, and legal considerations tell you that you ought not lie to the customers in that way, but it looks as though financial considerations tell you that you ought to go along with company policy, then what ought you to do? *So far, our analysis of "ought's" does not even allow this important question to arise.* We can say that from the point of view of the law, you ought not to lie to the customers, from the point of view of morality, you ought not to lie to the customers, from the point of view of company policy, you ought to lie to the customers, and so on. But so far, you cannot ask which point of view ought to win, unless you want to ask whether from the legal point of view the moral point of view ought to win, or whether it ought to win from the moral point of view, or from the company policy point of view. In other words, so far, all you can ever do is to ask about what ought to happen *from a specified point of view.*

Ordinarily, we give advice and speak of obligations without wanting to restrict our advice or obligation-setting merely to what ought to happen according to some one point of view. Rather, we want to be able to say things like this: "Even though company policy requires you to lie, and it is in your financial interest to do so, you really ought not lie to the customers". When we say things like this, we are *not* merely reporting on the fact that morality and legality say something different from company policy about what you ought to do. Rather, we are saying that *all things considered,* you ought not to lie to the customers. To say this is to say something more far-reaching and sweeping than merely to say that morality requires you not to lie to the customers. When you make an "all things considered" type of "ought" judgement, you are deciding which standards win.

I need a name for this kind of "ought" statement to distinguish it from the kind of "ought" which carries a reference to some specific point of view or set of standards. I will call that kind of "ought" a *relative* "ought" (because of its special relation to a set of standards), and I will call the "all things considered" sort of "ought" an *absolute* "ought".

Just as there can be both hypothetical and categorical *relative* "ought" statements (as described in the previous Section) there are both hypothetical and categorical *absolute* "ought" statements, such as

If you want to be happy, you ought, all things considered, to get married

and

You ought, all things considered, to stop using drugs.

We will consider in the next Section how one might go about trying to argue successfully in favor of some absolute "ought" statement. Our purpose at present is merely to clarify the various kinds of "ought" statements, since that clarification is essential to the proper evaluation of reasons given in support of an "ought" judgement.

Relative vs. absolute "ought's" and prima facie vs. deep disagreement

66.1 Devise three examples of deep disagreement about what one ought to do where the disagreement stems from controversy over the facts, and not from anything else. In each case specify in some detail what the evaluative standards are. The standards may be different for the different examples, if you wish.

66.2 Let's assume that you are going to operate *solely* on the basis of the following moral standard: Treat everyone fairly. Now you are confronted with the following factual situation: there is a high demand for people trained in accounting, and this has forced up salaries for beginning accountants, since there are not enough well-trained beginners to go around. Your university needs trained accountants to teach courses in the field, and must pay the going rate in order to attract any accounting teachers. This rate is considerably higher than the going rate for beginning faculty members in many other disciplines which your university also needs, such as the disciplines of history, philosophy, and political science. You have the power to decide what to pay beginning faculty in each area. You know that if you pay all of them the same rate as the beginning accountants you will not be able to hire nearly enough faculty to teach all the students. How would the fairness standard be applied in this case? Does it yield just one clear choice? If so, how? If not, what does that show about the fairness standard? (Refer back to the discussion of the third type of deep disagreement.)

66.3 Construct a set of moral principles (moral standards) which contain a hidden internal conflict. Try to make the conflict not too obvious. Show that the conflict exists by showing how application of your standards to a concrete case yields two or more conflicting results even when the standards are applied correctly.

66.4 A conversation is given below. It contains a number of "ought" claims, sometimes expressed without the use of the actual word "ought". You are to identify each "ought" claim, and then state whether it is a relative or an absolute "ought", and what reason you have for thinking so. In the case of each "ought" which you identify as a relative one, state what sort of standard is being appealed to by that "ought".

Dad: Well, have you decided what you're going to take up in college?

Son: Not for sure. I think I might like music or psychology.

Mom: If you studied music, what would you do with it? Teach?

Son: Naw, I couldn't stand listening to a bunch of brats murder all the stuff I like to hear. I'd just want to get into a band or something.

Dad: That sounds pretty risky to me. A lot of musicians starve. I think you should think about something more practical.

Son: Like what?

Dad: Real estate. There's a lot of money in real estate. You ought to be thinking about your future.

Mom: Your Dad is right. You owe it to us not to waste this opportunity we're giving you to go to college. That's something we never had.

Son: But I don't care anything about selling real estate. You shouldn't make me do something I don't want to do.

Dad: Listen. It's my money that's paying for you to go. You have to take our opinions into account.

Son: But not real estate. I couldn't stand that.

Mom: Oh, it doesn't have to be real estate. That was just a suggestion. But it should be something practical. Is psychology practical? What would you do with that? Teach?

Son: I don't know. I just like it. I guess I should find out.

Dad: What is psychology anyway? Is that where you go around telling everybody what they think? Is that it? I don't like that idea at all. It isn't right—trying to figure out people's secrets like that. It's none of their business. They shouldn't be allowed to do it.

Son: Dad! The mind is real interesting. I want to find out more about it. You shouldn't put something down that you don't know anything about.

Dad: I know all I want to know about that stuff.

Mom: Could you maybe be a doctor if you studied psychology? I like that idea. You must find out more about it.

67

Evaluating arguments about what we ought to do

Keeping in mind the techniques for clarifying the structure of "ought" propositions, we now turn to the main business of evaluating "ought" arguments no matter what form they take.

Let's use the two arguments below as examples for a while. Although these arguments are related, they also display some important differences.

(a)   You've gotten thirteen parking tickets now within the last two months; so you'd better go and pay the fines.

(b)   You got a parking ticket because you forgot to put money in the meter; so now you have to pay a fine, even though you didn't mean to over-park.

These arguments illustrate what is perhaps the most common and the most simple structure for arguments about what one ought to do, summarized as follows:

Premises describing some facts.
Conclusion about what someone ought to do with respect to or because of the above facts.

In order to evaluate arguments (a) and (b), we have to know what the conclusions mean—that is, what propositions are being expressed by the words used to state the conclusions. In order to decide what propositions are being expressed in these cases, we will need to use every clue we can get a hold of, including verbal and nonverbal hints of all kinds: tone of voice, surrounding sentences, general context, and how people generally talk. Many of these features are missing from our examples above, but we can nevertheless make some rational guesses about what typically could be meant.

Using some common sense, I think we might agree that in argument (a), the conclusion is probably not *merely* a legal "ought", and may not be a legal "ought" *at all*. The speaker does not seem to be saying merely that the law requires the fine to be paid, but rather the speaker seems to be saying that bad things will happen if the fine is not paid. There is no hint here of moral considerations or religious ones. It would seem to me, the "ought" in the conclusion can be expanded into something like this: "From the point of view of what is in your own best interests, to avoid being punished, you ought to pay the fines". Alternatively, it might be reasonable to take the conclusion as an "all things considered" kind of "ought", where the speaker has in mind that the prudential considerations win out over any other possible considerations that might be relevant.

Argument (b) seems quite different, giving us a pretty clear case of a legal "ought". The speaker does not weigh up the chances of getting arrested if the fine is not paid, but merely points out that once one has overparked, without putting money into the meter, then one is automatically liable to a fine, no matter how innocent or unintentional the overparking. This is merely a legal point about parking laws, and the conclusion might be restated accordingly: "Even though you didn't mean to overpark, from the point of view of the law, you are liable to a fine—that is, you are supposed to pay a fine." Now that we know what propositions make up arguments (a) and (b), we are ready to evaluate them. I will begin by concentrating on the question whether these arguments are *valid,* despite the fact that many, if not most, good "ought" arguments are only TABA's. We will find that by paying attention to validity first, we will be in a better position later to talk about tightness. And in order to think about validity, we will try to come up with counterexamples for "ought" arguments. It will turn out that there are certain standard kinds of counterexamples worth trying out on any "ought" argument, and that one can find standard kinds of weaknesses in such arguments by going through these counterexamples.

Arguments (a) and (b) are quite easily shown to be invalid; for instance, imagine a world in which the parking laws are written in such a way that failure to put money in the meter is not punishable by a fine. In such a world, neither argument's conclusion would be true, even though the premisses of both arguments were true. Thus, such a world serves as a counterexample to both argument (a) and argument (b).

The moral to this story is simple: normally, in order to be valid, arguments which conclude that certain persons ought to do certain things must contain premisses which do more than state the factual background. The premisses must also contain some kind of evaluative principle or standard by which the factual situation is judged in order to arrive at the "ought" conclusion.[1] Of course, in simple cases, one might leave out the needed premiss, assuming that everyone could easily fill it in. However, such assumptions are dangerous, for it may not actually be all that clear to everyone just what evaluative premisses are needed or assumed. The only safe way to insure the validity of the argument is to state the evaluative premiss explicitly—get it out in the open where it can be examined. In the case of argument (b) this is fairly easy. The missing premiss presumably amounts to something like this:

The law requires that when one overparks a fine is to be paid.

The addition of such a premiss blocks the proposed counterexample given above, and in fact seems to make the argument valid.

It is harder to fix argument (a). Argument (a) plus the above premiss still is not valid. And there is a simple reason for that. We decided argument (a) does not conclude that the law requires something—the conclusion is not merely a legal "ought". So it is not surprising that the addition of one premiss which states merely what the law requires would not make argument (a) valid.

Before trying to fix argument (a), it will be helpful to consider other kinds of problems which prevent "ought" arguments from being valid. Perhaps argument (a) suffers also from some of these added problems.

Argument (b), as amended, represents one of the very simplest types of "ought" argument. As described in the added premiss, the law is clear and offers no option. (I'm not saying that this premiss correctly describes the law in any actual jurisdiction; I'm just talking about validity.) In real life cases, such simplicity is rare. Argument (a) is more typical in that regard, for it seems likely that any realistic way of adding premisses to fill out argument (a) will give us *options* to consider. It may be possible, for instance, that you would be better off to skip town than to pay the fines, or that you might be better off to take a jail term instead (if you were very poor). At any rate, such options need to be considered and ruled out before one can arrive at the conclusion that you ought to pay the fine. This point can be brought out by thinking of counterexamples which describe alternative courses of action not ruled out by the premisses but ignored by the conclusion. I have already mentioned two such counterexamples above—skipping town or going to jail.

---

[1]This sort of point goes back to the work of Aristotle. Unfortunately, over 2,000 years doesn't seem to have been enough time to get the point across to the general public.

This, then, is the second type of counterexample worth looking for when you are confronting an "ought" argument. Argument (a) suffers from this problem, and would continue to suffer from it even if the premiss about the legal system's requiring a fine for overtime parking were added to it. The alternatives of skipping town or going to jail would still be open.

To make argument (a) valid, then, we would need to add premisses which rule out any other alternative but paying the fine. However, first let's go on to find out what other typical counterexamples to "ought" arguments exist. Argument (a) may suffer from further problems.

The third typical kind of counterexample to "ought" arguments consists of a story in which the course of action the conclusion advocates has some bad features not mentioned by the argument which would tend to show that one ought not follow that course of action after all. (Be careful here. The counterexample must not fight the premisses of the argument. If the premisses say that a particular course of action is best or is the only one available, there is no point in fighting that by trying to come up with unmentioned bad features of that course of action.) If we try to fix up argument (a), we are quite likely to run into this kind of problem. Suppose, for instance, that we enlarge argument (a) as follows:

> You've gotten thirteen parking tickets now within the last two months. That's enough for the police to seek a warrant for your arrest. If you're arrested, you'll have to spend some time in jail awaiting trial, or else you'll have to post bond. But in any case you'll be expected to stand trial. The fines imposed by the judge when you go to trial may well be higher than those you owe now. But if you pay the fines now, they'll leave you alone. So you ought to go and pay the fines.

Notice what has happened here. Nothing more than factual background has been added. In particular, no evaluative standards have been added. So the argument still has the first kind of problem; it tries to move from a description of the factual background to an "ought" conclusion without specifying any evaluative standards. The argument above does attempt, however, to deal with the second kind of counterexample, by giving some reasons for ruling out some of the alternatives such as the alternative of just ignoring the tickets. But it does not yet deal with all the alternatives, such as skipping town or going to jail instead of paying the fine. So the argument still has the second kind of problem too. It also has the third kind of problem, for it is possible that there are bad features of paying the fines and that these features would make it inadvisable to pay them.

To illustrate, the following counterexample would work. Suppose that all the premisses of the above version of argument (a) are true. But suppose in addition that you are quite poor and that if you pay the fines you will be completely penniless. Suppose also that you are disabled sufficiently that you cannot work; so the money used to pay the fines would come from your spouse's meager earnings. Suppose also the jail in town where you would have to go if you were arrested isn't too bad a place, and that your term would not be all that long. In such a case it might be that you would be better off to take the jail term if they arrest you rather than paying the fines.

The fourth type of problem which "ought" arguments may easily have is closely related to the third type just described. Many "ought" arguments, like the amended argument (a) above, work partly by comparing possible courses of action open to the persons to whom the arguments are addressed. When an "ought" argument works by comparing courses of action and trying to show that one of them is the best, it can fail to be valid in either of two ways: (1) It can leave out some *bad* features of the course of action *recommended* in its conclusion. (That's the third kind of problem, described above.) (2) Or, it can leave out some *good* features of the *rejected* courses of action. (That's the fourth kind of problem which I have in mind now.)

To bring out this fourth kind of problem within argument (a) as amended, I might add a bit to my last counterexample. I might add the supposition that your going to jail would relieve your spouse of having to support you, and that your family would be better off financially as a result. This possibility is not ruled out by the argument as amended. Thus, argument (a) as it stands is terribly incomplete. It deals with none of these issues.

There is one final kind of typical problem with "ought" arguments. The problem I have in mind arises when the premisses provide evidence for one kind of "ought" but the conclusion contains another kind. This problem exists in argument (a) when one adds to its original version just the premiss which describes the law regarding overtime parking in a metered zone. This added premiss counts as an evaluative premiss stating a standard sufficient to establish a legal "ought". But the conclusion of argument (a) is not a legal "ought". It is either an absolute "ought" or a prudential "ought". A specific limited type of standard expressed in a premiss can never successfully prove an absolute "ought" without the addition of a further premiss which states that there are no other relevant standards which are more important or which make any difference in the case at hand. Nor can a legal standard-setting premiss by itself establish a prudential "ought" conclusion.

I would be hiding something if I failed to mention that this last issue can cause considerable debate. The debate arises because some people think that once one has established a legal "ought", one has *automatically* established a moral "ought". If these people are right, then legal standard-setting premisses can be legitimately used to establish moral "ought" conclusions. This position can be expressed as the view that we automatically have a moral obligation to do what is legally obligatory. I personally don't see that *logic* guarantees the connection (if any) between legal and moral obligation. Thus, I don't see how one can move from legal obligation to moral obligation without further argument. Similarly, some people think that once one has established a religious obligation, that automatically establishes a moral obligation to do the same thing. For example, if it is thought to be sacrilegious to work on Sunday, then according to this point of view, it is also automatically immoral. Here again, I tend to resist these suggestions, preferring to hold that in order to move from one type of "ought" to another requires further premisses or argument.

Let's take stock of all the problems with "ought" arguments mentioned so far, and then see if we can fix argument (a) to avoid all these pitfalls and make it valid.

So far, we have come up with the following common types of counterexample for "ought" arguments:

(1) The argument may fail to specify any evaluative standard in its prem-
     isses, thereby allowing construction of numerous counterexamples
     employing evaluative standards different from those needed to arrive
     at the conclusion stated in the argument.

(2) The argument may fail to mention some alternative courses of action
     besides the one recommended by its conclusion—alternatives which
     can serve as the basis for a counterexample if the premises of the
     argument do not give enough information to allow us to rule out these
     alternatives. A valid argument need not actually mention all alternative
     courses of action, but it must contain enough premises to rule out all
     alternatives except the one recommended by its conclusion.

(3) If one can agree with the argument's premises and yet tell a story in
     which the course of action recommended by the conclusion has so
     many bad features that it is not the course of action which ought to be
     followed, then the argument is invalid.

(4) If one is evaluating an argument which proceeds in part by giving rea-
     sons for rejecting alternative courses of action so as to leave the au-
     dience with just one alternative, namely the alternative recommended
     in the conclusion, then consider whether the alternatives which have
     been ruled out are really as bad as they are made out to be in the
     argument. If you can tell a story which fits with the premises of the
     argument but which also brings out some unmentioned good features
     of the alternatives that have been ruled out, you may be able to come
     up with enough real or imagined good features to make one of those
     alternatives look better than the course of action recommended by the
     argument's conclusion. That will constitute a counterexample, for the
     argument must provide enough information to prevent the possibility
     that such good features can make such a difference.

(5) Sometimes it is possible to construct a counterexample because the
     premises of the argument contain evaluative principles or standards
     which support a conclusion containing an "ought" of one kind, such
     as a moral "ought", but the actual conclusion of the argument contains
     an "ought" of a different kind, such as a legal one. In a case such as
     this, the counterexample would consist of saying, for example, that
     although one ought, morally speaking, to do the thing mentioned by
     the argument, one ought not, legally speaking, to do that thing. When
     the conclusion contains an absolute "ought", then the premises must
     be strong enough to avoid this kind of counterexample. In such a case
     the premises are going to have to say something to the effect that no
     other considerations are important, or no other conflicting "ought's"
     exist, or that all other conflicting "ought's" are less important.

Although I would not want to claim that these problems are the only ones
ever arising in "ought" arguments, I do believe they are the main ones, and that
an "ought" argument which passes all these tests is likely to be valid. (Of
course, the standard techniques for proving arguments valid still apply here.
However, often such arguments have a structure which is too complex to be
represented adequately by our techniques, with the result that you will not be
able to actually prove them valid.)

What would argument (a) have to look like, then, to avoid all these problems? Here is one attempt:

> You've gotten thirteen parking tickets now within the last two months. That's a large enough number that if you don't pay the fines, they'll come looking for you to arrest you. They won't have any trouble finding you, if you stay in town, and you can't skip town because of your job. If they arrest you, you'll have to stand trial, and you're sure to be convicted and be assessed a larger fine than if you pay now. You're better off paying less now rather than more later. You're better off paying the fine than going to jail. There are no other alternatives, and no other relevant considerations. So, all things considered, you should pay the fines now.

Go through this argument to see if all the five types of counterexamples described above have been blocked. If not, think of a way to block them. But be careful. It may be tempting, for instance, to say that perhaps skipping town wouldn't be so bad after all, since maybe you could get a better job elsewhere. But such a temptation should be resisted. It would not give us a counterexample, because the argument simply says now that you *can't* skip town. That's all right to question if you're trying to assess the truth of the premisses, but not if you're trying to come up with a counterexample that would show the argument to be *invalid*.

We are ready now to move on to a consideration of arguments which contain "ought not" conclusions.

Let me use the following example of this type of argument:

> Our normal bail bond practices hit poor people much harder than people who are better off financially, since the poor can't make bail and are left to rot in jail until their trials, even though they haven't been proved guilty of any crime. So we should not keep our present system.

I will now go through each of the five types of counterexample already mentioned, to see how each type might be changed to apply to this sort of argument.

(1) Failure to specify any evaluative standard in the premisses still constitutes a major problem. This type of counterexample will work just as well for "ought not" arguments as for "ought" arguments. For example, the bail bond argument can be shown to be invalid by pointing out that it is possible that it is a *good* thing to treat the poor more harshly than the rich.

(2) The matter of alternatives which have been left out works somewhat differently for this type of argument. When the argument concludes that we ought not do something, it is irrelevant whether there are also other alternatives, not mentioned by the argument, that we ought to do, or that we also ought not to do. However, in order for it to be true that we *ought not* follow a certain course of action, there must be some alternative to following that course of action. If there is no alternative to the bail bond system, then we cannot conclude that we ought not keep it. Therefore, it is worth trying a counterexample which says that there is *no* viable alternative. (I'm afraid that in the case of my bail bond

argument, though, I can all too easily think of alternatives that might work; so it will be difficult to construct this type of counterexample for this particular argument.)

(3) With arguments containing "ought" conclusions, we worried about the possibility that the course of action recommended by the conclusion had bad features not mentioned by the argument. Now, with arguments containing "ought not" conclusions, we will have a different worry: does the course of action rejected by the conclusion have possible *good* features not mentioned by the argument? If there are enough of these features, that may be enough to make the conclusion wrong, and if these good features fit with the premises of the argument, the argument will be invalid. It may sound cruel to say so, but in the realm of the possible it seems we might conceive that the poor people are better off in jail, and that keeping them there before their trials is thus best. (Keep in mind here that we are only dealing with what is logically possible, and not necessarily with what is actual.)

(4) Suppose the bail bond argument is altered by adding a premiss which says that there is an alternative to keeping the poor in jail without being able to make bond. The alternative is to move up the trial dates for poor people who cannot make bond. That way, they won't have to spend many days in jail before their trials. The point of adding such a premiss is to make it appear that we ought not keep the present system, partly because there is a better one available. Thus, the bail bond argument as amended becomes an "ought not" argument which proceeds in part by giving reasons for recommending an alternative course of action so as to persuade the listener to give up on the course of action rejected by the conclusion. In such cases, it is worth trying for a counterexample which brings out unmentioned *bad* features of the alternatives mentioned by the premises. If these features are bad enough, and fit with the premises, then we have a counterexample. In the case of the altered bail bond argument, one might point out that moving up the trial dates would make it impossible in many cases for the defense attorneys representing the poor people to prepare their cases adequately, with the result that these people would receive less than appropriate legal defense at their trials. This would be more unfair to these people than keeping them in jail for a few weeks prior to their trials.

(5) The fifth kind of problem exists for "ought not" arguments in exactly the same form as it exists for "ought" arguments.

This completes our discussion of standard difficulties arising with both "ought" and "ought not" arguments. These difficulties have been presented as blocks to validity for such arguments. It remains to be seen how all this applies to the question of whether these arguments are tight.

First of all, let's consider the matter of background assumptions. The person who presented the bail bond argument in its original form left out any evaluative standard. Presumably, he would be likely to acknowledge some such standard as this: it isn't fair to treat poor people differently than rich people in our criminal justice system. This might then be thought of as a background assumption which needs to be added to the argument in order to make the statement of the argument more complete.

Sometimes, the addition of such background assumptions will make an argument *valid*. In such cases, we are *not* dealing with arguments that are tight or that are TABA's, and we merely have to observe the usual precautions about adding premises to arguments—see Section 59.

On other occasions, the addition of the missing assumptions serves only to make the argument *tight* rather than valid. In such cases, we are dealing with arguments that only make their conclusions *probable*.

Let's assume from now on that we are dealing with completely stated arguments, where all the background assumptions have been written out and included. How could such arguments be tight? That is, if we are trying to show that some course of action is one we ought to pursue, or one that we ought not pursue, how might it happen that we would have an argument that only made it probable that we were on the right track?

There are at least three ways in which probability enters into arguments of this sort. (1) Often we are not in a position to state the factual situation definitely. We may not know for sure that failure to pay the thirteenth parking ticket will lead to arrest. Rather, we will have to say that it *probably* will. We may not know whether moving up the trial dates would definitely lead to unprepared defense attorneys; we may have to say only that it probably would. (2) Somewhat less frequently, the evaluative standards themselves are probabilistic rather than definite. We may want to say that leaving poor people in jail when rich ones can get out is *probably* unfair, or that the law is structured in such a way that a fine for overtime parking is *probable*. (3) When weighing up various factors in an argument, some of which count against the conclusion and some of which count in its favor, it may well be that the weighing process cannot be stated with much precision, and that the weights assigned to each factor are only approximate, with the result that the outcome is only probable at best. Thus, if there are some reasons for skipping town and some for staying, some reasons for paying the fine and some for going to jail instead, it may be that we cannot say for sure just how all these weigh out, and that we can at best say that it is probably better to pay the fine.

I believe the introduction of probability into "ought" and "ought not" arguments does not make any significant difference to our list of primary types of problems which can arise in such arguments. The types of counterexamples we have gone through still are relevant even in probabilistic arguments, although they are no longer directed toward showing invalidity. They are now directed toward showing that the arguments are seriously incomplete—incomplete enough to undermine even the *probability* of the conclusion, given the premisses.

Evaluating arguments about what we ought to do

67.1 For each of the following arguments, make up counterexamples of each of the five standard types applicable to the argument. When a given type of counterexample does not seem to be possible for a given argument, state why you think it is not possible. Some of the arguments may be valid.

a. Your father wants you to be a doctor, and you have the ability; so you should try to become a doctor.

b. Your wife has been coming home late from work during the last few weeks, and she has been acting somewhat cold toward you. Occasionally, the phone rings and when you answer it the other party hangs up without saying anything. You found a pocket comb lying in her car yesterday, but it wasn't yours. So, you should believe that she is being unfaithful to you.

c. It isn't right to tell someone that you love them when you don't mean it and you are only trying to get them to have sex with you. But that's just what you did. So, you did something you should not have done.

d. Our normal bail bond practices unfairly discriminate against the poor, since they can't make bail and are left to rot in jail, sometimes for months, until their trials, even though they haven't been proved guilty of any crime. So, we should change to a system whereby poor people are allowed to be free until tried and convicted, without the payment of large bail bonds.

e. Musical works must have some kind of unity of form or theme in order to be good. If you introduce the proposed changes into that composition, you will break the continuity between the first three measures and the beginning of the second page. So, I don't think you should do it.

67.2 Make some suggestions about how each of the preceding arguments could be fixed to avoid any counterexamples you found for them.

67.3 If each of the arguments for which you found counterexamples were changed so as to claim only that its conclusion was probable, would that have made them any better?

Rules for Constructing Derivations

Horseshoe elimination (⊃E)	Negative horseshoe elimination (N ⊃ E)	Horseshoe introduction* (⊃I)	Triple bar elimination* (≡E)
			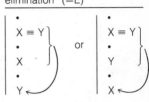

Horseshoe to wedge (⊃ T ∨)   Wedge to horseshoe (∨ T ⊃)   Triple bar introduction* (≡I)

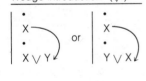

Wedge introduction* (∨I)   Wedge elimination* (∨E)   Redundant wedge elimination* (R ∨ E)

Ampersand to wedge (& T ∨)   Wedge to ampersand (∨ T &)   Repetition (R)

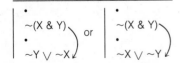

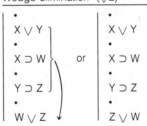

*indicates a basic rule of the system.

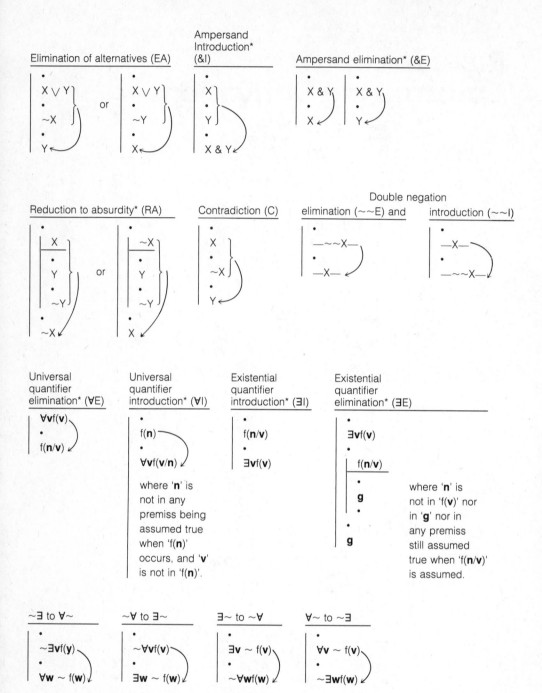

Restriction on all four of the above rules: '**w**' must not occur in 'f(**v**)' unless **w** is **y.**

*indicates a basic rule of the system.

Answers to Selected Exercises

2.1 ,Only (d) is clearly part of logic, for (d) describes support relations between propositions, while the other answers do not.

2.2a One possible cause for belief in God is a psychological need for security. The fact that belief in God satisfies this needs seems not to have anything to do with whether there really is a God.

3.1 Any serious campaign for equal pay for equal work would be destructive. The society must support in one way or another the role of the male as principal provider for the family. The present sexual constitution is cheaper than the alternatives. (Many other answers are possible.)

3.2 The evidence of all human history and anthropology as well as an increasing body of hormonal research is contrary to the claim that women have no more innate disposition to nurture children than do men. A man's body is full only of undefined energies. (Many other answers are possible.)

3.3 (b) is clearly not reasonably treated as an argument, for it constitutes an explanation of how something happened without providing any hint that this explanation is given in order to prove something.

4.2 Yes. An intermediate conclusion plays both roles.

4.3a (1) is a better answer than (2) or (3). (3) reverses the support role in the passage, because the survey results support the tie-up. Answer (2) claims the conclusion of the whole passage is the tie-up statement; however it seems more reasonable to see the tie-up statement as an intermediate conclusion, supporting the ultimate conclusion that the campaign to accept marijuana should be rejected.

4.4a Probably not. (1) does not seem intended as support for (2)—if anything, it's the other way around, with (2) supporting (1). (6) comes closest to supporting (2); however, (6) says nothing about the differentiation between North and South, or about racism.

4.6c P.   Pipes and pipe tobacco are heavily advertized in snobbish ads which run in magazines whose primary audience is upper middle class and upper class whites.

P.   Eighty-five percent of the American pipe smokers are in executive or professional occupations.

C.   Pipe smoking is an establishment habit.

P.   Pipe smoking is an establishment habit.
P.   The Chancellor smokes a pipe.

C.   The Chancellor has an establishment habit.

P.   The Chancellor has an establishment habit.
C.   He is not to be trusted.

5.1a   Key words: "so that", "hence". (Both indicate conclusions.)

P.   The external ears of the common mouse are supplied in an extraordinary manner with nerves.

C.   They no doubt serve as tactile organs.

P.   They no doubt serve as tactile organs.

C.   The length can hardly be quite unimportant.

5.1d   Key words: "since" (premiss indicator), possibly "because" (premiss indicator), and "therefore it follows that" (conclusion indicator).

P.   Whatever is inferior to the mind is weaker than the mind. (?)

C.   Whatever is inferior to the mind cannot make the mind a slave to lust.

P.   Whatever is inferior to the mind cannot make the mind a slave to lust.

P.   Whatever is equal or superior to the mind that possesses virtue and is in control does not make the mind a slave to lust.

C.   Nothing can make the mind a companion of desire except its own will and free choice.

It would be possible to look at this argument as a one-step argument by including the weakness clause in the first premiss of the second argument written above, and dropping the first argument entirely.

6.1   On the first conception, the two strings of words are different sentences, because they are located in different places.

6.2   On the second conception, the two strings of words count as just one sentence, because the words appear in each in the same order.

6.5   The train crushed the truck at 10 P.M.

8.3   Fallacious ad hominem argument: Jack wants the company to buy the cars, but his brother and sister are partners in a car dealership and Jack hopes the company will buy the cars from them, making him look good in their eyes. Obviously, then, we can't pay any attention to Jack.

Not fallacious: Mary thinks that we are better off buying rather than leasing, but the facts do not bear out her position. I have the figures here to show that the net cost to the company for leasing will be $220,000 per year for five years, while the comparable cost for buying the cars outright would amount to $350,000 per year for five years. So, we should reject Mary's plan and go ahead with consideration of the lease.

8.6   No. The fallacy is committed only by arguments; lies are not necessarily arguments. Moreover, only arguments containing conclusions of the right form can be ad hominem arguments.

9.1b   This is hard. Much depends on context. If it seemed that the mayor wanted the information leaked, or wanted to look good to his or her staff, the argument is fallacious. If you believe no one is in a good position to predict crime rates, because no one could have enough information, the argument will be judged fallacious. On the other hand, this argument seems generally better than the previous one, in Ex. 9.1a.

9.1d It would be reasonable to say that this argument is not fallacious, given truth-in-advertising laws and the likelihood that a scientist would not wish to lose a good reputation in the scientific community. However, if we found out that Dr. Schenk had been paid an enormous sum, or was about to retire from active scientific work, our opinion of the argument would perhaps change. This is not a clear case.

10.4 This argument is not of the proper form to commit the "you too" fallacy. In order to commit the fallacy, the speeder should have said that his speeding was justifiable because the other guy was behaving similarly.

10.5 Yes, it does seem that the politician commits the fallacy here, because he or she is arguing that bribery is acceptable merely on the grounds that many other politicians have accepted bribes in the past. The fact that lawlessness is common has nothing to do with the justifiability of lawlessness.

11.1c You should treat other people in the same way you would like them to treat you. Therefore, if you like people to drop in on you without an invitation, you should drop in on other people without an invitation.

11.2b One consequence which follows legitimately from this principle is that the doctor who gives a shot to someone who never harmed the doctor is acting immorally. Since this consequence is unacceptable, and the argument for it is not fallacious, we can know that there is something wrong with the principle.

12.2 Once the sample reaches the neighborhood of 1,200 to 1,500 it can be deemed quite likely representative, no matter how many people there are in the state all together.

12.3a Fallacious. Too small a sample.

12.3d Fallacious. The sample is obviously large enough, but it is not chosen properly to reveal natural abilities. This type of sample selection automatically excludes inferior women and superior men, and is thus biased.

14.1a Fallacious situation: I am not in a position to have heard if my brother has died. For example, I may have lost contact with him over the last ten years, and I have no relatives who would let me know of his death. Non-fallacious situation: I am in contact with my brother in a close enough way that I would be very likely to know if he had died. For example, he and I are living in the same house under normal conditions and he just stepped out into the kitchen to fix a sandwich two minutes ago. Nothing odd has been heard in the meantime.

14.3 Even though the given argument is fallacious, it would be equally fallacious to argue that since she has not proved herself trustworthy, she is untrustworthy. In other words, you don't have enough evidence to form a good argument for either conclusion, and how you should treat Leslie will depend on what other truths you know about how to treat people who have not yet been proved either trustworthy or untrustworthy.

15.1a Judy could help out financially or suggest going out to free events.

15.2   The argument from 15.1a which commits the fallacy is this one:

P.   "I will break up with John" is not true.

C.   "Things will go on as usual and I won't get to go out much" is true.

16.1b Perhaps the clearest example of a premiss not sufficiently independent of the conclusion is the claim buried in the middle of the argument to the effect that the fetus is a human life. Since the whole argument depends on seeing the fetus as a human being with the same moral rights as other people, this premiss will be viewed by pro-abortionists as question-begging.

17.1   At least *d, e,* and *f* are complex. Perhaps *c* can be viewed as complex. *D,* for example, presupposes that the garage will be built.

17.2a Presupposition: You found it necessary to flatter him. Fallacious context: Suppose that the argument is put forward in a context in which it is not at all clear there was any flattery going on, so that to assume there was flattery amounts to assuming you made a mistake. Nonfallacious context: Suppose it is clear to the audience that you did flatter him.

18.1   P1.   A six-year-old child is entitled to full legal and moral protection, just as adults are.

P2.   Anyone one day younger than a person who is entitled to full legal and moral protection is also entitled to it.

etc.

C.   A fetus is entitled to the same protection.

18.5   The only long chains shown by this Section to be fallacious are those meeting all the conditions for the slippery slope type of argument.

19.1h One can look at this argument as containing a fallacious appeal to the authority of the authors (not the instructor), for there is no reason to think that these authors are in any special position to know about the sexual characteristics of women.

19.1o This argument tries to use the lack of a convincing case for one side as evidence for the conclusion that the opposite side is right. Since engine durability would presumably be difficult to establish conclusively in six weeks, this lack of a convincing case doesn't mean much, and the argument commits the fallacy of arguing from ignorance.

21.1a Premiss: Hypoclycemia is a drop in blood sugar. Defines "hypoglycemia". Presumably a complete definition.

21.3   While some people may actually mean to say that the divine is identical with love, most people probably think there is more to God than love and so would not be trying to define "God", unless they are giving only a partial definition. However, a more likely candidate for a definition of "God" would say something like "the supreme being" or "the source of all things". Thus, it is likely that the statement "God is love" is intended merely to describe an important attribute of God, namely, His loving character.

21.4a An important but very hazy phrase in this argument is "war crimes". It stands in need of clarification before we can tell whether the actions of our leaders even come close to being war crimes. For instance, is it a crime to exercise bad judgment about how to conduct a war?

22.1a It is possible for people to value a big statue, such as the Statue of Liberty, but yet such an object would not count as money on the ordinary definition. This shows the proposed definition too broad by lexical standards.

22.2 An author might use 22.1a in an attempt to show that what we ordinarily think of as money is somewhat arbitrary, and that the important fact about money is the willingness of people to value it. Accordingly, if people valued dandelion flowers, they could be used as money, according to 22.1a.

22.5a The ordinary concept of causation seems to include the idea that causal relationships are constant only because of the causal connection and not because of mere coincidence. Thus, the proposed definition may be too broad, since it seems to include cases of coincidental relationships among those which count as necessary.

23.1a The first sentence is false if it is supposed to be a lexical definition, because it does not accurately report the meaning of "human".

23.1b Probably the author means to be concluding that only people who can care about others are deserving of the rights we normally think of as human rights. If so, his conclusion uses the word "human" in two ways—once as stipulatively defined, and once as normally defined, in the phrase "human rights". Not only is this confusing, but also it makes the argument look much better than it actually is. The conclusion sounded clearly true when worded in its original form, but when we understand what it actually says it is no longer clearly true, and no longer do the premises seem relevant to showing that it is true.

23.2a Perhaps "absolute" here means something like "allowing no exception or doubts or room for error". This sort of definition would fit with the rest of the argument, anyway.

23.3a If the center is right, the proposed definition is too narrow, since it might be possible to attempt to contain an opponent's influence without risky confrontations. The proposed definition is also too broad, for not all risky confrontations need have a place within a policy of attempting to contain an opponent's influence. For example, some risky confrontations might be motivated by a desire to unify the country, or might serve only to make someone look strong.

23.3b No. It might have been that the U.S. engaged in risky confrontations with the U.S.S.R. in an attempt to contain the influence of the U.S.S.R., and that the U.S.S.R. was an opponent.

24.1 Correct statements: a, b, e, f, i, and k.

24.2 Valid: b and d.

25.1a Suppose Angela would have run if guilty, and also that she did run. But suppose also that she was not guilty. She ran because she thought she would not ever get to have a fair trial since the evidence against her was strong.

25.1b Suppose that dishonesty and incompetence are both good reasons why any senator should be defeated if opposed by a good candidate. Suppose also that our senator is honest, incompetent, and opposed by a superior candidate. Then the senator should be defeated even though honest.

25.1c Suppose that although the men have not had any food or water for ten days they are still alive because they went into a state of suspended animation.

25.1d Valid.

26.1a All crows are birds, and that animal over there with the blue feathers is a bird; so it must be a crow. (Context: That animal is a blue bird.)

26.1b If you are now reading a novel, your eyes are open. Your eyes are open. So you must be reading a novel now.

27.1a Kyle ran for election AND JoAnn ran for election.
   b NOT (I like movies).
   c I'll go with you ONLY IF you promise to be quiet.
   d We'll have to give up UNLESS we get some action on this soon.
   f IF the county is allowed to keep the money, THEN NOT (there will be enough to run the town).

27.2 In *a, b,* and maybe *d.*

28.1a Replace "for" by "since" or "because" or "for the reason that".
   d The floods came AFTER it rained.

28.2b You'll get rich doing that kind of work IF AND ONLY IF you are able to keep it up for a while.
   c IF NOT (we operate) THEN the patient will die.
   f IF the barometer will fall THEN (that means) it will rain.

28.3a A AND B
   c A IF AND ONLY IF B (Sometimes, A IF B.)
   d IF NOT B THEN A (Other answers are possible.)
   f IF A THEN B

28.4 Several connectors, such as "because", cannot be replaced by preferred ones.

29.2a $\dfrac{\text{IF A THEN B}}{\text{NOT B}}$ **29.2b** $\dfrac{\text{A AND B}}{\text{C AND D}}$ **29.2c** $\dfrac{\text{(A AND B) AND C}}{\text{D}}$
   NOT A

30.1a

The driver fell asleep.	The driver was drunk.	The driver either fell asleep or he was drunk.
T	T	T
T	F	T
F	T	T
F	F	F

30.1d

I made it on time.	I missed the bus.	NOT (I made it on time) BECAUSE I missed the bus.
T	T	F
T	F	F
F	T	?
F	F	F

31.1h

A	B	A BECAUSE NOT B
T	T	F
T	F	?
F	T	F
F	F	F

31.1j

A	B	A UNLESS NOT B
T	T	?
T	F	?
F	T	F
F	F	?

31.3b

You are brave.	You deserve to win the crown.	You deserve to win the crown ONLY IF you are brave.
T	T	?
T	F	?
F	T	F
F	F	?

32.3

The explosion happened.	The fire broke out.	The roof collapsed.	The explosion happened BEFORE the fire broke out.	The roof collapsed AFTER the fire broke out.	The roof collapsed AFTER the explosion happened.
T	T	T	?	?	?
T	T	F	?	F	F
T	F	T	F	F	?
T	F	F	F	F	F
F	T	T	F	?	F
F	T	F	F	F	F
F	F	T	F	F	F
F	F	F	F	F	F

33.1a We got the full amount promised in the contract.	(conclusion) The contract was violated.	(premiss) NOT (we got the full amount promised in the contract).	(premiss) We got the full amount promised in the contract OR the contract was violated.
T	T	F	? or T
T	F	F	? or T
F	T	T	? or T
F	F	T	F

No bad lines, since there are no lines with true or questionable premisses when the conclusion is false.

332.a 33.1a is not T-invalid.   **33.2b** The argument is valid, in 33.1a.

34.1 Diagram for (1):

A	B	IF A THEN B (premiss)	NOT B (prem)	NOT A (conc)
T	T	?	F	F
T	F	F	T	F
F	T	?	F	T
F	F	?	T	T

No bad lines in this typical table.

34.2 Argument fitting (1): If Mary had been drinking heavily at the party, she would be drunk now. But she isn't drunk. So, we can see that she was not drinking heavily at the party.

34.3 Variation on (1):   **34.8** Valid: If today is Tuesday, then tomorrow is
IF NOT A THEN B   Wednesday. Today is not Tuesday. Thus, to-
NOT B   morrow is not Wednesday.
───────────
NOT NOT A

35.4 'A ⊃ B' is true when 'A' is false, no matter what value 'B' has. 'A & B' is false when 'A' is false, no matter what 'B' is.

35.5 Acceptable: *a, c, d, f, g, m, n, r, s* and *t*.

36.1a

A	B	~(A ∨ ~B)
T	T	F T F
T	F	F T T
F	T	T F F
F	F	F T T

37.1a A & B
b A ∨ B or
  ~A ⊃ B or
  ~B ⊃ A
c A
d A ⊃ B

e ~A & ~B or
  ~(A ∨ B)
f ~A & B
g A & ~B
h A ⊃ (B ∨ C)
i A & ~B

38.1 Valid: *a, b, c,* and *e*.

38.2a A ⊃ ~B   Table will show   **38.2e** A ⊃ (B & C)   The truth table test
    B   this valid.   B   is inconclusive. The
    ‾‾‾   ‾‾‾   diagram is invalid.
    ~A   A

  b A ⊃ B   Table will show   **f** A ⊃ B   The truth table test
    B ⊃ C   this valid.   ~A   is inconclusive.
    ‾‾‾‾‾   ‾‾‾
    A ⊃ C   ~B

39.1a

(pr)	(co)	(pr)
A	B	~B ⊃ ~A
T	Ⓣ	
T	F	T̸ Ⓕ F̸
Ⓕ	T	
Ⓕ	F	
(valid)		

39.2a Diagram: A ≡ B   Bad line occurs when
           B ⊃ C   'A' is T, 'B' is T,
           ~A ⊃ D   'C' is T, and 'D' is F.
           ‾‾‾‾‾
           D

  b Diagram: R ⊃ (N ∨ S)   Bad line occurs when
           N ⊃ H   'R', 'S', 'N', and
           ‾‾‾‾‾   'H' are all F.
           S ∨ H

40.1a Diagram: S ⊃ I   Bad line occurs when 'S' is F and 'I' is T. That sug-
           I   gests a story in which it is important for birth control
           ‾‾‾   information to be available for some other reason.
           S

  b Diagram: N ⊃ S   Ignoring the value of 'L', the bad line oc-
           W ⊃ ~H   curs when 'H' is T and 'N', 'S', and 'W' are
           (H ⊃ ~S) & ~L   all false. This suggests a counterexample
           ‾‾‾‾‾‾‾‾‾‾‾‾‾   in which we do not wait, and we do not
           ~H   have the release next week either. That is
                       the key. Have the release right away,
                       making everyone happy and avoiding
                       shorthandedness.

40.3a Key idea for counterexample: Tuition would not have to be raised if the legislature ignores the bitter complaints and increases funding for higher education.

40.3b Key ideas for counterexample: Man need not have any role to play in society.

(PREMISS)	(CONCLUSION)	(PREMISS)	(PREM)
41.1a It is cold outside. (= A)	The children should stay in. (= B)	IF it's cold outside, THEN the children should stay in.	A ⊃ B
T	T	?	T
T	F	F	F
F	T	?	T
F	F	?	T

Condition 1 is met automatically by the first premiss, since the table for the English sentence is identical to the table for 'A'. Comparison of the last two columns above reveals that the second premiss also meets condition 1. Condition 2 is trivially met because the English conclusion sentence has the same table as 'B'.

41.3 Since only *41.1a, b, d,* and *f* meet conditions 1 and 2, these are the only arguments for which there is a chance of a validity proof using our techniques. *f* and *d,* however, had invalid diagrams. So, only *a* and *b* are proved valid.

42.4

1	A ⊃ (B ⊃ C)	
2	B	
3	A	
4	B ⊃ C	1, 3, ⊃ E
5	C	2, 4, ⊃ E

42.8

1	C ⊃ (D & A)	
2	E	
3	E ⊃ ~(D & A)	
4	~(D & A)	2, 3, ⊃ E
5	~C	1, 4, N ⊃ E

43.3

1	(R & S) ⊃ E	
2	R	
3	A ≡ S	
4	A	
5	S	3, 4, ≡ E
6	R & S	2, 5, & I
7	E	1, 6, ⊃ E

43.6
1 | (A ⊃ B) & (~C ⊃ D)
2 | A
3 | B ≡ (D ⊃ ~C)
4 | ~C ⊃ D   1, & E
5 | A ⊃ B   1, & E
6 | B   2, 5, ⊃ E
7 | D ⊃ ~C   3, 6, ≡ E
8 | ~C ≡ D   4, 7, ≡ I

44.2
1 | E ⊃ (A ∨ B)
2 | A ⊃ (C & D)
3 | B ⊃ (C & D)
4 | E
5 | A ∨ B   1, 4, ⊃ E
6 | (C & D) ∨ (C & D)   5, 2, 3, ∨ E
7 | C & D   6, R ∨ E
8 | C   7, & E

44.3
1 | A ∨ B
2 | A ⊃ C
3 | B ⊃ C
4 | C ∨ C   1, 2, 3, ∨ E
5 | C   4, R ∨ E
6 | C ∨ D   5, ∨ I

44.7
1 | (B & E) ∨ A
2 | A ⊃ C
3 | B ≡ (C ∨ D)
4 | ~C
5 | ~A   2, 4, N ⊃ E
6 | B & E .   1, 5, EA
7 | B   6, & E
8 | C ∨ D   3, 7, ≡ E
9 | D   4, 8, EA

45.2a At line 7 we may not use line 5 because 5 does not count as being behind the vertical line closest to the left of 7. More informally, 3 was used to get 5, but 3 has been discharged before line 7 and thus one cannot assume 5 is true by the time one reaches line 7.

45.4
1 | (P ∨ R) ⊃ Q
2 | P
3 | P ∨ R   2, ∨ I
4 | Q   1, 3, ⊃ E
5 | P ⊃ Q   2-4, ⊃ I

45.6

1	P ⊃ R	
2	P	
3	Q	
4	R	1, 2, ⊃ E
5	Q ⊃ R	3-4, ⊃ I
6	P ⊃ (Q ⊃ R)	2-5, ⊃ I

46.3

1	A ⊃ B	
2	~B	
3	A	
4	B	1, 3, ⊃ E
5	~B	2, R
6	~A	3-4 & 5, RA

46.8

1	~B ⊃ ~A	
2	A	
3	~~A	2, ~~I
4	~~B	3, 1, N ⊃ E
5	B	4, ~~E
6	A ⊃ B	2-5, ⊃ I

46.11

1	~A & ~B	
2	A ∨ B	
3	A	
4	A	3, R*
5	A ⊃ A	3-4, ⊃ I
6	B	
7	~B	1, & E
8	A	6, 7, C
9	B ⊃ A	6-8, ⊃ I
10	A	2, 5, 9, ∨ E
11	~A	1, & E
12	~(A ∨ B)	2-10 & 11, RA

46.13b

1		*
2	~(A ∨ ~A)	
3	A	
4	A ∨ ~A	3, ∨ I
5	~(A ∨ ~A)	2, R
6	~A	3-4 & 5, RA
7	A ∨ ~A	6, ∨ I
8	A ∨ ~A	2-7, RA

*In the above derivations, these lines
could be omitted.

47.1 The argument from '~(A & B)' to 'A' is invalid. Our system should not allow invalid derivations. If the derivation is possible, the system is incorrectly designed.

47.3b

1	~A ∨ B	
2	~A ∨ ~~B	1, ~~I
3	~(A & ~B)	2, ∨T&

47.3i

1	B	
2	~A ∨ B	1, ∨ I
3	A ⊃ B	2, ∨ T ⊃

47.3n

1		
2	~(A ⊃ B)	
3	B	
4	~A ∨ B	3, ∨ I
5	A ⊃ B	4, ∨ T ⊃
6	~(A ⊃ B)	2, R
7	~B	3-5 & 6, RA
8	~(A ⊃ B) ⊃ ~B	2-7, ⊃ I

48.2b

1	U ⊃ I	
2	I ⊃ (~P & ~N)	
3	C ⊃ P	
4	U	
5	I	1, 4, ⊃ E
6	~P & ~N	2, 5, ⊃ E
7	~P	6, & E
8	~C	3, 7, N ⊃ E
9	~N	6, & E
10	~N & ~C	8, 9, & I
11	U ⊃ (~N & ~C)	4-10, ⊃ I
12	~U ∨ (~N & ~C)	11, ⊃ T ∨

48.3a Although the universe exhibits design, the designers were a committee, and there is not a God.

48.3b Suppose the tests do not accurately measure intelligence, and that non-whites are in fact more intelligent than whites.

48.5a

1	A ∨ (B & C)	
2	~B	
3	~B ∨ ~C	2, ∨ I
4	~(B & C)	3, ∨ T &
5	A	1, 4, EA

48.6 There are many possibilities. Here's one:

•
~X & ~Y
•
~(X ∨ Y)

49.2a ∀S's are R's
b ~(∃S's are R's)
c ∃S's are R's
d ∃S's are not R's

50.1 Correct: b, f, h, k, o, p, r

50.2a ∀x(Sx ⊃ Hx)
c Sj ⊃ Ej
e ∃x(Sx & Lx) & Hk
g ~∃x(Sx & Ex & Lx)
or ∀x((Lx & Ex) ⊃ ~Sx)

50.4 most, a few of, several, numerous, hardly any, a number of

51.1a   **51.1d**

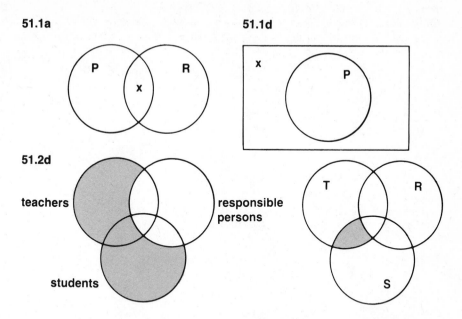

51.2d

Since the two diagrams differ, the symbolization is not adequate.

51.4a

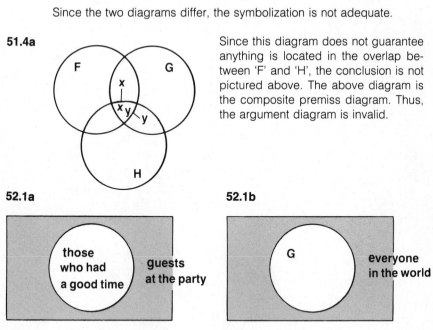

Since this diagram does not guarantee anything is located in the overlap between 'F' and 'H', the conclusion is not pictured above. The above diagram is the composite premiss diagram. Thus, the argument diagram is invalid.

52.1a   **52.1b**

52.3a Key: W( ):( ) is suspicious of the women
        M( ):( ) is a man   F( ):( ) works in the factory.
        S( ):( ) is suspicious of the women in the factory
In the first domain: ∀x((Mx & Fx) ⊃ Sx)
In the second domain: ∀x(Mx ⊃ Wx) or ∀x(Mx ⊃ Sx)

52.6a Likely to have existential import. Key: R( ):( ) is someone in this room. O( ):( ) should get out immediately. '∀x(Rx ⊃ Ox) & ∃xRx' with import. '∀x(Rx ⊃ Ox)' without import.

53.2a

1	∃x(Fx & Gx) ⊃ Ha	
2	∀xFx	
3	Fb ⊃ Gc	
4	Fb	2, ∀E
5	Gc	3, 4, ⊃ E
6	Fc	2, ∀E
7	Fc & Gc	5, 6, & I
8	∃x(Fx & Gx)	7, ∃I
9	Ha	1, 8, ⊃ E

53.2h

1	Fa	
2	∃xFx	1, ∃I
3	∃yFy	1, ∃I
4	∃xFx & ∃yFy	2, 3, & I

53.2k

1	∀x((Gx & Hx) ⊃ Cx)	
2	∀x(Bx ⊃ (Gx & Hx))	
3	~Ca	
4	(Ga & Ha) ⊃ Ca	1, ∀E
5	~(Ga & Ha)	3, 4, N ⊃ E
6	Ba ⊃ (Ga & Ha)	2, ∀E
7	~Ba	5, 6, N ⊃ E

54.1 Correct: *a, b, d* only.

54.3a

1	∀x(Cx ⊃ (Tx & Wx))	
2	∀x((Tx & Wx) ⊃ (Fx ∨ Lx))	
3	∀x((Fx ∨ Lx) ⊃ ~Dx)	
4	Ca	
	etc.	

54.3b

1	∀x((Cx & Dx) ⊃ Tx)
2	∀x((Cx & Tx) ⊃ Lx)
	etc.

54.3d Prem: ∀x((Cx & Tx) ⊃ ~Lx) & ∀x((Cx & Wx) ⊃ Lx)
Conc: ∀x(~Lx ⊃ (~Cx ∨ ~Wx))

54.4e

1	~∃x(Fx & Gx)	
2	∀x ~(Fx & Gx)	1, ~∃T∀~
3	~(Fa & Ga)	2, ∀E
4	~Fa ∨ ~Ga	3, &T∨

54.6c
1	∀xFx ∨ ∀xGx	
2	∀xFx	
3	Fa	2, ∀E
4	∀xFx ⊃ Fa	2-3, ⊃ I
5	∀xGx	
6	Ga	5, ∀E
7	∀XGx ⊃ Ga	5-6, ⊃ I
8	Fa ∨ Ga	1, 4, 7, ∨ E
9	∀x(Fx ∨ Gx)	8, ∀I

55.4a
1	∃y ~Fy	
2	∀x(~Fx ⊃ Gx)	
3	~Fa	
4	~Fa ⊃ Ga	2, ∀E
5	Ga	3, 4, ⊃ E
6	∃zGz	5, ∃I
7	∃zGz	1, 3-6, ∃E

55.4e
1	Fb	
2	∀x(Gx ⊃ ~Fx)	
3	~Gb ⊃ ∃xHx	
4	Gb ⊃ ~Fb	2, ∀E
5	~~Fb	1, ~~I
6	~Gb	4, 5, N ⊃ E
7	∃xHx	3, 6, ⊃ E

55.4g
1	∃x ~(Fx & Gx)	
2	~(Fa & Ga)	
3	~Fa ∨ ~Ga	2, &T∨
4	∃x(~Fx ∨ ~Gx)	3, ∃I
5	∃x(~Fx ∨ ~Gx)	1, 2-4, ∃E

55.4j
1	∀x(~Fx ∨ ~Gx)	
2	~Fa ∨ ~Ga	1, ∀E
3	~(Fa & Ga)	2, ∨T&
4	∀x~(Fx & Gx)	3, ∀I
5	~∃x(Fx & Gx)	4, ∀ ~T ~∃

55.5b 1 | Gf ⊃ ∀xGx
2 | Gf ⊃ Gl
3 | Gl ⊃ ∀xGx
4 | ∃x~Gx
5 | ~∀xGx   4, ∃ ~T ~∀
6 | ~Gf   1, 5, N ⊃ E
7 | ~Gl   3, 5, N ⊃ E
8 | ~Gf & ~Gl   6, 7, & I
9 | ∃x~Gx ⊃ (~Gf & ~Gl)   4-8, ⊃ I

55.5h 1 | ∃x(Cx & Sx)
2 | ∀x((Sx & Cx) ⊃ Nx)
3 | Ca & Sa
4 | (Sa & Ca) ⊃ Na   2, ∀ E
5 | Ca   3, & E
6 | Sa   3, & E
7 | Sa & Ca   5, 6, & I
8 | Na   4, 7, ⊃ E
9 | ∃xNx   8, ∃ I
10 | ∃xNx   1, 3–9, ∃ E

55.5i 1 | ∀x(Dx ⊃ ((Ix & Px) ⊃ ~Vx))
2 | Pa & Da
3 | ∀y(Py ⊃ Iy)
4 | (Da ⊃ ((Ia & Pa) ⊃ ~Va)   1, ∀E
5 | Da   2, & E
6 | (Ia & Pa) ⊃ ~Va   4, 5, ⊃ E
7 | Pa ⊃ Ia   3, ∀E
8 | Pa   2, & E
9 | Ia   7, 8, ⊃ E
10 | Ia & Pa   8, 9, & I
11 | ~Va   6, 10, ⊃ E
12 | ∀y(Py ⊃ Iy) ⊃ ~Va   3-11, ⊃ I
13 | (Pa & Da) ⊃ (∀y(Py ⊃ Iy) ⊃ ~Va)   2-12, ⊃ I
14 | ∀x((Px & Dx) ⊃ (∀y(Py ⊃ Iy) ⊃ ~Vx))   13, ∀I

56.1

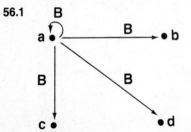

Here, *a* is the *x* which bears relation *B* to all things.

56.2a None. Neither '∀yBay' nor '∀xBxa' is true. Similarly, '∀yBby' and '∀xBxb' are false, as are the other instances of '∃x∀yBxy' and '∃y∀xBxy'. Thus, these two latter formulas are false. Also, '∀x∃yBxy' is false of *a*, as is '∀y∃xBxy', thus making these two formulas false.

56.3a ∀x∃yFxy
  b ∀x∃yFyx
  c ∃x∀yFyx
  d ∀x∀yFxy
  e ∃x∃yFxy
  f ~∃x∃yFxy
  g ~∃xFsx

56.4a False: True:

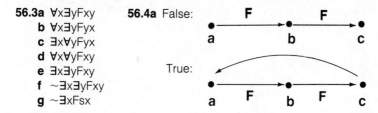

56.6c Same as *a*.
  e ∃x∃y(Ix & Iy & ~Sxy & ∀z(Iz ⊃ (Szx ∨ Szy)))
  g Same as *c*.
  h Without existential import: ∀x∀y∀z((Ix & Iy & Iz) ⊃ (Szx ∨ Szy))

57.1e

1	~∃x∀yFxy	
2	∀x~∀yFxy	1, ~∃T∀~
3	~∀yFay	2, ∀E
4	∃y~Fay	3, ~∀T∃~
5	∀x∃y~Fxy	4, ∀I

57.1h

1	∀x(∃yFy ⊃ Gx)	
2	∃xFy ⊃ Ga	1, ∀E
3	Fb	
4	∃yFy	3, ∃I
5	Ga	2, 4, ⊃ E
6	Fb ⊃ Ga	3-5, ⊃ I
7	∀y(Fy ⊃ Ga)	6, ∀I
8	∀x∀y(Fy ⊃ Gx)	7, ∀I

57.2a

1	∀x∀y∀z((Sxy & Syz) ⊃ Sxz)	
2	Sae & Sem	
3	∀y∀z((Say & Syz) ⊃ Saz))	1, ∀E
4	∀z((Sae & Sez) ⊃ Saz)	3, ∀E
5	(Sae & Sem) ⊃ Sam	4, ∀E
6	Sam	2, 5, ⊃ E

57.2d   1 | ∀x(Ax ⊃ Px)
      2 | | Oab & Ab
      3 | AB ⊃ Pb   1, ∀E
          | etc.
      7 | Oab & Pb   (from above)
      8 | ∃z(Oaz & Pz)   7, ∃I
      9 | (Oab & Ab) ⊃
             ∃z(Oaz & Pz)   2-8, ⊃ I
     10 | ∀y[(Oay & Ay) ⊃
             ∃z(Oaz & Pz)]   9, ∀I
     11 | ∀x∀y[(Oxy & Ay) ⊃
             ∃z(Oxz & Pz)]   10, ∀I

57.2e   Translation Key:
     Px: x is a problem in the US
     Hx: x is one of the people
     Bxyz: x blames y for z
     CXYZ: x believes person y has control over z
     p: the president

57.2i   1 | ∀x∀y(Bxy ⊃ Pxy)
      2 | ∀x∀y(Pxy ⊃ ~Ixy)
      3 | ∀xIxx
      4 | ∃xBxx
          | (Use RA to show that 4 is false.)

Translation Key: (Assume domain is events.)
Bxy: x brings about y
Pxy: x is prior to y in beginning
Ixy: x is identical with y

58.1   It is necessary that a triangle have three sides. Therefore, it is possible for triangles to have three sides.

58.2 Tensed logic.

1	∀x∀y((Px & Hy) ⊃ Bypx)	
2	∀x∀y∀z((Hx & Hy & Bxyz) ⊃ Cxyz)	
3	Hp	
4	Pa & Hb	
5	∀y((Pa & Hy) ⊃ Bypa))	1, ∀E
6	(Pa & Hb) ⊃ Bbpa	5, ∀E
7	Bbpa	4, 6, ⊃E
8	∀y∀z((Ha & Hy & Bbyz) ⊃ Cbyz)	2, ∀E
9	∀z((Hb & Hp & Bbpz) ⊃ Cbpz)	8, ∀E
10	(Hb & Hp & Bbpa) ⊃ Cbpa	9, ∀E
11	Hb	4, &E
12	Hb & Hp	3, 11, &I
13	Hb & Hp & Bbpa	7, 12, &I
14	Cbpa	10, 13, ⊃E
15	∃zCbza	14, ∃I
16	(Pa & Hb) ⊃ ∃zCbza	4-15, ⊃I
17	∀y((Pa & Hy) ⊃ ∃zCyza)	16, ∀I
18	∀x∀y((Px & Hy) ⊃ ∃zCyzx)	17, ∀I

59.2a One possible premiss: If John decides not to go to the party, he will see Linda this Friday, and if he sees her she will not be mad at him.

b Could add: We can't step up production of the new model, because we don't have enough capacity.

c Could add: If the regulations require most students to live in the dormitories unless they get a waiver, then regulations deny people the right to live where they want and should be revised.

59.4 Some premisses which could be added: Chaos should be avoided at all costs. The only alternative to abolishing the grading system is to keep the present one.

59.6a Suppose the counterexample turns on the idea that it is a good idea to have an unstable society, to bring about some reforms. This sort of example can be blocked by saying that it is undesirable to have instability. However, the addition of the mere statement that it is undesirable to have instability does not make the argument valid, for it may still be the case that we ought not to discriminate, even though discrimination prevents an undesirable consequence.

60.1 People will complain in any case.

60.9 Nixon was guilty of conspiracy to obstruct justice.

60.13 No, for A and E might both be true. No, for A might contradict B and E might contradict F. Yes, if A, B, and C are all true, then D is also true and NOT D is false; thus, E, F, and G cannot all be true. Similarly, the answer to the last question is also yes.

61.1 Correct: *b, d, i, m,* and *p.*

61.2 Suppose that people rarely perspire unless they are nervous and Mary is not nervous. This supposition is not ruled out by the given premiss, and thus that premiss, by itself, does not make the conclusion likely.

61.3 The argument of 61.2 seems quite reasonable because we are all aware of certain facts about people and under what conditions they perspire. We would reasonably take such facts as background assumptions here. For example, we could use the background assumption that almost everyone perspires on hot, humid days when working near the ground in the sun. This assumption when added to the argument would render the conclusion likely to be true and thus shows the original argument to be a TABA.

62.1 II from the first pair is more easily made into a TABA, because the subset of characteristics defining the group in its conclusion is larger than the subset in argument I. In the second pair, I is easier to think of as a TABA, because the conclusion in argument II refers to Idaho which has been described as different in a possibly relevant respect from all the other items in the argument—perhaps zinc is found only in the Rockies for all we know in argument II.

62.2a All businessmen (or maybe all businessmen appointed to the committee) are pushy and arrogant.
  b The two businessmen mentioned first are representative of the vast majority of businessmen, or at least of the vast majority of businessmen who get appointed to committees like this one. Whether such an assumption is likely to be true is an empirical question for which I do not have the evidence.

62.4 Alternative c is the only one which would give everyone in the neighborhood an equal chance to be chosen. The first alternative excludes those who lack telephones—likely to be a sizable group in a poor neighborhood. The second alternative excludes all those who live away from the center of the neighborhood.

62.7 A random sample might fail to be representative just by chance. A random sample of 1,000 U.S. citizens could contain more than 700 males, just by chance. A representative sample could fail to be random if it was chosen in some nonrandom way. For example, I might by chance happen to get a sample representative of the student body on my campus if I pick those who are enrolled in my logic class, even though certain students are systematically excluded from my class by their need to take courses which conflict with it in timing.

62.8 Even though a random sample can fail to be representative, it is unlikely to be drastically unrepresentative if it is large; so a large random sample can give us high probability of being approximately correct about the whole population.

63.1 There is a perfect direct correlation between success and the combination strategy of advertising in the newspaper while also setting a price lower than the competition. This correlation suggests a causal connection between success and that strategy.

63.3a The ancient division of labor between men and women according to which the men were assigned to hunt while the women were gatherers lies at the root of male domination over women in modern society.

63.3b Playing a hunter role is causally connected with male dominance in a society.

63.3c Assume premiss *b* above. Add the premiss that the early division of labor between the sexes established a tradition which never broke down, and the premiss that in ancient societies the men were assigned to hunt and the women to gather. These premisses lead to the conclusion that is described in *a*.

63.3d Probable assumptions: The situation today among these primitive peoples is similar in all relevant respects to the situation that existed among the ancient peoples. There are no other relevant differences between these societies that might explain the variations in male dominance.

63.3e Data about other primitive peoples living today in which the hunter role is associated with dominance would help. Data about ancient primitive peoples showing the same thing would help, more than data about current peoples. Data showing that all these people were different from one another in various ways that do not correlate with the degree of male dominance would help to rule out alternative hypotheses about the possible causes of male dominance.

63.3f If it were discovered that in all societies in which men are physically larger than women, men dominate, while in societies in which men are physically smaller than women (assuming there are some such societies) the women dominate, then the argument would be weakened.

64.1 The argument is not valid because the best explanation could be best and still be incorrect. The argument is not even tight, because there is no guarantee that the best explanation is even likely to be correct. (The last two tests of a good explanation are not designed to guarantee the truth of the explanation, and they can be important in forcing us to choose as best some explanation which is in fact not correct.)

64.2 Evidence from background theory: The influence of other people, and especially the influence of close friends, is quite often important in changing the attitudes a person holds. It is not likely that a person would change an attitude about something important in life without some reason for doing so. Evidence from facts: Bob's roommate attends church regularly. The roommate is known to talk persuasively to his friends in favor of Christianity. Bob's girlfriend attends church regularly.

64.3 One possible explanation: Something is wrong with the directions in the lab manual.

64.4 One possible argument, using the above explanation:

P.   There is something wrong with the directions in the lab manual.
P.   The experiment was performed according to the directions and no color change occurred even though the manual said one should occur.
_____
C.   These directions have not been in use for very many years.

This argument is a TABA, because it is reasonable to assume that the manual would have been revised if it contained an error like this. Thus, this argument tends to show that the proposed explanation does not fit well with the fact that the directions have been in use for years.

65.1a No "ought" obvious here, although one might suppose one is implied.
  b Categorical "ought".
  d Hypothetical "ought".

65.5a From the point of view of your own self-interest, you ought to attend Harvard.
  b Morally speaking, you ought not knock down old ladies.

66.2 Although one might think at first that fairness requires everyone during equally valuable work be paid the same, doing so will require the school to deny admission to qualified students or else create extra work for all the faculty who do stay on. So it is not clear to me what is fair. The fairness standard seems to be vague enough here to allow deep disagreement to arise over its application.

66.3 Principle #1: Always tell the truth. Principle #2: Never hurt anyone's feelings intentionally. Case: You are asked your opinion of a new sweater your friend has just bought at great cost. You think it's an ugly sweater. You know that your friend will be very hurt if you say what you think. By Principle #1, you should say what you think. By Principle #2, you should not.

67.1a Type 1 counterexample: Suppose the appropriate evaluative standard were "One ought to do things that one cannot do well, and above all avoid following your father's advice". Type 2 example: Perhaps your father would also like it if you became a pharmacist, and you have the ability in that field as well, but you got a scholarship to go to pharmacy school and not to go to medical school, thus making pharmacy more attractive as an option. Type 3 example: You have a great many debts and no way to pay them off without going to work now in some field far removed from medicine, and there is no way you will be able to go back into the medical field later; moreover, you have no desire to become a doctor. Type 4 examples seem to be inapplicable here since the argument does not proceed in part by rejecting alternatives. Type 5 examples are not applicable unless one takes the conclusion to be expressing an absolute "ought". Taking the conclusion in that sense, we can get a counterexample by supposing that some considerations other than those mentioned in the premisses are more important.

Index

Index

409